**21世纪高等学校规划教材｜计算机应用**

# 基于Java的综合课程设计

尉哲明 主编
冀素琴 郭珉 编著

清华大学出版社
北 京

## 内 容 简 介

本书涵盖基于 Java 的三个层次的课程设计：一是 Java 编程基础，包括 Java 面向对象技术、字符串、数组、集合类、异常处理及 Java I/O 流等，针对每种技术的两三个实用案例进行详细剖析，为后面的综合课程设计打下坚实基础；二是 Java 应用技术，根据 Java GUI、多媒体、线程等知识点精心设计一些实用的综合案例，并给出详细讲解，旨在帮助读者完成一个比较大的课程设计项目；三是基于 Java 的综合课程设计，将 Java 课程与信息技术类的其他课程结合起来，进行综合课程设计。

本书内容翔实，层次清晰，可以作为各专业的 Java 课程和高校信息技术类专业的综合课程设计的教材使用，也可以为 Java 的专业人员提供参考。

本书封面贴有清华大学出版社防伪标签，无标签者不得销售。
版权所有，侵权必究。举报：010-62782989，beiqinquan@tup.tsinghua.edu.cn。

图书在版编目(CIP)数据

基于 Java 的综合课程设计/尉哲明主编；冀素琴，郭珉编著. —北京：清华大学出版社，2014 (2021.8重印)
(21 世纪高等学校规划教材·计算机应用)
ISBN 978-7-302-36484-9

Ⅰ. ①基… Ⅱ. ①尉… ②冀… ③郭… Ⅲ. ①JAVA 语言－程序设计－高等学校－教材 Ⅳ. ①TP312

中国版本图书馆 CIP 数据核字(2014)第 099315 号

责任编辑：闫红梅　薛阳
封面设计：傅瑞学
责任校对：时翠兰
责任印制：丛怀宇

出版发行：清华大学出版社
    网　　址：http://www.tup.com.cn，http://www.wqbook.com
    地　　址：北京清华大学学研大厦 A 座　　邮　编：100084
    社 总 机：010-62770175　　邮　购：010-83470235
    投稿与读者服务：010-62776969，c-service@tup.tsinghua.edu.cn
    质量反馈：010-62772015，zhiliang@tup.tsinghua.edu.cn
    课件下载：http://www.tup.com.cn，010-83470236
印 装 者：三河市龙大印装有限公司
经　　销：全国新华书店
开　　本：185mm×260mm　　印　张：25.25　　字　数：612 千字
版　　次：2014 年 12 月第 1 版　　印　次：2021 年 8 月第 **7** 次印刷
印　　数：5201～5500
定　　价：59.50 元

产品编号：042193-03

# 出版说明

随着我国改革开放的进一步深化，高等教育也得到了快速发展，各地高校紧密结合地方经济建设发展需要，科学运用市场调节机制，加大了使用信息科学等现代科学技术提升、改造传统学科专业的投入力度，通过教育改革合理调整和配置了教育资源，优化了传统学科专业，积极为地方经济建设输送人才，为我国经济社会的快速、健康和可持续发展以及高等教育自身的改革发展做出了巨大贡献。但是，高等教育质量还需要进一步提高以适应经济社会发展的需要，不少高校的专业设置和结构不尽合理，教师队伍整体素质亟待提高，人才培养模式、教学内容和方法需要进一步转变，学生的实践能力和创新精神亟待加强。

教育部一直十分重视高等教育质量工作。2007年1月，教育部下发了《关于实施高等学校本科教学质量与教学改革工程的意见》，计划实施"高等学校本科教学质量与教学改革工程(简称'质量工程')"，通过专业结构调整、课程教材建设、实践教学改革、教学团队建设等多项内容，进一步深化高等学校教学改革，提高人才培养的能力和水平，更好地满足经济社会发展对高素质人才的需要。在贯彻和落实教育部"质量工程"的过程中，各地高校发挥师资力量强、办学经验丰富、教学资源充裕等优势，对其特色专业及特色课程(群)加以规划、整理和总结，更新教学内容、改革课程体系，建设了一大批内容新、体系新、方法新、手段新的特色课程。在此基础上，经教育部相关教学指导委员会专家的指导和建议，清华大学出版社在多个领域精选各高校的特色课程，分别规划出版系列教材，以配合"质量工程"的实施，满足各高校教学质量和教学改革的需要。

为了深入贯彻落实教育部《关于加强高等学校本科教学工作，提高教学质量的若干意见》精神，紧密配合教育部已经启动的"高等学校教学质量与教学改革工程精品课程建设工作"，在有关专家、教授的倡议和有关部门的大力支持下，我们组织并成立了"清华大学出版社教材编审委员会"(以下简称"编委会")，旨在配合教育部制定精品课程教材的出版规划，讨论并实施精品课程教材的编写与出版工作。"编委会"成员皆来自全国各类高等学校教学与科研第一线的骨干教师，其中许多教师为各校相关院、系主管教学的院长或系主任。

按照教育部的要求，"编委会"一致认为，精品课程的建设工作从开始就要坚持高标准、严要求，处于一个比较高的起点上；精品课程教材应该能够反映各高校教学改革与课程建设的需要，要有特色风格、有创新性(新体系、新内容、新手段、新思路，教材的内容体系有较高的科学创新、技术创新和理念创新的含量)、先进性(对原有的学科体系有实质性的改革和发展，顺应并符合21世纪教学发展的规律，代表并引领课程发展的趋势和方向)、示范性(教材所体现的课程体系具有较广泛的辐射性和示范性)和一定的前瞻性。教材由个人申报或各校推荐(通过所在高校的"编委会"成员推荐)，经"编委会"认真评审，最后由清华大学出版社审定出版。

目前，针对计算机类和电子信息类相关专业成立了两个"编委会"，即"清华大学出版社计算机教材编审委员会"和"清华大学出版社电子信息教材编审委员会"。推出的特色精品教材包括：

(1) 21世纪高等学校规划教材·计算机应用——高等学校各类专业，特别是非计算机专业的计算机应用类教材。

(2) 21世纪高等学校规划教材·计算机科学与技术——高等学校计算机相关专业的教材。

(3) 21世纪高等学校规划教材·电子信息——高等学校电子信息相关专业的教材。

(4) 21世纪高等学校规划教材·软件工程——高等学校软件工程相关专业的教材。

(5) 21世纪高等学校规划教材·信息管理与信息系统。

(6) 21世纪高等学校规划教材·财经管理与应用。

(7) 21世纪高等学校规划教材·电子商务。

(8) 21世纪高等学校规划教材·物联网。

清华大学出版社经过三十多年的努力，在教材尤其是计算机和电子信息类专业教材出版方面树立了权威品牌，为我国的高等教育事业做出了重要贡献。清华版教材形成了技术准确、内容严谨的独特风格，这种风格将延续并反映在特色精品教材的建设中。

清华大学出版社教材编审委员会

联系人：魏江江

E-mail：weijj@tup.tsinghua.edu.cn

目前，Java 技术被广泛应用于各种行业的信息处理系统，高校许多专业都开设了 Java 程序设计课程，且这些专业大都开设了综合课程设计。本书是作者经过长期的 Java 教学与科研实践的一个成果。

在 Java 课程中，教学过程一般被划分为两个大的阶段：

(1) Java 基础教学。在这一阶段中，重点完成 Java 面向对象程序设计的教学任务，将 Java 面向对象技术、字符串、数组、异常处理等基础知识细致、透彻地进行讲解，为后面综合应用技术的学习打下坚实的基础。

(2) Java 应用技术的教学。在这一阶段中，完成 Java 各种应用技术程序设计的教学任务，具体根据课时的多少，重点安排 Java 图形用户界面、Java I/O 流、Java 小应用程序设计、Java 多线程程序设计、Java 多媒体应用程序设计、Java 网络通信程序设计、Java 数据库连接程序设计等教学内容，其主要教学目的是增进学生学习 Java 的兴趣，切实培养和提高他们对 Java 技术的综合应用能力。

伴随两个阶段的 Java 教学，课程设计是一个重要且必不可少的环节。

在高校信息技术类专业开设的综合课程设计课程中，可以将 Java 技术作为开发工具，设计开发出管理信息系统、网络应用软件等多种类型的应用程序，有效结合 Java、数据库、网络、数据结构、软件工程等多门课程进行课程设计。

本书出版的宗旨就是为 Java 课程和基于 Java 的综合课程设计提供教材，主要涉及以下三方面的内容：

第 1 部分是 Java 基础课程设计，包括第 1～9 章。内容包括 Java 面向对象技术、字符串、数组、集合类、异常处理以及 Java 输入输出流，每一章都首先总结本章知识点，然后针对两三个小的课程设计题目，分别进行详细的分析、设计和实现。最后给出这些课程设计的拓展设计作业。

第 2 部分是 Java 应用技术，包括第 10～13 章。第 10～12 章分别介绍了 Java 图形用户界面设计、Java 多媒体应用程序设计、Java 多线程设计。每一章都利用案例详细介绍了如何使用 Java 语言来进行面向对象编程。每个案例都是针对各章相关的知识点精心设计的，将知识的讲解融入案例中，使读者能够从实践中来理解和巩固知识，每个案例都提供了相应的练习题目，读者可以继续思考并上机实践来进一步完善案例。最后，第 13 章通过一个拼图游戏的综合案例帮助读者巩固第 1～12 章所介绍的知识，因此本章内容不再采用第 10～12 章的形式对关键知识点进行分析和介绍，案例只是按照设计要求、总体设计、详细设计、案例练习题目的流程给出了具体步骤。第 2 部分的案例设计都是以面向对象的实际应用展开的，并希望能锻炼学生的计算思维能力，来提高学生运用计算机进行知识抽象、问题求解和形式化描述的能力，最终解决专业和生活中遇到的各种实际问题。

第 3 部分是基于 Java 的综合课程设计，包括第 14～17 章。第 14 章通过一个资料室图书管理信息系统的课程设计题目详细介绍了单机小型管理信息系统的设计和实现方法。该章内容可用于数据库课程的综合课程设计,或者将 Java 与数据库两门课程结合起来进行课程设计。第 15 章和第 16 章通过两个典型的网络程序案例,详细讲解了网络应用程序的设计和实现方法,这两章内容可用于网络课程的综合课程设计,或者将 Java 与网络两门课程结合起来进行课程设计。第 17 章通过一个用户登录系统的课程设计题目,讲解了如何设计和实现一个 B/S 模式的数据库、网络应用程序,该章内容可用于数据库、网络以及 Java 三门课程的综合课程设计。第 3 部分每一章都详细讲解了系统的需求分析、功能设计、实现思路、实现步骤、系统调试与软件发布方法,章末布置了综合课程设计的作业。读者可以参考作者的思路对该系统功能进行进一步的扩展和完善,或是重新设计和实现同一类的、新的系统软件。

本书第 1 部分由冀素琴编写,第 2 部分由郭珉编写,第 3 部分由尉哲明编写。书中全部代码由作者亲自编写,都在 JDK 1.7 运行环境下调试通过。读者可以从清华大学出版社网站上下载,仅供学习,不得以任何方式抄袭出版。

如有错误之处,敬请读者批评指正。

作者

尉哲明　冀素琴　郭珉

Yuzhem2@163.com

2014 年 9 月

# 第 1 部分 Java 编程基础

## 第 1 章 Java 开发环境 ………………………………………………………………… 3
1.1 本章知识点 ……………………………………………………………………… 3
1.2 设计 1 初识 Java ……………………………………………………………… 4

## 第 2 章 Java 语言基础 ………………………………………………………………… 11
2.1 本章知识点 ……………………………………………………………………… 11
2.2 设计 1 基本运算练习 ………………………………………………………… 12
2.3 设计 2 控制结构练习 ………………………………………………………… 13

## 第 3 章 类与对象 ……………………………………………………………………… 16
3.1 本章知识点 ……………………………………………………………………… 16
3.2 设计 1 对象的创建和使用 …………………………………………………… 17
3.3 设计 2 包的使用与访问控制 ………………………………………………… 20

## 第 4 章 继承、多态和封装 …………………………………………………………… 24
4.1 本章知识点 ……………………………………………………………………… 24
4.2 设计 1 继承性 ………………………………………………………………… 25
4.3 设计 2 多态性 ………………………………………………………………… 28
4.4 设计 3 封装性 ………………………………………………………………… 32

## 第 5 章 抽象类与接口 ………………………………………………………………… 35
5.1 本章知识点 ……………………………………………………………………… 35
5.2 设计 1 抽象类 ………………………………………………………………… 35
5.3 设计 2 接口 …………………………………………………………………… 38

## 第 6 章 数组和常用类 ………………………………………………………………… 42
6.1 本章知识点 ……………………………………………………………………… 42
6.2 设计 1 数组 …………………………………………………………………… 43
6.3 设计 2 字符串 ………………………………………………………………… 47

## 第 7 章　集合类 ································································· 50

- 7.1　本章知识点 ································································· 50
- 7.2　设计 1　List 接口及实现该接口的常用类的练习 ················· 52
- 7.3　设计 2　Set 接口及实现该接口的常用类的练习 ················· 55
- 7.4　设计 3　Map 接口及实现该接口的常用类的练习 ················· 57

## 第 8 章　异常 ································································· 60

- 8.1　本章知识点 ································································· 60
- 8.2　设计 1　异常的捕获 ····················································· 61
- 8.3　设计 2　异常的抛出及搜索 ············································ 63

## 第 9 章　文件与流 ······························································· 67

- 9.1　本章知识点 ································································· 67
- 9.2　设计 1　文件管理 ························································· 69
- 9.3　设计 2　常用流练习 ····················································· 72
- 9.4　设计 3　RandomAccessFile 类的应用 ······························ 76

# 第 2 部分　Java 应用技术

## 第 10 章　Java 图形用户界面设计 ············································ 81

- 10.1　案例：几何图形计算器 ················································ 81
  - 10.1.1　案例问题描述 ···················································· 81
  - 10.1.2　案例功能分析及演示 ·········································· 81
  - 10.1.3　案例总体设计 ···················································· 83
  - 10.1.4　案例代码实现 ···················································· 91
  - 10.1.5　案例练习题目 ·················································· 108
- 10.2　案例：饭店点菜 ························································ 109
  - 10.2.1　案例问题描述 ·················································· 109
  - 10.2.2　案例功能分析及演示 ········································ 109
  - 10.2.3　案例总体设计 ·················································· 111
  - 10.2.4　案例代码实现 ·················································· 117
  - 10.2.5　案例练习题目 ·················································· 133

## 第 11 章　Java 多媒体程序设计 ·············································· 134

- 11.1　案例：随机绘图与动画 ·············································· 134
  - 11.1.1　案例问题描述 ·················································· 134
  - 11.1.2　案例功能分析与演示 ········································ 134
  - 11.1.3　案例总体设计 ·················································· 136

11.1.4 案例代码实现 ………………………………………………………………… 141
　　　11.1.5 案例练习题目 ………………………………………………………………… 153
　11.2 案例：多媒体图片查看器 …………………………………………………………… 154
　　　11.2.1 案例问题描述 ………………………………………………………………… 154
　　　11.2.2 案例功能分析与演示 ………………………………………………………… 154
　　　11.2.3 案例总体设计 ………………………………………………………………… 157
　　　11.2.4 案例代码实现 ………………………………………………………………… 163
　　　11.2.5 案例练习题目 ………………………………………………………………… 180

## 第 12 章　Java 多线程程序设计 ………………………………………………………… 181

　12.1 案例：两按钮反向运动——使用 Thread 子类 ……………………………………… 181
　　　12.1.1 案例问题描述 ………………………………………………………………… 181
　　　12.1.2 案例功能分析与演示 ………………………………………………………… 181
　　　12.1.3 案例总体设计 ………………………………………………………………… 182
　　　12.1.4 案例代码实现 ………………………………………………………………… 184
　　　12.1.5 案例练习题目 ………………………………………………………………… 190
　12.2 案例：两按钮反向运动——使用 Runnable 接口 …………………………………… 191
　　　12.2.1 案例问题描述 ………………………………………………………………… 191
　　　12.2.2 案例功能分析与演示 ………………………………………………………… 191
　　　12.2.3 案例总体设计 ………………………………………………………………… 191
　　　12.2.4 案例代码实现 ………………………………………………………………… 193
　　　12.2.5 案例练习题目 ………………………………………………………………… 197
　12.3 案例：使用 Thread 类实现图像动画 ………………………………………………… 197
　　　12.3.1 案例问题描述 ………………………………………………………………… 197
　　　12.3.2 案例功能分析与演示 ………………………………………………………… 198
　　　12.3.3 案例总体设计 ………………………………………………………………… 200
　　　12.3.4 案例代码实现 ………………………………………………………………… 202
　　　12.3.5 案例练习题目 ………………………………………………………………… 207
　12.4 案例：线程同步——模拟跑步接力 ………………………………………………… 207
　　　12.4.1 案例问题描述 ………………………………………………………………… 207
　　　12.4.2 案例功能分析与演示 ………………………………………………………… 207
　　　12.4.3 案例总体设计 ………………………………………………………………… 208
　　　12.4.4 案例代码实现 ………………………………………………………………… 210
　　　12.4.5 案例练习题目 ………………………………………………………………… 215

## 第 13 章　综合案例：拼图游戏 …………………………………………………………… 217

　13.1 设计要求 ……………………………………………………………………………… 217
　13.2 总体设计 ……………………………………………………………………………… 218
　　　13.2.1 GameWindow.java …………………………………………………………… 219

13.2.2 PuzzlePanel.java ………………………………… 219
13.2.3 Cell.java ………………………………………… 219
13.2.4 ControlGamePanel.java …………………………… 219
13.2.5 SplitImage.java …………………………………… 219
13.2.6 MusicDialog.java ………………………………… 219
13.2.7 Player.java ……………………………………… 220
13.2.8 ResultRecordDialog.java ………………………… 220
13.3 详细设计 …………………………………………………… 220
13.3.1 GameWindow 类 …………………………………… 220
13.3.2 PuzzlePanel 类 …………………………………… 231
13.3.3 Cell 类 …………………………………………… 235
13.3.4 ControlGamePanel 类 …………………………… 237
13.3.5 SplitImage 类 …………………………………… 245
13.3.6 MusicDialog 类 ………………………………… 246
13.3.7 Player 类 ………………………………………… 250
13.3.8 ResultRecordDialog 类 ………………………… 252
13.4 案例练习题目 ……………………………………………… 254

# 第3部分 基于 Java 的综合课程设计

## 第14章 Java 与数据库：资料室图书管理系统 ……………………………… 259
14.1 资料室图书管理系统需求分析 ……………………………… 259
14.2 资料室图书管理系统设计 …………………………………… 259
14.2.1 数据库设计 ……………………………………… 259
14.2.2 系统功能设计 …………………………………… 260
14.3 资料室图书管理系统实现思路 ……………………………… 261
14.4 资料室图书管理系统实现 …………………………………… 262
14.4.1 建立数据库表 …………………………………… 262
14.4.2 登录功能的实现 ………………………………… 263
14.4.3 主界面类 BooksManager 的实现 ………………… 267
14.4.4 录入图书功能的实现 …………………………… 271
14.4.5 检索图书功能的实现 …………………………… 277
14.4.6 修改图书功能的实现 …………………………… 287
14.4.7 删除图书功能的实现 …………………………… 291
14.4.8 借书功能的实现 ………………………………… 294
14.4.9 查看借书记录功能的实现 ……………………… 298
14.4.10 还书功能的实现 ………………………………… 300
14.4.11 查看还书记录功能的实现 ……………………… 306
14.5 资料室图书管理系统调试与软件发布 ……………………… 307

        14.5.1 系统调试 ………………………………………………………… 307
        14.5.2 软件发布 ………………………………………………………… 309
    14.6 综合课程设计作业 ……………………………………………………… 309
        14.6.1 资料室图书管理信息系统扩展 …………………………………… 309
        14.6.2 综合课程设计题目 ……………………………………………… 309

## 第 15 章 Java 与网络：P2P 聊天系统 …………………………………………… 310

    15.1 P2P 聊天系统需求分析 ………………………………………………… 310
    15.2 P2P 聊天系统设计 ……………………………………………………… 310
        15.2.1 信息服务器功能设计 …………………………………………… 310
        15.2.2 P2P 聊天端设计 ………………………………………………… 311
    15.3 P2P 聊天系统实现思路 ………………………………………………… 318
        15.3.1 传输协议选择 …………………………………………………… 318
        15.3.2 P2P 端与信息服务器的应用协议 ………………………………… 319
    15.4 P2P 聊天系统实现 ……………………………………………………… 319
        15.4.1 Request 类和 Response 类 ……………………………………… 319
        15.4.2 信息服务器的实现 ……………………………………………… 322
        15.4.3 P2P 聊天端的实现 ……………………………………………… 327
    15.5 P2P 聊天系统调试与软件发布 ………………………………………… 351
        15.5.1 系统调试 ………………………………………………………… 351
        15.5.2 软件发布 ………………………………………………………… 351
    15.6 综合课程设计作业 ……………………………………………………… 352
        15.6.1 P2P 聊天系统扩展 ……………………………………………… 352
        15.6.2 综合课程设计题目 ……………………………………………… 352

## 第 16 章 Java 与网络：Web 服务器与浏览器 …………………………………… 353

    16.1 Web 服务器与浏览器需求分析 ………………………………………… 353
        16.1.1 Web 服务器需求分析 …………………………………………… 353
        16.1.2 浏览器需求分析 ………………………………………………… 353
    16.2 Web 服务器与浏览器系统设计 ………………………………………… 353
        16.2.1 Web 服务器功能设计 …………………………………………… 353
        16.2.2 浏览器功能设计 ………………………………………………… 355
    16.3 Web 服务器与浏览器系统实现思路 …………………………………… 355
        16.3.1 传输协议选择 …………………………………………………… 355
        16.3.2 浏览器与 Web 服务器的应用协议 ……………………………… 355
        16.3.3 增加"生成配置文件"功能 ……………………………………… 355
    16.4 Web 服务器与浏览器系统实现 ………………………………………… 356
        16.4.1 应用协议的实现 ………………………………………………… 356
        16.4.2 Web 服务器的实现 ……………………………………………… 360

16.4.3　浏览器的实现 ································ 364
　　　16.4.4　"生成配置文件"功能实现 ···················· 370
16.5　Web 服务器与浏览器系统调试与软件发布 ················ 372
　　　16.5.1　系统调试 ···································· 372
　　　16.5.2　软件发布 ···································· 372
16.6　综合课程设计作业 ······································ 373

# 第 17 章　Java 与网络、数据库：基于 B/S 的用户登录管理系统 ··· 374

17.1　基于 B/S 的用户登录管理系统需求分析 ·················· 374
17.2　基于 B/S 的用户登录管理系统设计 ······················ 374
　　　17.2.1　数据库设计 ···································· 374
　　　17.2.2　登录服务器功能设计 ···························· 374
　　　17.2.3　客户端功能设计 ································ 375
17.3　基于 B/S 的用户登录管理系统实现思路 ·················· 375
　　　17.3.1　系统实现采用分层结构模型 ······················ 375
　　　17.3.2　客户端与服务器的应用协议 ······················ 376
　　　17.3.3　客户端的实现思路 ······························ 376
17.4　基于 B/S 的用户登录管理系统实现 ······················ 376
　　　17.4.1　建立数据库表和数据源 ·························· 376
　　　17.4.2　应用协议的实现 ································ 377
　　　17.4.3　登录服务器的实现 ······························ 379
　　　17.4.4　JApplet 的实现——LoginApplet 类 ··············· 386
17.5　基于 B/S 的用户登录管理系统调试与软件发布 ············ 391
　　　17.5.1　系统调试 ······································ 391
　　　17.5.2　软件发布 ······································ 391
17.6　综合课程设计作业 ······································ 392

# 第1部分

# Java编程基础

第 1 章　Java 开发环境
第 2 章　Java 语言基础
第 3 章　类与对象
第 4 章　继承、多态和封装
第 5 章　抽象类与接口
第 6 章　数组和常用类
第 7 章　集合类
第 8 章　异常
第 9 章　文件与流

# 第1章 Java开发环境

## 1.1 本章知识点

### 1. JDK

JDK(Java Development Kit)是 Oracle 公司推出的针对 Java 开发人员发布的免费软件开发工具包,是整个 Java 的核心,主要包括 Java 的类库、编译 Java 源代码的编译器、执行 Java 字节码的解释器、运行时环境,还有其他一些有用的工具。

### 2. JRE

JRE 是 Java Runtime Environment 的缩写,是 Java 程序的运行环境。JRE 的内部有一个 Java 虚拟机器(Java Virtual Machine,JVM)以及一些标准的类别函数库(Class Library)。它面向 Java 程序的使用者,而不是开发者。

### 3. UltraEdit

JDK 提供了 Java 程序的命令行编译和运行方式,没有提供一个集成开发环境(Integrated Development Environment,IDE)。从初学者角度来看,最好不要使用强大的 IDE,采用 JDK 开发 Java 程序能够很快理解程序中各部分代码之间的关系,有利于理解 Java 面向对象的设计思想。UltraEdit 是一套小巧且功能全面、灵活好用的文本编辑器,尤其适合于 Java 源代码的编写。它可以取代记事本,内建英文单词检查,它可以对 Java 的关键词进行识别,可以着色 Java 核心类中的部分类名,可以区分变量和字符串等。

### 4. Java API

API(Application Programming Interface)是应用程序编程接口。Java API 是为程序开发人员提供的、可直接调用的类库。Java 初学者最好将其下载到本机,在学习 Java 语言时要经常查看该文档。

### 5. Eclipse

Eclipse 最初是由 IBM 公司开发的软件产品,2001 年 11 月发布第一个版本,后来作为一个开源项目捐献给了开源组织。Eclipse 是一个优秀的集成开发环境,深受广大开发人员的青睐,应用非常广泛。Java语言学习到中后期时可以Eclipse为开发平台。

## 1.2 设计1 初识Java

### 1. 设计目的

(1) 掌握JDK的安装过程以及环境变量的配置。
(2) 掌握UltraEdit工具的安装。
(3) 掌握Java应用程序的编译和运行命令。
(4) 了解Java API docs的使用。
(5) 熟悉Eclipse集成开发环境(本部分可以在学期中后期进行)。

### 2. 设计要求

(1) 尝试进行JDK的安装,注意设置JDK的安装路径,并进行环境变量JAVA_HOME、Path和ClassPath的配置。

(2) 尝试进行UltraEdit的安装。

(3) 下面是一个简单的Java程序,功能是使用标准输出System.out的println()方法输出字符串。请为程序命名,并编译运行程序,观察程序的运行结果。

```
public class HelloWorld{
    public static void main(String [] args){
        System.out.println("Hi,你好,我要开始学习Java了。");
    }
}
```

### 3. 设计步骤

1) JDK的下载、安装

JDK安装程序的最新版本可以从网站http://www.oracle.com下载。在下载时要注意自己计算机的操作系统类型。下载的安装程序应当与自己计算机的操作系统相匹配,而且版本一般选择最新的。本书使用了JDK 1.6版本,文件名为"jdk-6u35-windows i586.exe"。双击该文件进行JDK的安装,安装的默认目录为"C:\Program Files\Java\jdk1.6.0_35\"。本教材中JDK的安装目录为"D:\Java\"。

2) JDK的环境配置

JDK安装完毕后,还需要进行系统环境变量JAVA_HOME、Path和ClassPath的配置,以便在Java程序编译和运行时,由系统自动定位Java类库和调试、运行等工具。环境变量JAVA_HOME指明JDK的安装路径,环境变量Path中包含JDK的Java开发工具(bin子目录)的路径,而ClassPath的值为Java类库的路径和程序需要使用的类的路径。

环境变量的配置可以在Windows图形界面中直接配置,具体方法是右击"我的电脑"图标,在弹出的快捷菜单中选择"属性"命令,打开"系统属性"对话框。打开"高级"选项卡,在出现的界面中单击"环境变量"按钮,弹出"环境变量"对话框,如图1-1所示。

在"系统变量"框中单击"新建"按钮,出现"新建系统变量"对话框,在"变量名"一栏的文本框中输入 JAVA_HOME,在"变量值"一栏的文本框中输入"d:\java"(即 JDK 安装的主目录),如图 1-2 所示。单击"确定"按钮,返回图 1-1。

图 1-1 "环境变量"对话框

图 1-2 "新建系统变量"对话框 1

在"系统变量"框中选中 Path,单击"编辑"按钮(如果没有 Path 变量,则单击"新建"按钮),出现"编辑系统变量"对话框,在"变量值"一栏的文本框中,在原有值的末尾加入";%JAVA_HOME%\bin",如图 1-3 所示。其中,分号用来分隔原来的路径和新加入的 Java 运行路径。单击"确定"按钮,返回图 1-1。

在"系统变量"框中单击"新建"按钮,出现"新建系统变量"对话框,在"变量名"一栏的文本框中输入 ClassPath,在"变量值"一栏的文本框中输入".;%JAVA_HOME%\lib",如图 1-4 所示。其中"."表示当前目录。单击"确定"按钮,返回图 1-1。

图 1-3 "编辑系统变量"对话框

图 1-4 "新建系统变量"对话框 2

三个环境变量设置完成后,选择开始菜单的"开始"|"运行",输入 cmd 命令,进入 DOS 窗口。在命令行提示符后输入 java 或 javac 并回车,如果出现其用法参数提示信息,则安装正确,如图 1-5 所示。

3) UltraEdit 下载、安装

UltraEdit 的下载、安装很简单,从相关网站下载 UltraEdit-32 软件,双击安装程序包,按照安装向导,分步单击"下一步"按钮就可完成安装。

4) Java 源程序的编辑、编译和运行

双击安装好的应用程序图标,在编辑窗口中输入 HelloWorld.java 源程序,如图 1-6 所示。将其存放到指定目录,例如 d:\example。

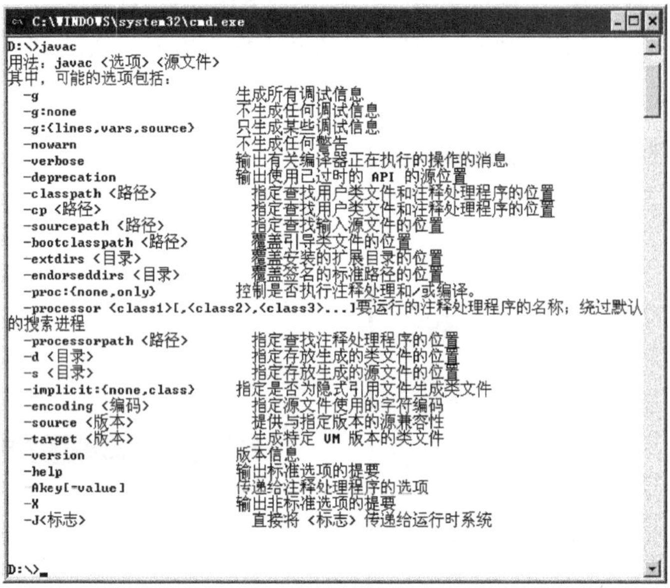

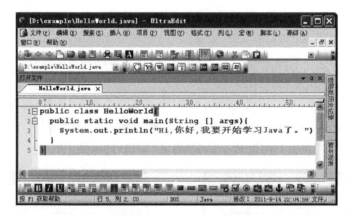

图1-5　JDK参数提示信息

图1-6　Java源程序的编辑

在DOS窗口中,通过控制台命令cd进入Java源程序HelloWorld.java所在的目录。然后通过编译命令javac HelloWorld.java来编译源程序。注意,被编译的文件名的后缀一定是".java",而且一定要写。

在编译完成后,可以进入执行阶段。通过命令java HelloWorld运行编译生成的".class"字节码文件,产生运行结果。这里需要注意的是,在输入运行程序的命令中,程序名HelloWorld一定不能含有任何后缀。整个过程如图1-7所示。

5) Java API

从网站"http://www.oracle.com"下载文件jdk-6u30-apidocs.zip并将其解压,打开其中文件夹api中的index.html文件(在IE浏览器中就可以打开),进入Java在线文档的主页面,如图1-8所示。这个主页面分成三个区域,在左上角的区域内,可以选择所有的类或者某个特定的包(Java语言中的包是一些类或接口的集合);在左下角的区域内,可以选择

图 1-7　Java 程序的编译、运行

具体的类或接口。这些类和接口是按字母排序的,因此查找很方便。在右边的区域内显示具体包或类或接口的详细说明。例如,通过这个页面,可以很方便地找到类 String 的详细说明,如图 1-8 所示。

图 1-8　在线帮助文档主页面

6) Eclipse 集成开发环境

Eclipse 可以在官方网站 http://www.eclipse.org 中下载其最新版本,下载后直接解压缩即可使用。Eclipse 无须安装,对资源要求低,简单易用而且免费。

双击 Eclipse 图标,就可以启动并运行 Eclipse 开发环境。

启动 Eclipse 即可进入如图 1-9 所示的界面,在该界面中选择 File→New→Java Project,新建一个项目。

在 Project name 中输入项目名,如图 1-10 所示,例如 FirstProject,单击 Finish 按钮关闭对话框,这样一个名为 FirstProject 的新项目就建成了。

接下来新建包。包是有效管理项目中很多个类的方式。操作如图 1-11 所示。

类似地,在包中创建类。例如在包 firstPac 中创建类,在 Name 中输入 HelloWorld,勾选 public static void main(String[] args)复选框,让 Eclipse 创建 main 方法,然后单击 Finish 按钮关闭对话框,如图 1-12 所示。

新建类后,即可在编辑器里编写相应代码,并运行写好的类,在该界面中选择 Run→Run,就可以看到程序的运行结果,如图 1-13 所示。

这样,在 Eclipse 环境中,编写和运行一个 HelloWorld 的 Java 应用程序就完成了。

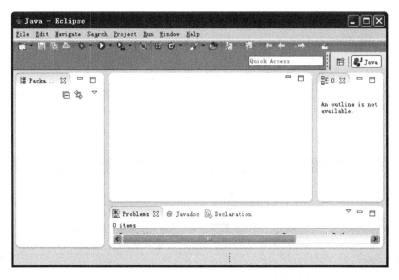

图 1-9　Eclipse 开发平台的界面

图 1-10　新建工程

图 1-11　新建"包"

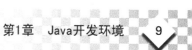

图 1-12　新建"类"

图 1-13　编辑程序并输出结果

# 第 2 章 Java语言基础

## 2.1 本章知识点

本章学习 Java 基础知识,包括 Java 基本数据类型、运算符以及流程控制语句。

### 1. 基本数据类型

Java 语言中,数据类型分为基本数据类型和引用数据类型,基本数据类型包括整型、浮点型、逻辑型和字符型;引用数据类型包括类、数组、枚举和接口。

1) 整型

Java 定义了 4 种表示整数的整型:字节型(byte)、短整型(short)、整型(int)、长整型(long)。每种整型的数据都是带符号位的,一个整数隐含为整型。

2) 浮点型

Java 定义了两种表示浮点数的浮点型:单精度浮点型(float)和双精度浮点型(double)。一个浮点数隐含为双精度浮点型。

3) 布尔型

布尔型用来表示逻辑值,它只有真(true)和假(false)两个值,这两个值不能转换成数字表示形式。

4) 字符型

字符型(char)用来存储字符,一个字符用一个 16 位的 Unicode 码来表示。

**注意**:基本数据类型按精度级别由低到高的顺序是:byte → short → int → long → float → double。将级别低的变量值赋给级别高的变量时,系统自动完成数据类型的转换;将级别高的变量值赋给级别低的变量时,必须使用显式类型转换运算。

### 2. 运算符与表达式

(1) 算术运算符:+、-、*、/、%、++、--。

(2) 关系运算符:>、<、>=、<=、==、!=。

(3) 逻辑运算符:&&、||、&、|、!、^。

(4) 位运算符:&、|、~、^、>>、<<、>>>。

(5) 赋值类运算符:=、+=、-=、*=、/=、%=、&=、|=、>>=、<<=、

>>>=。
(6) 条件运算符：?:。
(7) 其他运算符：[ ]、( )、new、instanceof、.。

### 3. 流程控制语句

按照程序的执行流程,程序的控制结构分为 3 种：顺序结构、分支结构和循环结构。
Java 提供了如下控制流程的语句。
分支语句：if,switch。
循环语句：for,while,do…while。
跳转语句：break,continue。

## 2.2 设计 1 基本运算练习

### 1. 设计目的

(1) 掌握基本数据类型及其之间的相互转换。
(2) 掌握 Java 语言的运算类型。

### 2. 设计要求

分析以下程序,得到该程序的运行结果。
程序清单 TypeConvert.java：

```java
public class TypeConvert {
    public static void main(String arg[]) {
        int i1 = 12;
        int i2 = 65;
        char c1 = 'A';
        float f1 = (float)((i1 + i2) * 1.0);
        double d1 = (i1 + c1) * 1.0;
        System.out.println("f1 = " + f1 + "; d1 = " + d1);
        byte b1 = 34;
        byte b2 = 78;
        byte b3 = (byte)(b1 + b2);
        System.out.println("b3 = " + b3);
        double d2 = 1e200;
        float f2 = (float)d2;
        System.out.println("f2 = " + f2);
        float f3 = 1.23f;
        long l1 = 123;
        long l2 = 30000000000L;
        float f4 = l1 + l2 + f3;
        long l3 = (long)f4;
        System.out.println("l3 = " + l3);
    }
}
```

### 3. 设计步骤

(1) 打开 UltraEdit 编辑软件,新建一个名为 TypeConvert.java 的文件,将其保存在 d:\example 目录下,在编辑窗口输入上述程序。

(2) 编译并运行程序,观察运行结果,如图 2-1 所示。

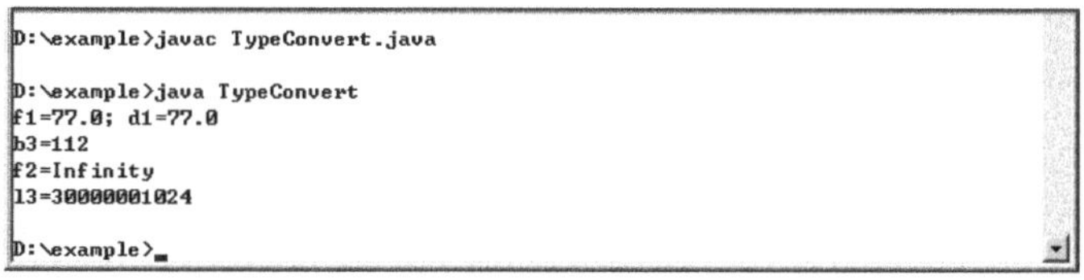

```
D:\example>javac TypeConvert.java

D:\example>java TypeConvert
f1=77.0; d1=77.0
b3=112
f2=Infinity
l3=30000001024

D:\example>
```

图 2-1 程序 TypeConvert.java 的运行结果

### 4. 拓展设计

针对上述程序,思考并设计以下问题:

(1) 将语句 float f1 = (float)((i1+i2)*1.0);中强制类型转换去掉,程序编译时提示怎样的错误?

(2) 将语句 byte b3 = (byte)(b1+b2);中强制类型转换去掉,程序编译时提示怎样的错误?

(3) 在语句 byte b1 = 34;中,34 是一个整型值,为什么前面不加强制转换符(int)?

(4) 在语句 float f2 = (float)d2;中,强制转换符(float)能去掉吗?为什么?

(5) 在 main()方法中追加语句 System.out.println((f3==1.23) && (i1<i2));,程序运行结果如何?

### 5. 设计提示

(1) float 类型强制转换成 long 类型时,将损失数据精度,即舍去小数部分。

(2) 计算机在表示浮点数以及浮点数运算时均存在误差,因此,一般建议不直接比较两个浮点数是否相等。

## 2.3 设计 2 控制结构练习

### 1. 设计目的

(1) 理解程序设计三种结构:顺序、选择和循环。
(2) 掌握 break、continue 的用法。
(3) 掌握 Java 语言的控制语句,并且可以根据具体情况灵活使用它们。

### 2. 设计要求

分析以下程序,得到该程序的运行结果。
程序清单 ForWhile.java:

```java
public class ForWhile{
    public static void main(String[] args) {
        for (int i = 101; i < 200; i += 2){
            boolean f = true;
            int j = 2;
            while(j < i){
                if (i % j == 0){
                    f = false;
                    break;
                }
                j++;
            }
            if (!f) {continue;}
            System.out.print(" " + i);
        }
    }
}
```

**3. 设计步骤**

(1) 打开 UltraEdit 编辑软件,新建一个名为 ForWhile.java 的文件,将其保存在 d:\example 目录下,在编辑窗口输入上述程序。

(2) 编译并运行程序,观察运行结果,该程序功能为求 100～200 之间的素数,程序运行结果如图 2-2 所示。

```
D:\example>javac ForWhile.java

D:\example>java ForWhile
 101 103 107 109 113 127 131 137 139 149 151 157 163 167 173 179 181 191 193 197
 199
D:\example>_
```

图 2-2　程序 ForWhile.java 的编译运行结果

**4. 拓展设计**

(1) for 语句和 while 语句是等价的,试用嵌套的 for 循环语句改写上述程序。

(2) 将程序的运行结果改为以每行 5 个数据的形式输出。

(3) 如果去掉 if (!f) {continue;} 语句,程序运行结果将如何?

(4) 将程序中的 while 语句改用 do…while 语句实现。

(5) 增加语句,统计运行结果中个位分别是 1、3、7、9 的素数的个数。

(6) 打印出 100～999 之间所有的"水仙花数",所谓"水仙花数"是指一个三位数,其各位数字立方和等于该数本身。例如 153 是一个"水仙花数",因为 $153 = 1^3 + 5^3 + 3^3$。

**5. 设计提示**

(1) for 语句和 while 语句的等价关系表示如下:

for (初始化表达式; 条件表达式; 更新表达式)
　　语句或语句块

改写为：

```
初始化表达式;
while (条件表达式){
    语句或语句块
        更新表达式;
}
```

(2) 对于拓展设计(5)，可以通过 switch 语句来统计运行结果中个位分别是 1、3、7、9 的素数的个数。在原程序的 for 循环语句前定义 4 个整型变量 a、b、c 和 d，用来统计个位分别是 1、3、7、9 的素数的个数，将语句 switch (i%10){

```
case 1: a++;break;
case 3: b++;break;
case 7: c++;break;
case 9: d++;break;}
```

追加到 System.out.print(" " ＋ i);语句之后，在 for 循环语句结束后执行输出语句

```
System.out.println("个位是 1 的有"+a+"个; ");
System.out.println("个位是 3 的有"+b+"个; ");
System.out.println("个位是 7 的有"+c+"个; ");
System.out.println("个位是 9 的有"+d+"个; ");
```

(3) 对于拓展设计(6)，可以利用 for 循环控制 100～999 之间的数，每个数分解出个位、十位、百位。然后求这三个数的立方和，并判断该和是否与原数相等。参考程序如下：

```
class DaffodilNum   {
    public static void main(String[] args) {
        int a = 0,b = 0,c = 0;
        for(int num = 100;i<=999;i++){
            a = num / 100;
            b = num % 100 / 10;
            c = num % 10;
            if(Math.pow(a,3) + Math.pow(b,3) + Math.pow(c,3) == num){
                System.out.println(num);
            }
        }
    }
}
```

# 第3章 类与对象

##  3.1 本章知识点

### 1. 类和对象

类是用来创建对象的模板,是一组相似的对象实体共性的抽象。它包含被创建对象的属性和功能,是 Java 语言程序设计的基本单位,Java 的源文件就是由若干个形式上相互独立的类构成的。类中包含两大成员,其一是成员变量,也称为类的属性,不同的属性值使得同一类中的对象相互区别。其二是成员方法,用来体现类的功能,是类中用来完成某个任务的一组相关的代码。对象是类的实例化,或者说具体化。

创建一个对象,就是创建一个类的实例,通过 new 运算符和类的构造方法来实现。类的构造方法不同于成员方法,其名称与类名完全相同,且没有返回值。

类定义时,类体中一般先定义成员变量,接着编写构造方法,最后是成员方法。类定义中可以缺省构造方法,这时系统会为该类定义一个不含任何参数的构造方法。一旦在类中定义了构造方法,系统不会再创建这个默认的不含参数的构造方法。

### 2. 变量

Java 语言中的变量按作用域分为成员变量和局部变量两种,成员变量指类的静态变量和实例变量,局部变量指方法中定义的变量和参数。成员变量的作用域是整个类,而局部变量的作用域仅限于方法,即从方法的左括号开始直到右括号结束。

当方法中的局部变量(包括参数)与成员变量同名时,则成员变量被隐藏,如果要将该成员变量显露出来,则需在该变量前加上修饰符 this。

成员变量中的静态变量也称为类变量,是与类相关联的数据变量,实例变量仅仅与对象相关联。不同对象的实例变量将分配不同的内存空间,而所有对象的类变量都分配相同的内存,也就是说,一个类的所有对象共享该类的类变量。

### 3. 包

Java 将程序中的类以包的形式组织起来,便于管理,而且可以减少命名冲突。用户可以通过 package 语句自定义包或使用 import 声明导入其他的包,从而直接使用其中的类。

## 3.2 设计1 对象的创建和使用

**1. 设计目的**

（1）理解类的定义。
（2）掌握对象的声明。
（3）学会使用构造方法创建对象。
（4）掌握对象的使用。

**2. 设计要求**

（1）定义一个 Cat 类，该类有三个属性，分别是名字、年龄、颜色。定义构造方法以初始化类的属性，并定义方法获取 Cat 对象的属性信息。另外，Cat 类还可以有 speak()方法。
（2）定义一个测试类，对 Cat 类进行实例化，并调用相应方法来进行功能测试。

**3. 设计步骤**

1）分析

类 Cat 的三个属性：名字、年龄和颜色，可分别定义为 String、int、String 类型，构造方法需要带有三个参数，分别用于接收名字、年龄和颜色。定义获取 Cat 对象的属性信息的成员方法时可以使用 return 语句，而方法的返回值类型为相应的属性类型。成员方法 speak()中可以有一条输出语句，用来在屏幕显示猫的叫声。

测试类中只定义 main()方法，首先定义并构造一个 Cat 对象，然后调用相应的成员方法以实现功能测试。

2）编写程序

程序清单 CatTest.java：

```java
class Cat {
    String name;
    int age;
    String colour;
    Cat(String a_name, int a_age ,String a_colour) {
        name = a_name;
        age = a_age;
        colour = a_colour;
    }
    public String getName(){
        return name;
    }
    public int getAge(){
        return age;
    }
    public String getColour(){
        return colour;
```

```
        }
        public void speak(){
            System.out.println("喵喵…");
        }
    }
    public class CatTest{
        public static void main(String arg[]){
            Cat mimi = new Cat("mimi",3,"white");
            Cat jiafei = new Cat("jiafei",5,"yellow");
            System.out.println(mimi.getName() + " is " + mimi.getColour());
            System.out.println(jiafei.name + " is " + jiafei.getAge());
            mimi.speak();
            System.out.println("听,有声音");
        }
    }
```

3）编译并运行程序

结果如图 3-1 所示。

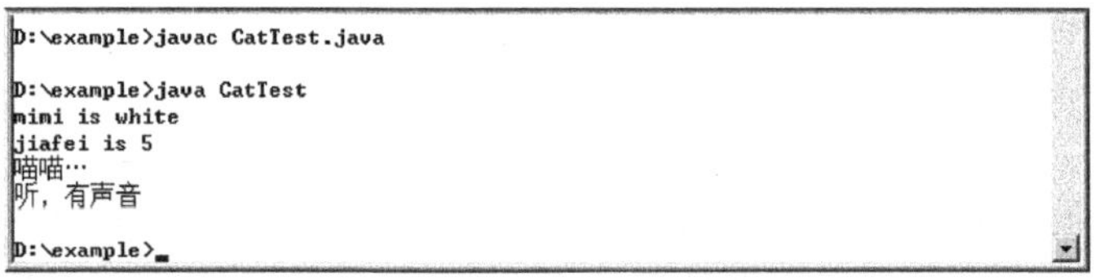

```
D:\example>javac CatTest.java

D:\example>java CatTest
mimi is white
jiafei is 5
喵喵…
听, 有声音

D:\example>
```

图 3-1  程序 CatTest.java 的编译运行结果

### 4．拓展设计

（1）在原程序基础上,增加"体重"属性,并增加获取该属性的方法。

（2）将语句 Cat mimi = new Cat("mimi",3,"white");改写为 Cat mimi = new Cat();,程序编译时将会出现什么情况？试分析之。

（3）去掉构造方法,程序编译能通过吗？为什么？

（4）将 speak()方法的声明部分改为 public static void speak(),并且将语句 mimi.speak();改为 Cat.speak(),程序能否编译、运行？

（5）将构造方法中的参数依次改名为 name、age、colour 后,程序编译能否通过？运行时会出现什么情况？应如何修改程序？

（6）按下列要求完成程序：

定义一个类点(Point),具有两个属性 x 和 y,其构造方法带有两个参数,四个成员方法分别用来设置和获取两个属性值。

定义一个类圆(Circle),具有两个属性 o 和 r,其中 o 是 Point 类型。两个构造方法：一个带有两个参数,分别给 o 和 r 传值；另一个构造方法为一个参数,用来给 r 传值,默认圆心在原点处。六个成员方法：其中四个分别用来设置和获取两个属性值,一个方法用来求圆的面积,最后一个方法用来判断一个点(Point 对象)是否在圆(Circle 对象)内。

定义一个测试类 Test,包括一个 main 方法,用来测试满足上述要求的程序。

### 5. 设计提示

(1) 增加"体重"属性：double weight;。

增加获取"体重"属性的方法：

```
public double getWeight(){
    return weight;
}
```

(2) 将语句 Cat mimi = new Cat("mimi",3,"white");改写为 Cat mimi = new Cat();，程序编译时将会出现如图 3-2 所示的情况。

```
D:\example>javac CatTest.java
CatTest.java:25: 找不到符号
符号： 构造函数 Cat()
位置： 类 Cat
        Cat mimi = new Cat();
                   ^
1 错误

D:\example>
```

图 3-2　拓展设计(2)的编译结果

(3) 去掉构造方法，程序编译将不能通过。

(4) 将 speak()方法的声明部分改为 public static void speak()，并且将语句 mimi.speak();改为 Cat.speak();，程序能编译、运行。因为此时的方法为静态方法，静态方法可以通过类名.方法名来引用。

(5) 将构造方法中的参数依次改名为 name、age、colour 后，程序编译能通过，但运行时会出现如图 3-3 所示的结果。因为构造方法中的参数与成员变量同名，将会在构造方法中隐藏成员变量，达不到给成员变量赋值的目的，所以应使用 this 关键字。

```
D:\example>javac CatTest.java

D:\example>java CatTest
null is null
null is 0
喵喵…
听，有声音

D:\example>
```

图 3-3　拓展设计(5)的编译运行结果

(6) 拓展设计(6)参考程序如下：

```
class Point {
    double x;
    double y;
    Point(double x1, double y1) {
        x = x1;
        y = y1;
    }
    double getX() { return x; }
    double getY() { return y; }
    void setX(double i) { x = i; }
    void setY(double i) { y = i; }
}
```

```java
class Circle {
    Point o;
    double r;
    Circle(Point p, double r) {
        o = p;
        this.r = r;
    }
    Circle(double r) {
        o = new Point(0.0, 0.0);
        this.r = r;
    }
    void setO(double x, double y) {
        o.setX(x);
        o.setY(y);
    }
    Point getO() { return o; }
    double getR() { return r;}
    void setR(double r) { this.r = r;}
    double area() {   return 3.14 * r * r;}
    boolean contains(Point p) {
        double x = p.getX() - o.getX();
        double y = p.getY() - o.getY();
        if(x * x + y * y > r * r)
          return false;
        else return true;
    }
}
public class Test {
    public static void main(String args[]) {
        Circle c1 = new Circle(5.0);
        Circle c2 = new Circle(new Point(3.0,2.0), 2.0);
        System.out.println("c1:(" + c1.getO().getX() + ","
            + c1.getO().getY() + ")," + c1.getR());
        System.out.println("c2:(" + c2.getO().getX()
            + "," + c2.getO().getY() + ")," + c2.getR());
        System.out.println("c1 的面积是: " + c1.area());
        System.out.println("c2 的面积是: " + c2.area());
        Point p1 = new Point(3.2, 2.8);
        System.out.println(c1.contains(p1));
        System.out.println(c1.contains(new Point(10.9,13.0)));
    }
}
```

## 3.3 设计 2　包的使用与访问控制

### 1. 设计目的

（1）理解 Java 包的组织结构。
（2）学会编写带有包结构的程序。

（3）掌握包结构下的成员访问控制。

**2．设计要求**

编写三个类，类名分别为 Cat、Mouse、CatMouse，其中类 Cat 放在包 packCat 中，而类 Mouse 则放在包 packMouse 中，包 packCat 和包 packMouse 放在同一个目录 d:\example 下。类 CatMouse 置于 d:\example 下。

**3．设计步骤**

（1）首先在 d:\example 目录下创建 Mouse.java 程序。编译该程序,将生成的 Mouse.class 文件放在 d:\example\pcakMouse 目录下。

程序清单 Mouse.java：

```java
package packMouse;
public class Mouse{
    public String name;
    public Mouse(String str){
        name = str;
    }
    public void run(){
        System.out.println("I can run!");
    }
}
```

（2）在 d:\example 目录下创建 Cat.java 程序。编译该程序,将生成的 Cat.class 文件放在 d:\example\pcakCat 目录下。

程序清单 Cat.java：

```java
package packCat;
import packMouse.*;
public class Cat {
    public String name;
    public Cat(String a_name) {
        name = a_name;
    }
    public void catchMouse(Mouse m){
        m.run();
        System.out.println("haha...I catch a mouse!");
    }
}
```

（3）在 d:\example 目录下创建一个名叫 CatMouse.java 的程序。在编译该程序之前，将当前目录下的 Cat.java 和 Mouse.java 文件移走。编译 CatMouse.java 生成 CatMouse.class 文件放在 d:\example 目录下。

程序清单 CatMouse.java：

```java
import packCat.*;
import packMouse.*;
```

```
public class CatMouse{
    public static void main(String args[]){
        Cat tom = new Cat("Tom");
        Mouse jack = new Mouse("Jack");
        jack.run();
        tom.catchMouse(jack);
    }
}
```

（4）在 DOS 窗口下运行程序，结果如图 3-4 所示。

```
D:\example>javac CatMouse.java

D:\example>java CatMouse
I can run!
I can run!
haha…I catch a mouse!

D:\example>
```

图 3-4  程序 CatMouse.java 的编译运行结果

### 4．拓展设计

（1）针对上述设计中的第（3）步，如果不把 d:\example 目录下的 Cat.java 和 Mouse.java 文件移走，将会出现什么情况？

（2）试对 Cat.java 程序做如下修改：

```
package packCat;
public class Cat {
    public String name;
    public Cat(String a_name) {
        name = a_name;
    }
    public void catchMouse(packMouse.Mouse m){
        m.run();
        System.out.println("haha…I catch a mouse!");
    }
}
```

然后编译运行，观察结果是否正确。

（3）在 d:\example 目录下使用命令 md 创建两个子目录 sourse 和 dest，将上述设计中三个源文件置于 d:\example\sourse 目录下。进入该目录，使用命令 javac -d d:\example\dest Cat.java Mouse.java 编译 Cat.java 和 Mouse.java 文件，可以看到在 dest 子目录下生成了两个子目录 packCat 和 packMouse，分别存放着 Cat.class 和 Mouse.class 两个类文件。此时，在 d:\example\sourse 目录下将 Cat.java 和 Mouse.java 文件移走，编译 CatMouse.java 文件，将会出现什么情况？

（4）在拓展设计（3）中，已经在 d:\example\dest 下生成了两个子目录 packCat 和 packMouse，分别存放着 Cat.class 和 Mouse.class 两个类文件。将这两个类文件打包，在 d:\example\dest 下使用命令 jar -cvf test.jar *.* 将会生成一个 test.jar 文件，将该文件及其所在路径即 d:\example\dest\test.jar 加入到环境变量 classpath 中，编译并运行以下

程序：

```java
import packCat.*;
public class Dog{
    void makeFriend(Cat c){
        System.out.println("my friend is " + c.name);
    }
    public static void main(String args[]){
        Cat c = new Cat("jiafei");
        Dog d = new Dog();
        d.makeFriend(c);
    }
}
```

**5. 设计提示**

（1）用 import 语句可以导入所需的类，如果不想使用 import 语句，那么在使用包中的类时就必须带上完整的包路径。

（2）在拓展设计（3）中，将会出现编译异常，找不到包 packCat 和 packMouse，这时一定切记要修改环境变量 classpath，增加路径"d:\example\dest"。

（3）使用 jar 命令将类打包生成.jar 文件，其他程序就可以引入 jar 包中的类并使用。

# 第4章 继承、多态和封装

 **4.1 本章知识点**

### 1. 继承

面向对象技术的三大特性分别是继承性、多态性和封装性。继承性是软件重用的一种形式,通过继承,一个子类可以具有它的父类的属性和方法。

子类除了继承其父类的属性和方法外,还可对其父类进行扩展,创建子类自己的属性和方法。一个类可能是另一个类的子类,但同时又可能成为其他类的父类,这样一来,就形成了一种类的层次结构。

**注意**:Java 只支持单重继承(指一个类只能有一个直接父类),而不支持多重继承。

子类可以继承父类的所有成员,然而构造方法却不能被继承。因此,子类要有自己的构造方法,子类的构造方法分为两部分,一部分用于构造父类的成员,另一部分用于构造子类自己的成员。对父类部分的构造要使用 super 关键字。

### 2. 多态

多态性也是实现软件可重用性的手段之一。多态性包括两种类型:静态多态性和动态多态性。

静态多态性指在同一个类中同名方法在功能上的重载(overload)。在方法声明的形式上要求同名的方法具有不同的参数列表。

**注意**:

(1) 这里的方法可以是成员方法也可以是构造方法。

(2) 不同的参数列表指的是方法的参数个数不同、参数的数据类型不同或者参数的数据类型排列顺序不同。

动态多态性指在子类和父类的类体中均定义了具有基本相同声明的非静态成员方法。这时也称为子类的成员方法对其父类基本相同声明的成员方法的重写(override)。

**注意**:

(1) 基本相同声明的成员方法要求子类的成员方法和其父类对应的成员方法具有相同的方法名,相同的参数个数,对应参数的类型也相同,而且子类的成员方法应当比其父类对应原成员方法具有相同或更广的访问控制方式。

(2)方法重写与方法重载是不同的,重载是发生在一个类中有多个同名但参数不同的方法时,而重写是发生在当子类和父类的方法具有完全相同的名字、参数和返回类型时,从而子类将父类中的方法屏蔽掉。

### 3. 封装

在Java语言中,封装性是通过访问控制来实现的,如果一个类不是内部类,则该类的访问控制方式有两种:公共模式(public)和默认模式(default)。类的成员的访问控制方式有4种:公共模式(public)、保护模式(protected)、默认模式(default)和私有模式(private)。表4-1给出了类成员的访问控制模式及其允许访问范围,其中空白的单元格表示不允许访问。

表 4-1 类成员的访问控制模式及其允许访问范围

| 修饰符 | 同一个类内 | 同一个包内 | 子类 | 所有类 |
| --- | --- | --- | --- | --- |
| public | 允许访问 | 允许访问 | 允许访问 | 允许访问 |
| protected | 允许访问 | 允许访问 | 允许访问 | |
| default | 允许访问 | 允许访问 | | |
| private | 允许访问 | | | |

## 4.2 设计1 继承性

### 1. 设计目的

(1)理解继承的概念。
(2)学习创建子类对象。
(3)掌握成员变量的继承与隐藏。
(4)掌握方法重写的使用。

### 2. 设计要求

编写一个Java程序,该程序中有三个类:类Person是一个父类,类Student和类Teacher是类Person的子类,类Person中包含有姓名、年龄等属性,以及显示信息等方法。类Student除了具有类Person的属性外,还有自己的属性如成绩,同时有自己的显示成绩的方法。类Teacher除了具有类Person的属性外,还有自己的属性如工龄、工资等,同时还有自己的显示工龄、获取工资等方法。最后程序中应包含一个主类来使用上述三个类并显示它们的信息。

### 3. 设计步骤

1)分析

类Student和类Teacher作为类Person的子类,有自己特有的属性和方法,所以在定义时需通过extends语句继承父类Person,而在类体中只定义自己特有的属性和方法。

2) 编写程序

程序清单 TestPerson.java：

```java
class Person {
    String name;
    int age;
    void showInfo() {
        System.out.println("name:" + name);
        System.out.println("age:" + age);
    }
    void setAge(int age) {
        this.age = age;
    }
    String getName(){
        return name;
     }
}
class Student extends Person {
    int score;
    Student(String name, int age, int score){
        this.name = name;
        this.age = age;
        this.score = score;
    }
     void showScore() {
        System.out.println("score:" + score);
    }
}
class Teacher extends Person {
    int workYear;
    int salary;
    Teacher(String name, int age, int workYear, int salary){
        this.name = name;
        this.age = age;
        this.workYear = workYear;
        this.salary = salary;
    }
    int getSalary() {
        return salary;
    }
    void showWorkYear() {
        System.out.println("workYear:" + workYear);
    }
}
public class TestPerson {
    public static void main(String arg[]){
        Person p = new Person();
        Student s = new Student("Mary",18,88);
        Teacher t = new Teacher("Lee",40,13,5000);
        s.showInfo();
```

```
            s.showScore();
            t.showInfo();
            System.out.println("Teacher's salary is " + t.getSalary());
            t.showWorkYear();
        }
    }
```

3）编译并运行程序

结果如图 4-1 所示。

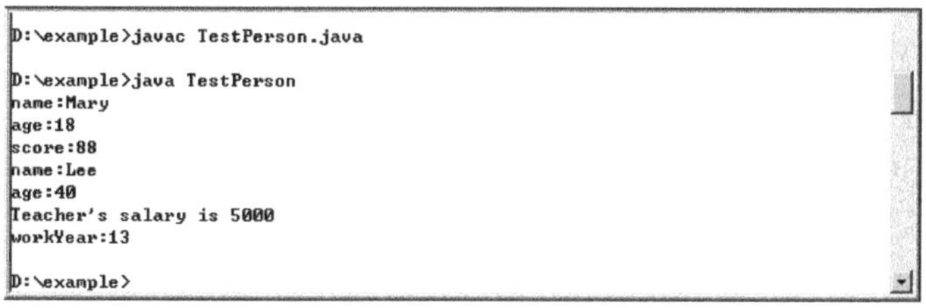

```
D:\example>javac TestPerson.java

D:\example>java TestPerson
name:Mary
age:18
score:88
name:Lee
age:40
Teacher's salary is 5000
workYear:13

D:\example>
```

图 4-1 程序 TestPerson.java 的编译运行结果

### 4．拓展设计

（1）在原程序 TestPerson.java 的基础上，去掉类 Student 中的构造方法，同时将 main()方法中的语句 Student s = new Student("Mary",18,88);改为 Student s = new Student();，程序编译会通过吗？为什么？

（2）在原程序 TestPerson.java 的基础上，类 Teacher 中的构造方法中增加 super();，并将其作为构造方法的第一条语句，程序编译会通过吗？为什么？

（3）在原程序 TestPerson.java 的基础上，类 Person 中增加一个构造方法：

```
Person(String name){
    this.name = name;
}
```

程序编译时将会出现什么情况？

（4）在原程序 TestPerson.java 的基础上，类 Teacher 中的构造方法中增加语句：

```
super();
```

并将其作为构造方法的第一条语句，类 Person 中增加一个构造方法

```
Person(String name){
    this.name = name;
}
```

程序编译时将会出现什么情况？

（5）在原程序 TestPerson.java 的基础上，进行如下操作：

① 类 Person 中增加成员变量：

```
String job;
```

② 同时在其成员方法 showInfo()中增加语句：

`System.out.println("job:" + job);`

③ 类 Teacher 中增加成员变量：

`String job = "Teacher";`

④ main()方法中第 7 行插入语句：

`System.out.println("Teacher's job is " + t.job);`

⑤ 编译并运行程序，其结果如图 4-2 所示。

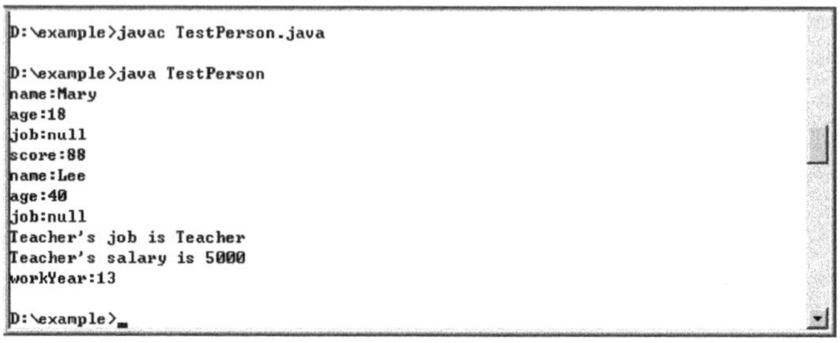

```
D:\example>javac TestPerson.java

D:\example>java TestPerson
name:Mary
age:18
job:null
score:88
name:Lee
age:40
job:null
Teacher's job is Teacher
Teacher's salary is 5000
workYear:13

D:\example>
```

图 4-2 拓展设计(5)的运行结果

观察运行结果的第 7、8 行，仔细体会子类对父类中同名的成员变量的隐藏。

（6）在类 Teacher 中增加两个构造方法，一个为无参的构造方法，另一个为只有一个参数（为成员变量 name 传值）的构造方法。

（7）在类 Student 和类 Teacher 中分别重写父类的方法 showInfo()，以实现各类中所有属性信息的输出。

### 5. 设计提示

（1）在类的继承体系中，子类的构造过程一定会调用其父类的构造方法。因此，如果子类的构造方法中第一条语句显式调用父类的构造方法，则父类中必须有相应的构造方法与之匹配，否则编译将出错。

（2）子类可以定义与父类同名的成员变量，这样，父类中同名的成员变量将会在子类中被隐藏。

（3）如果子类可以继承父类的方法，子类就有权利重写这个方法，子类通过重写父类的方法可以改变方法的具体行为。

（4）子类在重写父类的方法时，一定要保证方法的名字、类型、参数个数和类型与父类的相应方法完全相同，只有这样，父类中被重写的方法才会在子类中被隐藏。

## 4.3 设计 2 多态性

### 1. 设计目的

（1）掌握多态性在继承中的运用。

(2) 理解动态绑定的含义。

(3) 掌握上转型(upcasting)对象的使用。

**2．设计要求**

编写一个 Java 程序,该程序中有三个类:类 Animal 是一个父类,包含有名字属性,以及 move()、speak()等方法。类 Bird 和类 Dog 是类 Animal 的子类,类 Bird 除了具有类 Animal 的属性外,还有自己的属性如羽毛颜色,同时有自己的 move()、speak()方法。类 Dog 除了具有类 Animal 的属性外,还有自己的属性如皮毛颜色,同时还有自己的 move()、speak()方法。最后程序中应包含一个主类来使用上述三个类并显示它们的信息。

**3．设计步骤**

1) 分析

类 Bird 和类 Dog 作为类 Animal 的子类,除有自己的属性外,还有不同于父类的 move()、speak()方法,所以在定义时需重写这两个方法。

2) 编写程序

程序清单 TestPolymoph.java:

```java
class Animal {
    String name;
    Animal(String name) {
        this.name = name;
    }
    public void speak(){
        System.out.println("叫声…");
    }
    public void move(){
        System.out.println("I can move");
    }
}
class Dog extends Animal {
    String furColor;
    Dog(String name,String furColor) {
        super(name);
        this.furColor = furColor;
    }
    public void speak() {
        System.out.println("狗叫声:汪汪…");
    }
    public void move(){
        System.out.println("I can run quickly.");
    }
}
class Bird extends Animal {
    String featherColor;
    Bird(String name,String featherColor) {
        super(name);
```

```java
            this.featherColor = featherColor;
        }
        public void speak() {
            System.out.println("鸟叫声:叽叽喳喳…");
        }
        public void move(){
            System.out.println("I can fly freely in the sky.");
        }
    }
    class Boy {
        String name;
        Animal pet;
        Boy(String name,Animal pet) {
            this.name = name;
            this.pet = pet;
        }
        public void trainPet(){
            System.out.println("I'm " + name + ",I'm trainning " + pet.name);
            pet.speak();
            pet.move();
        }
    }
    public class TestPolymoph {
        public static void main(String args[]){
            Bird bird = new Bird("littltbird","colorful");
            Dog dog = new Dog("littledog","black");
            Boy boy1 = new Boy("dongdong",dog);
            Boy boy2 = new Boy("nannan",bird);
            boy1.trainPet();
            boy2.trainPet();
        }
    }
```

3）编译并运行程序

结果如图 4-3 所示。

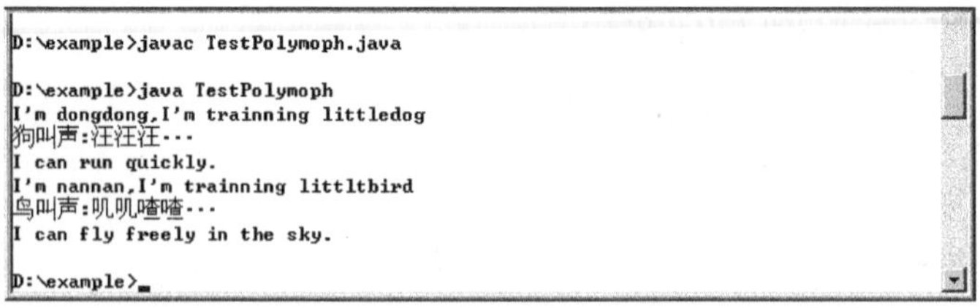

```
D:\example>javac TestPolymoph.java

D:\example>java TestPolymoph
I'm dongdong,I'm trainning littledog
狗叫声:汪汪汪…
I can run quickly.
I'm nannan,I'm trainning littltbird
鸟叫声:叽叽喳喳…
I can fly freely in the sky.

D:\example>
```

图 4-3 程序 TestPolymoph.java 的编译运行结果

### 4．拓展设计

（1）类 Bird 如果不重写 speak()方法和 move()方法,程序编译能通过吗？运行结果会如何？

（2）将原程序文件 TestPolymoph.java 中的各个类以独立的文件形式存储,并进行编译、运行。

(3) 在拓展设计(2)的基础上,类 Animal 中增加一个方法:

```
void printName(){
    System.out.println("My name is " + name);
}
```

类 TestPolymoph 的 main()方法中增加语句:

```
bird.printName();
dog.printName();
```

编译 Animal.java 文件和 TestPolymoph.java 文件,运行程序。

(4) 在拓展设计(3)的基础上,新建一个文件,创建一个 Cat 类,它也继承自类 Animal,同样重写类 Animal 的 speak()方法和 move()方法。在 main()方法中实例化一个 Cat 对象,将其作为 boy1 的 pet。编译 Cat.java 文件和 TestPolymoph.java 文件,运行程序。体会多态性带来的灵活性和程序的可扩展性,它使我们能够将发生改变的东西与没有发生改变的东西分开。

(5) 在原程序 TestPolymoph.java 的基础上,在 main()中增加语句:

```
Animal a = dog;
System.out.println("My name is " + a.name);
System.out.println("My name is " + a.furColor);
```

程序编译时会出现什么错误?分析并解释原因。

(6) 在原程序 TestPolymoph.java 的基础上,在 main()中增加语句:

```
Animal a = dog;
bird = (Bird)a;
System.out.println( "My name is " + bird.name);
```

程序编译时会通过吗?运行时会出现什么情况?

### 5. 设计提示

(1) 在 Java 中父类定义一个通用的方法,其不同的实现则由它的不同的子类来完成。当通过对象来调用这样一个方法时,Java 会依据运行时该对象的实际类型来决定用方法的哪一个版本来执行,也即动态地绑定一个方法来执行。而动态绑定的前提是 Java 允许将子类对象赋给父类类型的变量。

利用多态性,可以将一个方法的形参设为父类类型,而将来传递给该方法的实参既可以是父类对象也可以是其子类对象。具体执行时,Java 会自动根据实际对象所属的类型来选择方法执行,因此多态性增强了编程的灵活性和系统的可扩展性。

(2) 在拓展设计(5)中,要想输出 Animal 类型的变量 a 所指向的 Dog 对象的 furColor 属性,考虑使用如下语句:

```
if(a instanceof Dog){
    Dog d = (Dog)a;
    System.out.println("My name is " + d.furColor);
}
```

## 4.4 设计 3 封装性

**1. 设计目的**

（1）巩固带有包结构的程序。
（2）巩固 Java 语言的继承性。
（3）掌握访问控制的应用。

**2. 设计要求**

定义一个类 A，该类中包含 4 个具有不同访问控制权限的成员变量和 4 个不同访问控制权限的成员方法。设计实验分别验证类的两种访问控制权限和类成员的 4 种访问控制权限。

**3. 设计步骤**

（1）按要求定义类 A。

程序清单 A.java：

```java
package A;
public class A {
    private String privateVar = "private";
            String defaultVar = "default";
    protected String protectedVar = "protected";
    public String publicVar = "public";
    private void privateMethod(){
        System.out.println(privateVar);
    }
    void defaultMethod() {
        System.out.println(defaultVar);
    }
    protected void protectedMethod(){
        System.out.println(protectedVar);
    }
    public void publicMethod(){
        System.out.println(publicVar);
    }
}
```

（2）在类 A 中增加 main()方法，以验证具有不同访问控制权限的各成员在本类中的应用。（注意编译生成的类 A.class 的路径）

```java
public static void main(String [] args){
    A aa = new A();
    aa.privateMethod();
    aa.defaultMethod();
    aa.protectedMethod();
```

```
        aa.publicMethod();
        System.out.println(aa.privateVar);
        System.out.println(aa.defaultVar);
        System.out.println(aa.protectedVar);
        System.out.println(aa.publicVar);
    }
```

(3) 在步骤(1)的基础上,在同一个包中新建一个类 Test,类中仅有步骤(2)中的main()方法,观察程序编译结果,如图 4-4 所示。

图 4-4　设计步骤(3)的编译结果

(4) 将类 Test 置于包 B 中,观察程序的编译结果,如图 4-5 所示。

图 4-5　设计步骤(4)的编译结果

(5) 将类 Test 作为类 A 的子类置于包 B 中,并将 main()方法中的第一条语句改为:

```
Test aa = new Test();
```

观察程序编译结果,如图 4-6 所示。

### 4. 拓展设计

(1) 修改上述设计步骤(3)、(4)、(5)中编译发现的错误,继续运行程序得到正确结果。

(2) 将类 A 的访问控制修饰符 public 去掉,进行设计内容的步骤(3)、(4),观察编译结果。

### 5. 设计提示

如何确定成员应当具有的访问控制模式？这里给出一个基于程序安全性的设计方案。其具体方法是先要确定允许访问该成员的类有哪些,这些类与该成员所在的类是什么关系,

```
D:\example>javac Test.java
Test.java:6: 找不到符号
符号： 方法 privateMethod()
位置： 类 B.Test
                aa.privateMethod();
                  ^
Test.java:7: 找不到符号
符号： 方法 defaultMethod()
位置： 类 B.Test
                aa.defaultMethod();
                  ^
Test.java:10: privateVar 可以在 A.A 中访问 private
            System.out.println(aa.privateVar);
                                 ^
Test.java:11: defaultVar 在 A.A 中不是公共的；无法从外部软件包中对其进行访问
            System.out.println(aa.defaultVar);
                                 ^
4 错误

D:\example>_
```

图 4-6　设计步骤(5)的编译结果

如是否具有继承关系,或者是否在同一个包内。然后,确定能够刚好满足允许访问方式的访问模式。如果采用默认模式就能满足所需要的访问方式,则不应当采用保护模式。这样才能在访问模式方面最大限度地保证程序的安全性和鲁棒性。另外,一般将类的构造方法的访问控制方式设置成公共模式,将有特殊限制的成员域的访问控制方式设置成私有模式。

# 第 5 章 抽象类与接口

## 5.1 本章知识点

### 1. 抽象类

抽象类是类体中包含有抽象方法的类,抽象类只定义它的所有子类共享的通用格式,而其实现的细节由具体子类来完成。抽象类在声明时需要 abstract 修饰符。

抽象类中的抽象方法由子类在继承时实现,当一个子类继承一个抽象类时,它必须实现父类中所有的抽象方法,否则这个子类也必须被标识为 abstract 类型。

程序中定义抽象类的目的是为一类对象建立抽象的模型,在同类对象所对应的类体系中,抽象类往往在顶层,这样能使类的设计变得清晰,同时抽象类也为类的体系提供通用的接口。定义了抽象类后,就可以利用 Java 的多态机制,通过抽象类中的通用接口处理类体系中的所有类。

### 2. 接口

Java 中的接口使抽象类的概念更深入一层,Java 通过接口来实现多重继承。接口是一组抽象方法和常量的集合。接口中只定义需要做些什么,而不定义如何实现它,具体实现是由实现该接口的类来完成的。一个类实现接口时必须实现接口中所有的抽象方法,否则,只要有一个抽象方法未实现,那么该类就是抽象类,并且不能实例化。

## 5.2 设计 1 抽象类

### 1. 设计目的

(1) 学会使用抽象类和接口。
(2) 巩固 Java 语言的多态性和继承性。

### 2. 设计要求

试编写程序:动物园里有猫、狗和鸟三种不同的动物共 10 只,现在动物园要举行一个晚会,这些动物都要表演不同的才艺。它们的数量比例和出场的顺序都是随机的。

## 3. 设计步骤

### 1) 分析

不同的动物表演不同的才艺,这可以通过 Java 语言的多态性来实现。首先,定义一个抽象的动物类,在此类中定义两个抽象方法,一个用来显示动物的名称,一个是该动物要表演的才艺。让猫、狗和鸟三种动物都继承自这个动物类。其次,由于它们的数量和出场的顺序都是随机的,所以考虑使用 java.util 包中的 Random 类来实现。

### 2) 编写程序

程序清单 Zoo.java:

```java
import java.util.*;
abstract class Animal {
    private String name;
    abstract void showInfo();
    abstract void act();
}
class Cat extends Animal {
    void showInfo() {
        System.out.print("我是一只猫,");
    }
    void act(){
        System.out.println("我将表演猫步!");
    }
}
class Dog extends Animal {
    void showInfo() {
        System.out.print("我是一只狗,");
    }
    void act(){
        System.out.println("我将表演双脚站立行走!");
    }
}

class Bird extends Animal {
    void showInfo() {
        System.out.print("我是一只鸟,");
    }
    void act(){
        System.out.println("我将表演唱歌!");
    }
}
public class Zoo{
    public static void main(String args[]){
        Animal dongwu;
        for (int i = 0; i < 10; i++){
            Random rand = new Random();
            switch(rand.nextInt(3)){
                case 0:
```

```
                dongwu = new Cat();
                break;
            case 1:
                dongwu = new Dog();
                break;
            case 2:
                dongwu = new Bird();
                break;
            default:
                dongwu = null;
        }
        dongwu.showInfo();
        dongwu.act();
    }
}
```

3) 编译并运行程序

结果如图 5-1 所示。

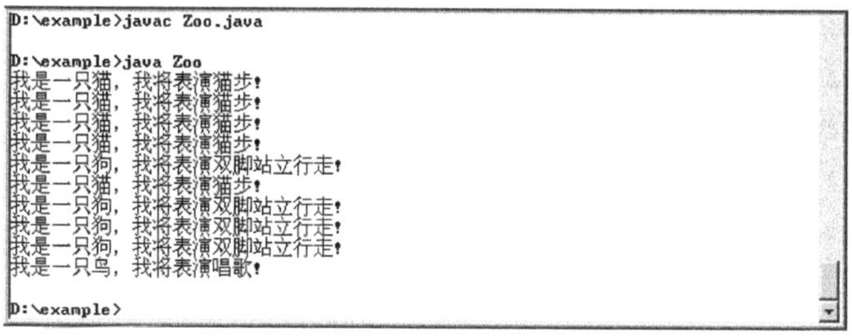

图 5-1　程序 Zoo.java 的编译运行结果

### 4．拓展设计

(1) 试将类 Zoo 的 main()方法中第一条语句改为：

`Animal dongwu = new Amimal();`

程序编译时将会出现什么错误？

(2) 新增一个 Pig 类，它也继承自 Animal 类，由于其无才艺表演，所以没有重写父类的 act()方法，其类定义语句如下：

```
class Pig extends Animal {
    void showInfo() {
        System.out.print("我是一头猪,");
    }
}
```

程序编译能通过吗？

(3) 将 main()方法中的 switch 语句的 default 部分

`default:dongwu = null;`

去掉,程序编译时将会出现什么错误?为什么?

(4) 将抽象类 Animal 的定义改为用接口定义,即

```
interface Animal {
    void showInfo();
    void act();
}
```

程序的其他部分应做何修改才能得到原运行结果?

### 5. 设计提示

(1) 抽象类是不能实例化的。

(2) 在拓展设计(2)中,Pig 类必须重写父类的 act() 方法,否则它也应该是一个抽象类。

(3) 在拓展设计(3)中,变量 dongwu 是一个局部变量,编译器确定局部变量没有经过初始化时,将产生编译错误。

(4) 拓展设计(4)中,由于接口 Animal 中声明的方法其访问控制权限默认是 public,所以在实现接口 Animal 的各类,如类 Cat、类 Dog、类 Bird 中,在实现 Animal 中定义的两个方法 showInfo()和 act()时,要在声明中增加 public,否则这些类对接口中方法的实现将缩小了访问权限,会在编译时出错。

## 5.3 设计2 接口

### 1. 设计目的

(1) 学会使用接口。
(2) 了解接口的继承体系。
(3) 巩固 Java 语言的多态性和继承性。

### 2. 设计要求

试编写程序:学校中有教师和学生两类人,教师每月都有工资收入,学生每年要付学费。而研究生既是教师又是学生,因此他们既要付学费,每月还会有收入。研究生又分为硕士研究生和博士研究生。将教师、学生和研究生都设计为接口,而硕士、博士研究生为实现研究生接口的两个类。编写一个主类,表示有一个研究生子女的家庭,如果该子女经济状态好,则可以自食其力,否则需要家庭对其资助。

### 3. 设计步骤

1) 分析

该问题中需要设计三个接口,Teacher 接口中有 getSalary 方法,Student 接口中有 payFee 方法,而 Graduate 接口继承自 Teacher、Student 接口,并且有反映其经济状况的 status 方法。类硕士生实现 Graduate 接口,有 name、fee、salary 属性,类博士生也实

Graduate 接口，有 name、fee、salary、projectFunds 属性。类 Family 有一个 Graduate 类型的 child 属性，且有一个 isLoad 方法，该方法判断 child 的经济状况，以决定是否需要给予资助。最后在 main 方法中实例化两个 Family。

2）编写程序

程序清单 Family.java：

```
interface Student{
    double payFee();
}
interface Teacher{
    double getSalary();
}
interface Graduate extends Student,Teacher{
    boolean status();
}
class 硕士生 implements Graduate{
    String name;
    double fee;
    double salary;
    硕士生(String name,double fee,double salary){
        this.name = name;
        this.fee = fee;
        this.salary = salary;
    }
    public double payFee(){
        return fee;
    }
    public double getSalary(){
        return salary;
    }
    public boolean  status(){
        if (salary * 12 < fee)
            return true;
        else
            return false;
    }
}
class 博士生 implements Graduate{
    String name;
    double fee;
    double salary;
    double projectFunds;
    博士生(String name,double fee,double salary,double projectFunds){
        this.name = name;
        this.fee = fee;
        this.salary = salary;
        this.projectFunds = projectFunds;
    }
    public double payFee(){
        return fee;
```

```java
        }
        public double getSalary(){
            return salary;
        }
        public double getProjFunds(){
            return projectFunds;
        }
        public boolean  status(){
            if (projectFunds + salary * 12 < fee)
                return true;
            else
                return false;
        }
}
class Family {
    Graduate child;
    Family(Graduate graduate){
        child = graduate;
    }
    public void isLoad(){
        if (child.status())
            System.out.println("需要父母支助,或其他兼职!");
        else
            System.out.println("孩子能自食其力了!");
    }
    public static void main(String [] args){
        Graduate s1 = new 硕士生("李力",6500, 500);
        Graduate s2 = new 博士生("梅莉",8800,2000,60000);
        Family a = new Family(s1);
        Family b = new Family(s2);
        a.isLoad();
        b.isLoad();
    }
}
```

3) 编译并运行程序

其结果如图 5-2 所示。

```
D:\example>javac Family.java

D:\example>java Family
需要父母资助,或其他兼职!
孩子能自食其力了!

D:\example>
```

图 5-2　程序 Family.java 的编译运行结果

### 4. 拓展设计

(1) 请将 Graduate 接口改为抽象类,还需要修改哪些地方,程序才能正常运行?
(2) 在原程序 Family.java 中,Teacher 接口和 Student 接口能改为抽象类吗?
(3) 在原程序 Family.java 中,child 属性能改为 Student 接口类型吗?

(4) 仔细体会类 Family 中定义 child 属性为 Graduate 类型的优势。

## 5. 设计提示

(1) 将 Graduate 接口改为抽象类,要修改三个地方,程序才能正常运行。

① 抽象类 Graduate 的定义改为:

```
abstract class Graduate implements Student,Teacher{
    abstract boolean status();
}
```

② 类硕士生的声明改为:

class 硕士生 extends Graduate

③ 类博士生的声明改为:

class 博士生 extends Graduate

(2) 拓展设计(2)中,若将 Teacher 接口和 Student 接口改为抽象类,则 Graduate 接口将无法实现。

(3) 拓展设计(3)中,不能将 child 属性改为 Student 接口类型。否则在编译时会出现如图 5-3 所示的错误。

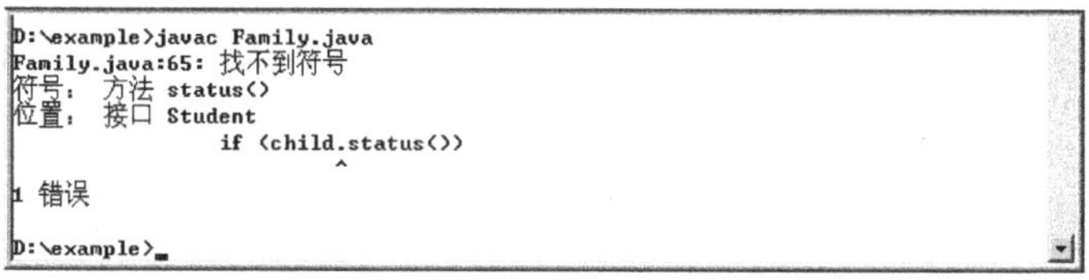

图 5-3 拓展设计(3)的编译结果

(4) 接口的特点在于只定义能做什么,而不定义怎么去做。在本设计中,硕士生和博士生分别以自己的方式实现了 Graduate 接口,当 child 定义为 Graduate 类型时,它可以指向任何实现了 Graduate 接口的对象。否则需要定义两个类,一个为 Family1 类,该类具有硕士生类的 child 属性;另一个为 Family2 类,该类具有博士生类的 child 属性。

# 第6章 数组和常用类

## 6.1 本章知识点

### 1. 数组

Java 语言通过数组来处理相同数据类型的多个数据。

Java 中为所有数组设置了一个表示数组元素个数的特性变量 length，它作为数组的一部分存储起来。Java 用该变量在运行时进行数组越界检查，应用程序中也可以访问该变量获取数组的长度。

在 Java 语言的 java.lang.System 类中提供了一种进行数组复制的方法：

public static void arraycopy(Object src, int srcPos, Object dest, int destPos, int length)

通过将原数组中 srcPos 位置开始的 length 个元素复制到目标数组的 destPos 开始的位置。

### 2. 字符串

字符串是 Java 语言中必不可少的数据结构。与其他语言不同，它并不是基本数据类型，而是由字符串的类来实现的。常用的进行字符串处理的类有 java.lang.String 和 java.lang.StringBuffer。

1) String 类

String 类是一个 final 类型的类，即该类是一个最终类，不能修改，也不能被继承。

String 的常用构造方法如下。

String()：创建一个空的 String 对象。

String(char [] value)：通过给定的字符数组 value，创建一个新的 String 对象。

String(char [] value, int offset, int count)：在字符数组 value 中，从 offset 开始，count 个字符序列创建新的 String 对象。

String(String original)：通过给定的字符串常量 original 创建一个新的 String 对象。

String(StringBuffer buffer)：利用已经存在的 StringBuffer 类对象，创建一个新的 String 对象。

String 常用的方法如下。

String 类可以实现字符串内容是否一致的比较,常用的方法有 equalsIgnoreCase(String anotherString)、equals(Object anObject)、compareTo(String anotherString)、compareToIgnoreCase(String str)。

String 类常用重载的 indexOf()和 lastIndexOf()方法来实现子串或字符的查找。

String 类常用 toLowerCase()、toUpperCase()方法来实现字符串的大小写转换。

String 类常用重载的 valueOf()方法来将特定类型的数据转换成 String 类对象。

String 类常用 length()方法返回该 String 对象中包含的字符的个数。

2) StringBuffer 类

与 String 类不同,StringBuffer 类可以创建和处理动态的字符串信息。二者相同的是,它们都是最终类,不能派生子类,以防用户修改其功能。

StringBuffer 类的构造方法如下。

StringBuffer():创建一个新的空的 StringBuffer 类的对象,其容量初值设置为 16 个字符。

StringBuffer(int length):创建一个新的空的 StringBuffer 类的对象,其容量初值设置为 length 个字符。

StringBuffer(String str):创建一个新的空的 StringBuffer 类的对象,其内容为 Str 的内容,容量初值设置为 str.length()+16 个字符。

StringBuffer 的一些常用方法如下。

重载的 append 方法用来将指定的内容追加到当前的字符串中。

重载的 insert 方法用来将指定的内容插入到当前的字符串中。

public char charAt(int imdex):返回该 StringBuffer 对象内容的第 index 个字符。

public void setCharAt(int index, char ch):将原 StringBuffer 对象的第 index 个字符替换成字符 ch。

public void getChars(int srcBegin, int srcEnd, char[] dst, int dstBegin):将该 StringBuffer 对象从 srcBegin 字符开始至 srcEnd 字符结束(不含该字符)的所有字符放到目标字符数组的 dstBegin 起始位置处。

public StringBuffer reverse():返回原 StringBuffer 对象内容反转后新的 StringBuffer 对象。

public int length():返回该 StringBuffer 对象中字符串的长度。

## 6.2 设计 1 数组

### 1. 设计目的

(1) 掌握数组的定义和使用方法。

(2) 熟悉数组的排序、查找等算法。

(3) 学会使用 Java 语言提供的数组复制方法。

## 2. 设计要求

有 100 个小孩按照 1~100 编号围成一个圈做游戏,从第 1 个小孩开始从 1 报数,数到 3 的小孩退出游戏,后面的小孩接着重新从 1 开始报数,数到 3 的小孩再退出,如此循环,直到圈中只有一个小孩为止。设计要求输出最后这个小孩的编号。

## 3. 设计步骤

1) 分析

做游戏的 100 个小孩,编号分别为 1~100,可以考虑用整型数组来存放其编号,初始时所有小孩都在圈中,数到 3 时退出,所以数组元素初始值为其相应编号,当有小孩退出时,其值变为 0。

游戏循环进行,直到圈中只有一个小孩,所以可以用 while 循环语句实现,条件为剩余小孩数 leftCount 不等于 1。

最后扫描整个数组,将不为 0 的数组元素值输出,此即为最后一个小孩的编号。

2) 编写程序

程序清单 Baoshu.java:

```java
public class BaoShu {
    public static void main(String[] args) {
        int [] child = new int[100];
        for(int i = 0; i < child.length; i++) {
            child[i] = i + 1;
        }
        int leftCount = 100;
        int countNum = 0;
        int index = 0;
        while(leftCount != 1) {
            if(child[index] != 0) {
                countNum ++;
                if(countNum == 3) {
                    countNum = 0;
                    child[index] = 0;
                    leftCount -- ;
                }
            }
            index ++;
            if(index == child.length) {
                index = 0;
            }
        }
        for(int i = 0; i < child.length; i++) {
            if(child[i] != 0) {
                System.out.println("最后一个小孩的编号是: " + child[i]);
            }
        }
    }
}
```

}

3）编译并运行程序

结果如图 6-1 所示。

图 6-1　程序 BaoShu.java 的编译运行结果

**4．拓展设计**

（1）如果将程序中的以下语句去掉

```
index ++;
if(index == child.length) {
    index = 0;
}
```

程序编译、运行将会出现什么情况？

（2）如果要求输出这 100 个小孩从游戏中退出的顺序，程序该如何修改？

（3）随机产生 20 个 0～100 之间的随机整数，用冒泡排序法将其按从小到大的顺序排列，并对排序前后的数组按照每行 10 个数的方式输出。

（4）在拓展设计（3）的基础上，采用折半查找法查找给定的数在排序后的数组中的位置。如果查到该数，输出信息：X 在数组的第 Y 个位置上的程序。其中，X 代表给定的数，Y 代表该数在数组中的位置（下标）。

（5）试编写一程序，定义两个二维数组，将其中一个数组初始化，用 Java 语言在 java.lang.System 类中提供的数组复制方法将已经初始化的数组复制到另一数组中，改变其中任一数组的元素值，看另一个数组中对应的元素有什么变化。

**5．设计提示**

（1）拓展设计（1）中，程序编译时将没有错误，但在运行时会陷入死循环。体会上述语句的作用。

（2）拓展设计（2）中，只需在每个小孩退出时，即在 while 循环中

```
child[index] = 0;
```

语句前增加一条输出语句即可。

（3）拓展设计（3）中，产生 20 个 0～100 之间的随机整数，可以通过以下语句实现：

```
int a[] = new int[20];
for (int k = 0;k < a.length;k++){
    a[k] = (int)(100 * Math.random());
}
```

而冒泡排序法可以通过如下方法实现：

```java
public static void bubble(int a[]){
    int count = a.length, i;
    for(i = 0;i < count;i++){
        for(int j = count - 1;j > i;j-- ){
            if(a[j]< a[j - 1]){
                int temp = a[j];
                a[j] = a[j - 1];
                a[j - 1] = temp;
            }
        }
    }
}
```

(4)拓展设计(4)中的折半查找法可参考以下代码：

```java
public static int bisearch(int a[], int n){
    int low = 0;
    int high = a.length - 1;
    int mid = (low + high)/2;
    if(n > a[high]||n < a[low])
        return -1;
    while(low <= high){
        if(a[low] == n) return low;
        else if(a[high] == n) return high;
        else if(a[mid] == n) return mid;
        else{
            if(n > a[mid]){
                low = mid + 1;
                high = high - 1;
            }
            else{
                high = mid - 1;
                low = low + 1;
            }
        }
    }
    return -1;
}
```

(5)拓展设计(5)参考程序如下：

```java
public class TestArrayCopy {
    public static void main(String args[]) {
        int[][] srcArray = {{1,2},{3,4,5},{6,7}};
        int[][] destArray = new int[3][];
        System.arraycopy(srcArray,0,destArray,0,srcArray.length);
        srcArray[2][1] = 100;
        for(int i = 0;i < destArray.length;i++){
            for(int j = 0;j < destArray[i].length;j++){
                System.out.print(destArray[i][j] + "  ");
            }
            System.out.println();
```

                }
            }
        }

## 6.3 设计 2  字符串

**1. 设计目的**

(1) 了解 String 类和 StringBuffer 类,掌握其构造方法和常用方法。
(2) 对字符串进行有效的处理。
(3) 学会利用 String 类、StringBuffer 类处理实际问题。

**2. 设计要求**

编写一个程序,要求能够将 15 位的身份证号码升到 18 位,或者验证所给的 18 位身份证号码是否有效。提示:18 位的身份证号码由 17 位数字本体码和 1 位校验码组成,其中前 6 位为数字地址码,表示编码对象常住户口所在县(市、旗、区)的行政区划代码,中间 8 位为数字出生日期码,4 位表示年,2 位表示月,2 位表示日,最后 4 位为 3 位数字顺序码和 1 位数字校验码,顺序码表示在同一地址码所标识的区域范围内,对同年、同月、同日出生的人编写的顺序号,奇数分配给男性,偶数分配给女性,校验码的计算方法如下。

(1) 17 位数字本体码加权求和公式:

$$S = sum(a_i * w_i), i=0,1,\cdots,16$$

$a_i$:表示第 i 个位置上的身份证号码数字值。
$w_i$:表示第 i 个位置上的加权因子,其值依次为 7、9、10、5、8、4、2、1、6、3、7、9、10、5、8、4、2。

(2) 计算模,即求 S 除以 11 的余数:

$$Y = mod(S,11)$$

(3) 通过模得到对应的校验码:

Y:          0  1  2  3  4  5  6  7  8  9  10
校验码:     1  0  x  9  8  7  6  5  4  3  2

**3. 设计步骤**

1) 分析

可以使用 String 类处理身份证,在计算校验码的过程中,用到的加权因子可以采用 int 型数组存放,而校验码可以用 char 型数组存放。

2) 编写程序

程序清单 TestID.java:

```
public class TestID {
    String ID;
    final int[] weight = {7,9,10,5,8,4,2,1,6,3,7,9,10,5,8,4,2};
```

```java
        final char[] checkCode = {'1','0','x','9','8','7','6','5','4','3','2'};
        int[] temp = new int[17];
        public char getCheckCode(String id){
            int pos = 0,sum = 0;
            if (id.length() == 18){
                id = id.substring(0,17);
            }
            if (id.length() == 17){
                for(int i = 0;i < 17;i++){
                    char c = id.charAt(i);
                    temp[i] = c - '0';
                }
                for (int i = 0;i < 17;i++)
                    sum = sum + weight[i] * temp[i];
                pos = sum % 11;
            }
            return checkCode[pos];
        }
        public String upTo18(String oldID){
            String newID = oldID.substring(0,6);
            newID = newID + "19";
            newID = newID + oldID.substring(6,15);
            newID = newID + getCheckCode(newID);
            return newID;
        }
        public boolean verifyCheckCode(String id){
            char ch = id.charAt(17);
            if(ch == getCheckCode(id))
                return true;
            else
                return false;
        }
        public static void main(String args[]) {
            TestID tid = new TestID();
          int len = args[0].length();
            if(len == 15)
                System.out.println("新的身份证号码是: " + tid.upTo18(args[0]));
            else if(len == 18){
                    if(tid.verifyCheckCode(args[0]))
                        System.out.println("这是一个正确的身份证号");
                    else
                        System.out.println("这是一个错误的身份证号");
                }
                else
                    System.out.println("对不起,您的输入有误,请重新输入!");
        }
    }
```

3) 编译并运行程序

结果如图 6-2 所示。

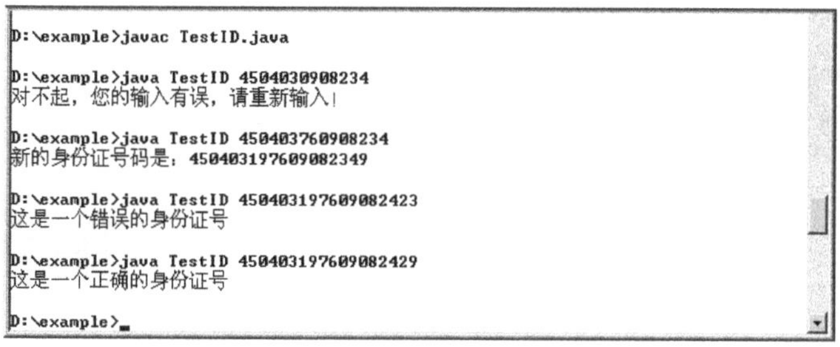

图 6-2 程序 TestID.java 的编译运行结果

**4. 拓展设计**

(1) 校验码的验证仅仅是身份证号码正确性验证的一个方面,试增加方法以验证身份证号码中间 8 位表示出生年、月、日的有效性。

(2) 编写一个应用程序,统计给定的一字符串中单词的个数。

(3) 编写一个应用程序,判断一个给定的整数的位数。

(4) 如果一个字符串去掉除字母和数字之外的所有字符后,正读和逆读都一样,则这个字符串就是一个回文。编写一个应用程序,对输入的字符串测试其是否为回文。

**5. 设计提示**

(1) 拓展设计(2)中可以考虑使用类 StringTokenizer,该类提供的方法 int countTokens() 会返回 StringTokenizer 对象中被分割后子字符串的个数。

(2) 拓展设计(3)中可以通过以下语句实现:

```
int i = 654378;
String s = String.valueOf(i);
System.out.println("i的位数为: " + s.length());
```

(3) 拓展设计(4)中,根据回文的定义,首先定义一个方法,去掉字符串中除字母和数字之外的其他字符,该方法中可以使用 String 类的 chatAt(int index)方法,以及 Character 类的静态方法 isLetterOrDigit(char c)。

接着,定义一个方法,将上述方法取得的整个字符串翻转,这一步可以使用 StringBuffer 类提供的 reverse()方法来辅助实现。

最后,定义一个方法,比较翻转前后的两个字符串,如果相等则读入的字符串就是回文,否则不是回文。在比较两个字符串时,要忽略字母的大小写,因此,可以使用 String 类的 equalsIgnoreCase(String anotherString)方法。

# 第7章 集合类

## 7.1 本章知识点

集合类是 Java 中一组很实用的类,位于 java.util 包中。Java 提供的核心集合接口如图 7-1 所示,它们形成了两个独立的树状结构。

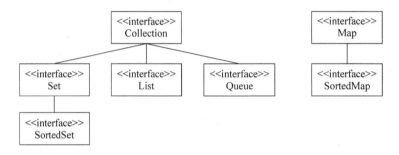

图 7-1 Java 集合类 API 的核心接口

**1. Collection 接口**

Collection 接口是集合接口树的根,它定义了集合操作的通用 API。要尽可能以常规方式处理一组元素时,就使用这一接口,其中提供的常用操作如下所示。

1) 基本操作

boolean add(Object o):将对象 o 添加给集合。

boolean remove(Object o):如果集合中有与 o 相匹配的对象,则删除对象 o。

int size():返回当前集合中元素的数量。

boolean isEmpty():判断集合中是否有任何元素。

boolean contains(Object o):查找集合中是否含有对象 o。

Iterator iterator():返回一个迭代器,用来访问集合中的各个元素。

2) 组操作

boolean containsAll(Collection c):查找集合中是否含有集合 c 中的所有元素。

boolean addAll(Collection c):将集合 c 中所有元素添加给该集合。

void clear():删除集合中的所有元素。

void removeAll(Collection c):从集合中删除集合 c 中的所有元素。

void retainAll(Collection c)：从集合中删除集合 c 中不包含的元素。

3) Collection 转换为 Object 数组

Object[] toArray()：返回一个内含集合所有元素的 array。

Object[] toArray(Object[] a)：返回一个内含集合所有元素的 array。运行期返回的 array 和参数 a 的型别相同，需要转换为正确型别。

### 2．Set 接口

Set 接口是 Collection 接口的子接口，其中不能包含重复的元素。Set 接口没有引入新方法，所以 Set 就是一个 Collection，只不过其行为不同。JDK 中提供了实现 Set 接口的三个实用类：HashSet 类、TreeSet 类和 LinkedHashSet 类。

### 3．List 接口

List 接口也是 Collection 接口的子接口，允许有重复的元素，是一个有序的集合。JDK 中提供了实现 List 接口的三个实用类：ArrayList 类、LinkedList 类和 Vector 类。该接口不但能够对列表的一部分进行处理，还添加了面向位置的操作。

(1) 面向位置的操作包括插入某个元素或 Collection 的功能，还包括获取、除去或更改元素的功能。在 List 中搜索元素可以从列表的头部或尾部开始，如果找到元素，还将报告元素所在的位置：

void add(int index, Object element)：在指定位置 index 上添加元素 element。

boolean addAll(int index, Collection c)：将集合 c 的所有元素添加到指定位置 index。

Object get(int index)：返回 List 中指定位置的元素。

int indexOf(Object o)：返回第一个出现元素 o 的位置，否则返回－1。

int lastIndexOf(Object o)：返回最后一个出现元素 o 的位置，否则返回－1。

Object remove(int index)：删除指定位置上的元素。

Object set(int index, Object element)：用元素 element 取代位置 index 上的元素，并且返回旧的元素。

(2) List 接口不但以位置序列迭代地遍历整个列表，还能处理集合的子集：

ListIterator listIterator()：返回一个列表迭代器，用来访问列表中的元素。

ListIterator listIterator(int index)：返回一个列表迭代器，用来从指定位置 index 开始访问列表中的元素。

List subList(int fromIndex, int toIndex)：返回从指定位置 fromIndex（包含）到 toIndex（不包含）范围中各个元素的列表视图。

### 4．Queue 接口

Queue 接口是存放等待处理的数据的集合，其中的元素一般采用 FIFO（First In First Out，先进先出）的顺序，也有以元素的值进行排序的优先队列。

### 5．Map 接口

Map 接口用来实现键值到值的映射。Map 中不能包含重复的键值，每个键值最多只能映射到一个值。JDK 中提供了实现 Map 接口的实用类，包括 Hashmap 类、HashTable 类、TreeMap 类和 WeekHashMap 类等。Map 接口提供的方法如下所示。

1）添加、删除操作

Object put(Object key, Object value)：将互相关联的一个关键字与一个值放入该映像。如果该关键字已经存在，那么与此关键字相关的新值将取代旧值。方法返回关键字的旧值，如果关键字原先并不存在，则返回 null。

Object remove(Object key)：从映像中删除与 key 相关的映射。

void putAll(Map t)：将来自特定映像的所有元素添加给该映像。

void clear()：从映像中删除所有映射。

2）查询操作

Object get(Object key)：获得与关键字 key 相关的值，并且返回与关键字 key 相关的对象，如果没有在该映像中找到该关键字，则返回 null。

boolean containsKey(Object key)：判断映像中是否存在关键字 key。

boolean containsValue(Object value)：判断映像中是否存在值 value。

int size()：返回当前映像中映射的数量。

boolean isEmpty()：判断映像中是否有任何映射。

3）视图操作

处理映像中键/值对组：

Set keySet()：返回映像中所有关键字的视图集。

Collection values()：返回映像中所有值的视图集。

Set entrySet()：返回 Map.Entry 对象的视图集，即映像中的关键字/值对。

### 6．SortedSet 接口

SortedSet 接口是一个特殊的 Set 接口，它保持元素的有序顺序。SortedSet 接口为 Set 的视图（子集）和它的两端（即头和尾）提供了访问方法。

添加到 SortedSet 实现类的元素必须实现 Comparable 接口，否则必须给它的构造函数提供一个 Comparator 接口的实现。TreeSet 类是它的唯一实现。

### 7．SortedMap 接口

SortedMap 接口是一个特殊的 Map 接口，它用来保持键的有序顺序。除了排序是作用于 Map 的键以外，处理 SortedMap 和处理 SortedSet 一样。

添加到 SortedMap 实现类的元素必须实现 Comparable 接口，否则必须给它的构造函数提供一个 Comparator 接口的实现。TreeMap 类是它的唯一实现。

## 7.2 设计 1　List 接口及实现该接口的常用类的练习

### 1．设计目的

（1）理解集合、泛型的概念。

(2) 掌握 List 接口及实现该接口的常用类的方法的应用。
(3) 掌握 Collections 类的应用。
(4) 学会应用集合类解决实际问题。

**2. 设计要求**

编写一个应用程序，要求：
(1) 能将用户从键盘输入的学生对象(包括学号、姓名和成绩)按成绩进行排序，并将排序结果输出到屏幕上。
(2) 实现查找指定学生的成绩。

**3. 设计步骤**

1) 分析

要将学生信息按成绩进行排序，可以定义学生类，且该类必须实现 java.lang 包中的 Comparable 接口，将每个学生对象置于 List 中，利用 java.util 包中的 Collections 类的静态方法 sort 进行排序。当要查找指定学生的成绩时，需要遍历整个 List，这里要用到 List 的 iterator 方法。

2) 编写程序

程序清单 TestList.java：

```java
import java.util.*;
class Student implements Comparable<Student>{
    String name;
    int id;
    int score;
    Student(int id,String name,int score){
        this.id = id;
        this.name = name;
        this.score = score;
    }
    public int compareTo(Student s){
        int d = this.score - s.score;
        return d;
    }
    public String toString(){
        return id + " " + name + " " + score;
    }
}
public class TestList{
    void append(List<Student> list){
        Scanner scanner = new Scanner(System.in);
        System.out.print("请输入学生人数：");
        int num = scanner.nextInt();
        for(int i = 1;i <= num;i++){
            System.out.println("请输入第" + i + "个学生的信息");
            System.out.print("学号：");
            int id = scanner.nextInt();
```

```java
            System.out.print("姓名：");
            String name = scanner.next();
            System.out.print("成绩：");
            int score = scanner.nextInt();
            list.add(new Student(id,name,score));
        }
        System.out.println("排序前：" + list);
        Collections.sort(list);
        System.out.println("排序后：" + list);
    }
    void search(List<Student> list){
        Scanner scanner = new Scanner(System.in);
        System.out.print("请输入学生姓名：");
        String searName = scanner.next();
        Iterator<Student> it = list.iterator();
        boolean find = false;
        while(it.hasNext()){
            Student stu = it.next();
            if(stu.name.equals(searName)){
                System.out.print("该学生的学号：" + stu.id);
                System.out.println(",成绩：" + stu.score);
                find = true;
            }
        }
        if(!find) System.out.println("查无此人！");
    }
    public static void main(String [] args){
        TestList tl = new TestList();
        List<Student> list = new LinkedList<Student>();
        tl.append(list);
        tl.search(list);
    }
}
```

3）编译并运行程序

其结果如图 7-2 所示。

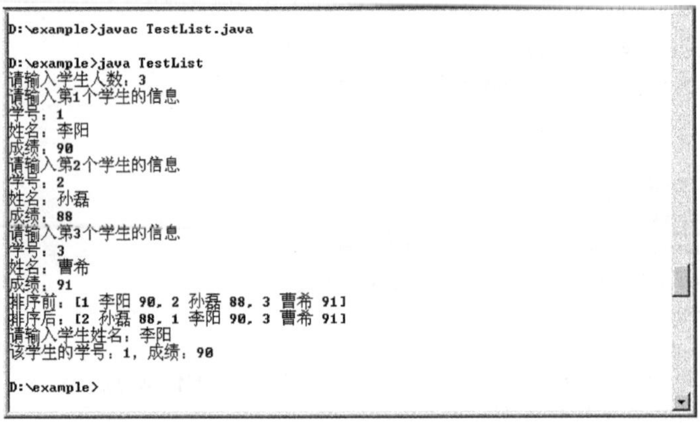

图 7-2　程序 TestList.java 的编译运行结果

**4．拓展设计**

（1）请实现对指定学生信息的删除操作，即用 List 的 remove 方法删除某个学生对象。
（2）请实现对从键盘输入的学生信息，输出该学生的名次。
（3）请在原程序的基础上进行修改，以输出成绩最好与最差的学生信息。

**5．设计提示**

（1）实现对指定学生信息的删除操作，学生类还应该实现 equals 方法和 hashCode 方法，参考代码如下：

```
public boolean equals(Object o){
    if(o instanceof Student){
        Student s = (Student)o;
        return((id == s.id)&&(name.equals(s.name))&&(score == s.score));
    }
    return super.equals(o);
}
public int hashCode(){
    return name.hashCode();
}
```

（2）对于拓展设计（2），可以使用 java.util 包中的 Collections 类的 binarySearch 方法实现对指定学生对象的查询操作。
（3）由于原程序 TestList.java 采用的是 Linkedlist 类，而该类提供了一些处理列表两端元素的方法，所以可以使用其 getFirst 方法和 getLast 方法作用于已排序的链表，以输出成绩最差与最好的学生信息。

## 7.3 设计2 Set 接口及实现该接口的常用类的练习

**1．设计目的**

（1）掌握 Set 接口及实现该接口的常用类的方法的应用。
（2）理解 HashSet、LinkedHashSet 类中元素的不同存放顺序。
（3）掌握 TreeSet 类的作用。

**2．设计要求**

编写程序，以验证 HashSet、LinkedHashSet、TreeSet 类中元素的不同存放顺序。

**3．设计步骤**

1）分析

为验证 HashSet、LinkedHashSet、TreeSet 类中元素的不同存放顺序，可以随机地产生一些数据，将其放入 HashSet、LinkedHashSet 中，使用 TreeSet 类对其进行排序输出。

2)编写程序

程序清单 TestSet.java:

```java
import java.util.*;
public class TestSet {
    public static void main(String args[]) {
        Set<Integer> set1 = new HashSet<Integer>();
        Set<Integer> set2 = new LinkedHashSet<Integer>();
        for(int i = 0;i<5;i++){
            int number = (int)(Math.random() * 100);
            set1.add(new Integer(number));
            set2.add(new Integer(number));
            System.out.println("第" + I + "次随机数产生为:" + number);
        }
        System.out.println("未排序前HashSet:" + set1);
        System.out.println("未排序前LinkedHashSet: " + set2);
        System.out.println("排序后set1 : " + new TreeSet<Integer>(set1));
        System.out.println("排序后set2 : " + new TreeSet<Integer>(set2));
    }
}
```

3)编译并运行程序

其结果如图 7-3 所示。注意在该次设计中,恰好产生了两个相同的随机数,由于 Set 中的对象不能重复,所以在结果中只有互不相等的 4 个数据。

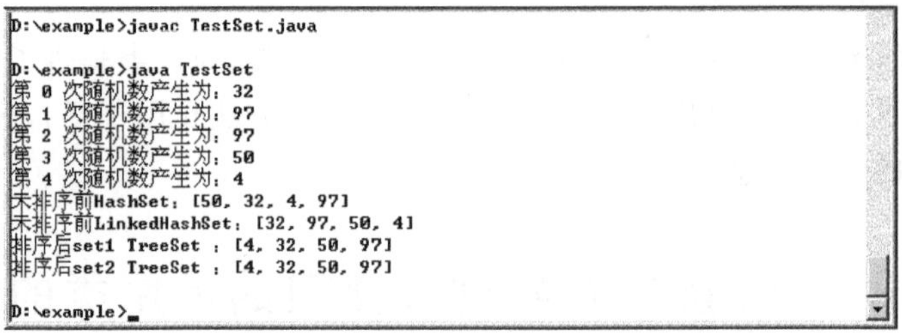

图 7-3  程序 TestSet.java 的编译运行结果

### 4. 拓展设计

(1)请用 TreeSet 类实现 7.2 节设计 1 的设计要求。

(2)请问能否用 java.util 包中的 Collections 类的静态方法 binarySearch 在树集中查找指定的对象?

### 5. 设计提示

(1)当使用构造方法 TreeSet()创建树集后,再用 add 方法增加节点时,节点会按其存放的数据的"大小"顺序一层一层地依次排列,在同一层中的节点从左到右按从小到大递增排列,因此,当直接输出树集时,其中的元素已经排序完成。但要注意节点中存放的对象必须是实现 java.lang 包中的 Comparable 接口的类的实例。

(2)由于 java.util 包中的 Collections 类的静态方法 binarySearch 只能实现在 List 中

查找给定对象,所以不能在树集中查找指定的对象。

## 7.4 设计 3 Map 接口及实现该接口的常用类的练习

### 1. 设计目的

(1) 掌握 Map 接口及实现该接口的常用类的方法的应用。
(2) 掌握 Java 对基本类型数据提供的自动装箱(autoboxing)和自动拆箱(autounboxing)功能。
(3) 学会应用集合类解决实际问题。

### 2. 设计要求

编写一个应用程序,以验证 Java.util 包中 Random 类产生随机数的随机性。

### 3. 设计步骤

1) 分析

要验证 Random 类产生随机数的随机性,可以通过若干次的实验以统计产生的不同随机数的次数,所以可以使用 HashMap 类来实现。其中键为随机数,值为随机数出现的次数。

2) 编写程序

程序清单 TestMap.java:

```java
import java.util.*;
public class TestMap {
    public static void main(String args[]) {
        Map<Integer,Integer> m = new HashMap<Integer,Integer>();
        int freq = 0;
        int k = 0;
        Random rand = new Random();
        for(int i = 0;i<300;i++){
            switch ( rand.nextInt(3) ){
                case 0: k = 0; break;
                case 1: k = 1; break;
                case 2: k = 2; break;
            }
            if (m.get(k) == null){
                freq = 1;
            }else{
                freq = m.get(k).intValue() + 1;
            }
            m.put(k,freq);
        }
        System.out.println(m.size());
        System.out.println(m);
    }
}
```

3) 编译并运行程序

其结果如图 7-4 所示。

图 7-4　程序 TestMap.java 的编译运行结果

### 4. 拓展设计

请编写程序实现对 Map 的遍历。

### 5. 设计提示

要对 Map 进行遍历，可以采用不同的方法，如下参考程序 TestIteratorMap.java 列出 4 种：

(1) 通过 Map.values() 遍历所有的 value，但是不能遍历键 key。这种方法比较少用，因为我们大多数时候都是同时需要 key 和 value 的。

(2) 通过 Map.keySet() 遍历 key 和 value。这是许多人最喜欢的一种方式，因为代码最少，看起来最简单，通过遍历 keySet，再将 key 所对应的 value 查询出来，这里有一个二次取值的过程，所以并不推荐。

(3) 通过 Map.entrySet() 使用 iterator 遍历 key 和 value。

(4) 通过 Map.entrySet() 遍历 key 和 value。

后两种原理是相同的，都是通过遍历 Map.Entry 的方式，将 Entry 中的 key 和 value 打印出来，第(4)种代码看起来比较整洁，是推荐写法。

综上所述，如果 Map 里面内容比较少，其实采用哪种方式都可以，第(2)种和第(4)种相对简洁一些；但是一旦容量非常大时，更推荐采用第(4)种方式，相比于第(2)种将极大地提高程序性能。

程序清单 TestIteratorMap.java：

```java
import java.util.HashMap;
import java.util.Iterator;
import java.util.Map;
public class TestIteratorMap {
    public static void main(String[] args){
        Map<String, String> map = new HashMap<String, String>();
        map.put("one", "first");
        map.put("two", "second");
        map.put("three", "third");
        System.out.println("通过 Map.values()遍历所有的 value:");
        for (String v : map.values()){
            System.out.println("value = " + v);
        }
        System.out.println("通过 Map.keySet 遍历 key 和 value:");
```

```java
        for (String key : map.keySet()) {
            System.out.println("key = " + key + " value = " + map.get(key));
        }
        System.out.println("通过Map.entrySet用iterator遍历key和value:");
        Iterator<Map.Entry<String, String>> it = map.entrySet().iterator();
        while (it.hasNext()){
            Map.Entry<String, String> entry = it.next();
            System.out.print ("key = " + entry.getKey());
            System.out.println(" value = " + entry.getValue());
        }
        System.out.println("通过Map.entrySet遍历key和value:");
        for (Map.Entry<String, String> entry : map.entrySet()){
            System.out.print ("key = " + entry.getKey());
            System.out.println(" value = " + entry.getValue());
        }
    }
}
```

程序 TestIteratorMap.java 编译运行结果如图 7-5 所示。

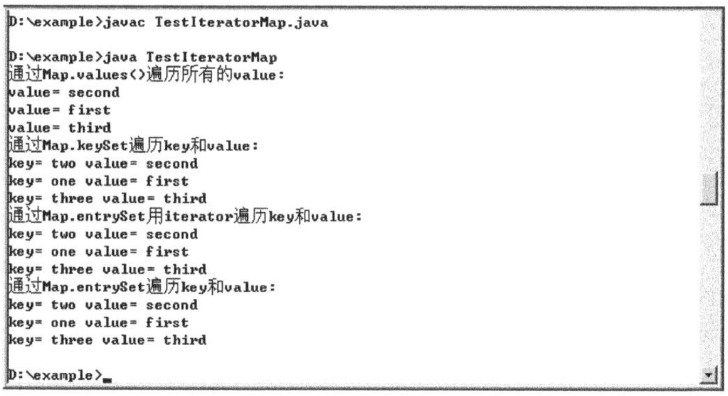

图 7-5　程序 TestIteratorMap.java 的编译运行结果

# 第8章 异常

## 8.1 本章知识点

在 Java 中，所有的异常都是用类来表示的，这些异常都继承自类 Throwable。类 Throwable 有两个直接子类 Error 和 Exception。Error 类型的错误与 Java 虚拟机本身有关，Exception 类型的错误是程序产生的错误，如除数为 0，数组越界等。图 8-1 展示了 Java 中的异常类体系。

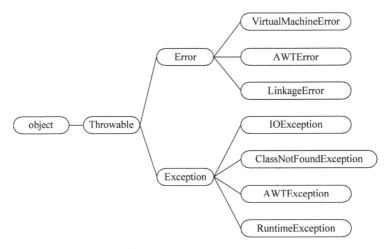

图 8-1 异常类体系

Java 要求如果程序中调用的方法有可能产生某种类型的异常，那么调用该方法的程序必须采取相应动作处理异常。异常处理具体有以下两种方式。

捕获并处理异常：用 try-catch-finally 语句组来完成。一个 try 语句块可以同时搭配多个 catch 语句使用，此时一定要注意 catch 语句的先后顺序。如果要同时捕获父类和子类的异常，就必须将捕获子类异常的 catch 语句放在捕获父类异常的 catch 语句的前面，否则捕获子类异常的 catch 语句将不可到达。

将方法中产生的异常抛出：用 throw 语句来实现。如果被抛出的异常在调用程序中未被处理，则该异常将被沿着方法的调用关系继续上抛，直到被处理。如果一个异常返回到 main()方法，并且在 main()中还未被处理，则该异常将把程序非正常地终止。

为处理特定的异常类,可以通过继承 Exception 类或其子类来创建自己的异常类。自定义异常类一般包含两个构造方法,一个是无参的构造方法,另一个构造方法则以字符串的形式接收一个定制的异常消息,并将该消息传递给父类构造方法。

## 8.2 设计 1 异常的捕获

### 1. 设计目的

(1) 了解 Java 异常处理的作用。
(2) 了解有异常处理与无异常处理的差别。
(3) 理解系统异常处理的机制。
(4) 掌握 try-catch-finally 语句的用法。

### 2. 设计要求

有如下程序 TestExcep.java,其中定义了一个类 TestExcep,在类中定义了一个静态方法 initArray,该方法通过循环语句实现数组初始化。main 方法调用类 TestExcep 的 initArray 方法对数组 arr 执行初始化,并通过循环语句将数组 arr 的所有元素输出。要求:
(1) 编译并运行该程序,观察其结果。
(2) 在原程序的基础上,增加异常处理语句,使程序运行得到如图 8-2 所示的结果。

图 8-2 增加异常处理后的运行结果

程序清单 TestExcep.java:

```java
public class TestExcep{
    static int[] initArray(int[] temp) {
        for(int i = 0;i <= temp.length;i++){
            temp[i] = 10/i;
        }
        return temp;
    }
    public static void main(String [] args){
        int[] arr = new int[10];
        arr = initArray(arr);
        for(int i = 0;i < arr.length;i++){
            System.out.print(arr[i] + ",");
        }
        System.out.println();
    }
}
```

### 3. 设计步骤

(1) 编辑原程序 TestExcep.java,编译并运行,其结果如图 8-3 所示。

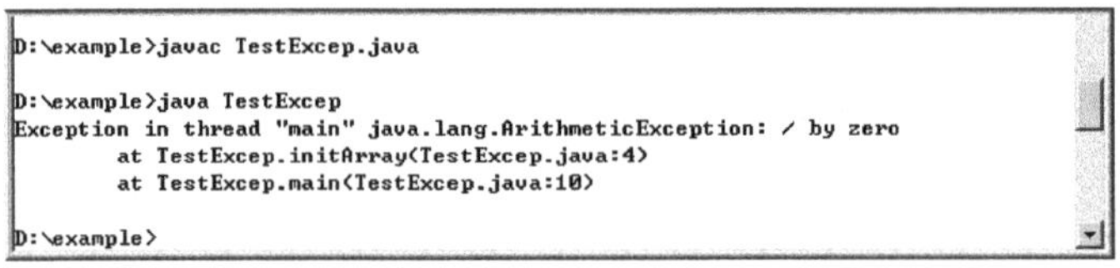

```
D:\example>javac TestExcep.java

D:\example>java TestExcep
Exception in thread "main" java.lang.ArithmeticException: / by zero
        at TestExcep.initArray(TestExcep.java:4)
        at TestExcep.main(TestExcep.java:10)

D:\example>
```

图 8-3　程序 TestExcep.java 的编译运行结果

(2) 分析步骤(1)的结果可知,语句

　　temp[i] = 10/i;

在 i 等于 0 时发生错误。仔细阅读程序,发现 for 循环的终止条件

　　i <= temp.length;

也有误。

根据设计要求,只能通过增加异常处理语句来得到如图 8-3 所示的运行结果。则程序中的 initArray 方法应做如下调整:

```java
static int[] initArray(int[] temp) {
    try{
        for(int i = 0;i <= temp.length;i++){
            temp[i] = 10/i;
        }
    }catch(Exception e){
        System.out.println("发现异常:整数被 0 除、数组下标越界");
        System.out.println("系统帮你修改了!请看结果:");
        temp[0] = 20;
        for(int j = 1;j < temp.length;j++){
            temp[j] = 10/j;
        }
    }
    return temp;
}
```

### 4. 拓展设计

(1) 在原程序的基础上,将 initArray 方法修改为:

```java
static int[] initArray(int[] temp) {
    try{
        for(int i = 0;i <= temp.length;i++){
            temp[i] = 10/i;
        }
    }catch(Exception e){
        System.out.println("发现异常:整数被 0 除.");
```

```
            System.out.println("系统帮你修改了!请看结果: ");
            temp[0] = 20;
        }
        return temp;
    }
```

编译并运行程序,观察运行结果,并解释此结果的原因。

(2) 通过 try-catch 语句分别捕获原程序中的两种异常。

### 5. 设计提示

当 try 语句块有异常发生时,中断 try 语句块剩余的语句的执行,并产生该异常所对应的实例对象,而且 try 语句块的剩余语句一般将不会被执行。所以拓展设计(1)会出现如图 8-4 所示的结果。

```
D:\example>javac TestExcep.java

D:\example>java TestExcep
发现异常: 整数被0除!
系统帮你修改了! 请看结果:
20,0,0,0,0,0,0,0,0,0,

D:\example>_
```

图 8-4 拓展设计(1)的运行结果

## 8.3 设计 2 异常的抛出及搜索

### 1. 设计目的

(1) 理解异常处理的搜索机制。

(2) 掌握自定义异常类的创建和使用。

(3) 学会使用 throw 语句抛出异常。

### 2. 设计要求

创建一个类 TestEx,该类中包含三个方法: method1、method2、method3,这些方法形成三级调用,主方法 main 调用 method1,method1 调用 method2, method2 调用 method3,而 method3 会抛出一个异常,该异常为自定义异常。编写程序,在 main 方法中捕获和处理 method3 抛出的异常。

### 3. 设计步骤

1) 分析

类 TestEx 中 method3 方法要抛出异常,所以应使用 throw 语句; 要在 main 方法中捕获和处理 method3 抛出的异常,则需要用到 try-catch 语句; 由于 method1、method2 和 method3 三个方法形成了逐级调用,因此 method1 和 method2 也应将异常抛出。此外,还要自定义一个异常类。

2) 编写程序 TestEx.java

程序清单 TestEx.java：

```java
public class TestEx {
    static void method1() throws MyException {
        System.out.println("method1 is calling");
        method2();
    }
    static void method2() throws MyException{
        System.out.println("method2 is calling");
        method3();
    }
    static void method3() throws MyException {
        System.out.println("method3 is calling");
        throw new MyException();
    }
    public static void main(String args[]) {
        try {
            System.out.println("main is running");
            method1();
        } catch (MyException e) {
            e.printStackTrace();
        }
    }
}
class MyException extends Exception{
    public MyException(){
        System.out.println("我是一个异常,我被抛出了!");
    }
}
```

3) 编译并运行程序

结果如图 8-5 所示。

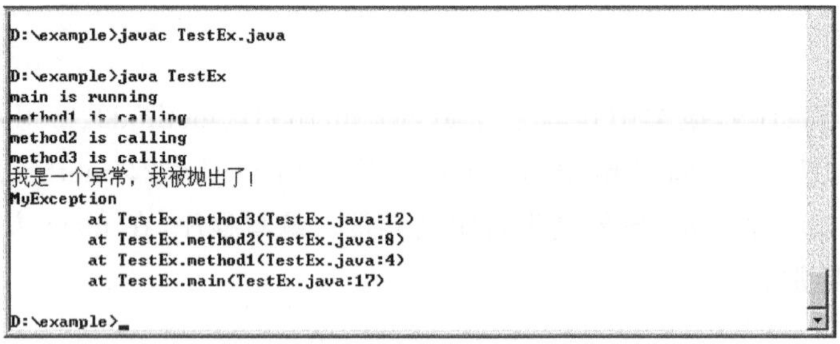

```
D:\example>javac TestEx.java

D:\example>java TestEx
main is running
method1 is calling
method2 is calling
method3 is calling
我是一个异常,我被抛出了!
MyException
        at TestEx.method3(TestEx.java:12)
        at TestEx.method2(TestEx.java:8)
        at TestEx.method1(TestEx.java:4)
        at TestEx.main(TestEx.java:17)

D:\example>
```

图 8-5　程序 TestEx.java 的编译运行结果

### 4. 拓展设计

（1）请将 method2 方法修改为如下代码，即方法中对 method3 的调用增加了一组 try-catch 语句。编译运行程序，观察并分析运行结果。

```
static void method2()throws MyException  {
    System.out.println("method2 is calling");
    try{
        method3();
    }catch(MyException e){
        System.out.println("method2 caught Exception ");
    }
}
```

（2）请问拓展设计（1）中方法 method2 的声明语句可以去掉 throws MyException 部分吗？为什么？那么 method1 的声明语句可以去掉 throws MyException 部分吗？请解释原因。

（3）请在拓展设计（1）的基础上，在 method2 方法和 main()方法中分别增加如下两个 finally 语句：

```
finally{
    System.out.println("method2's finally  executed");
}
finally{
    System.out.println("main's finally  executed");
}
```

编译、运行程序，观察并分析运行结果。

### 5．设计提示

（1）拓展设计（1）的运行结果如图 8-6 所示。在程序执行过程中，method3 中发生异常，程序便开始沿着调用顺序依次在 method3、method2、method1、main 中搜索异常处理器 catch 块，一旦找到了对应的 catch 块，便停止搜索，并做相应的处理，程序的剩余部分将会像没有异常发生一样，所以说一个异常一旦被捕获就不会再自动向外抛出了。

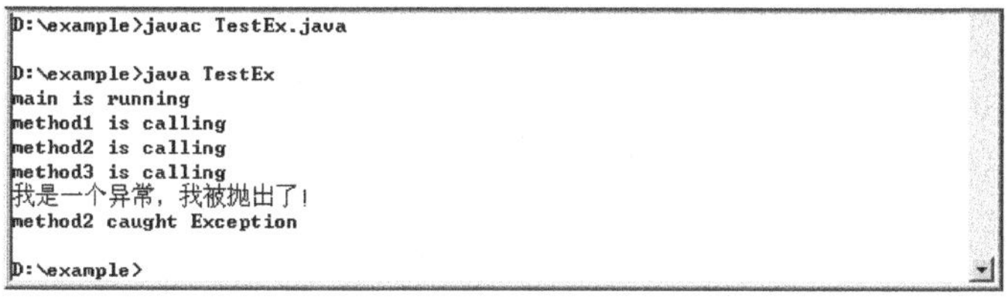

图 8-6　异常被 method1 捕获

（2）拓展设计（2）中 method2 方法声明语句中的 throws MyException 部分可以去掉，因为方法体中调用 method3 方法产生的异常已通过 try-catch 语句捕获，且方法体中没有新的 throw 语句，即不会产生 MyException 异常，所以去掉 method2 方法声明语句中的 throws MyException 部分不会有任何影响。

method1 方法中声明语句中的 throws MyException 部分不可以去掉，虽然 method1 方法中不会产生 MyException 异常，但是在 main()方法中有语句：

```
try {
    System.out.println("main is running");
```

```
        method1();
    } catch (MyException e) {
        e.printStackTrace();
    }
```

这就意味着 try 语句块中一定要有 MyException 异常抛出，否则会出现如图 8-7 所示的结果。

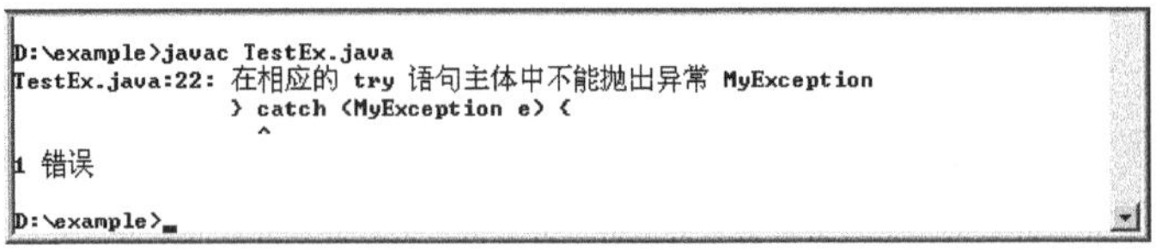

图 8-7　去掉 method1 方法声明中 throws MyException 部分后的运行结果

（3）在异常处理机制中，不论异常是否发生，在 try-catch-finally 语句中的 finally 语句块都要被执行，以完成一些必要的工作，如关闭之前打开的文件等。因此，拓展设计(2)将是如图 8-8 所示的运行结果。

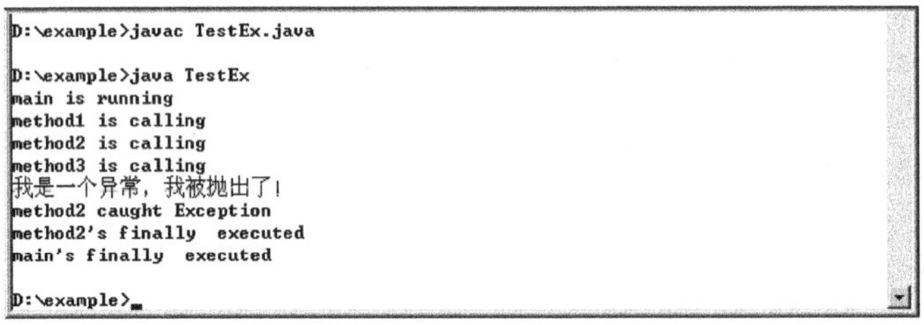

图 8-8　增加 finally 语句的运行结果

# 第9章 文件与流

## 9.1 本章知识点

文件系统的处理是各种语言中的重要内容,通过文件可以处理大量的数据,实现数据保存,为以后的使用和操作提供方便。在 Java 语言中文件和流不能分割,文件是流的一个分支,对文件的处理实际上就是对输入输出流的处理。Java 开发环境提供了 java.io 包,该包包括了文件类及一系列实现输入输出处理的流。本章介绍其中一些重要的类以及如何利用这些类进行输入输出的处理。

### 1. 文件

在 Java 语言中,File 类是对真实文件的抽象,它通过文件名列表来描述一个文件对象的属性,例如:权限、时间、日期和目录路径等。通过 File 类提供的方法,可以获得文件的描述信息。

1) 常用的构造方法
- public File(File parent, String child)
- public File(String pathname)
- public File(String parent, String child)

**注意**:这些方法并不能生成真正的文件。

2) 文件类的操作

File 类提供了一系列操作文件名称的方法:getName(),getPath(),get AbsolutePath(),getParent(),renameTo(),compareTo(),isAbsolute()。

File 类提供了一系列测试文件属性的方法:exist(),isFile(),isDirectory(),canRead(),canWrite(),setReadOnly(),length()。

File 类还提供了一系列有关文件的操作方法:creatNewFile(),mkdir(),mkdirs(),delete(),deleteOnExit(),list()。

### 2. Java 的 I/O 流

Java 语言是通过数据流来实现数据的输入和输出的。不论数据的类型还有所处的位置有何不同,程序均以流的方式进行数据处理。

程序读数据：建立输入流→依次读取信息→关闭流。

程序写数据：建立输出流→依次写入信息→关闭流。

数据流类按照数据类型可以分成两种继承层次：基于字节的数据流类和基于字符的数据流类，有时也简称为字节流和字符流。

1) 字节流类

Java 语言中，字节流有两大基本类：抽象类 java.io.InputStream 和 java.io.OutputStream，它们是所有基于字节的数据流类的父类。类 InputStream 是字节输入流，该类中包含了一套所有字节输入流都需要的方法，可以实现最基本的输入流读取数据的功能。类 OutputStream 是字节输出流，包含所有字节输出流都要使用的方法。

字节输入流的类层次如图 9-1 所示。字节输出流的类层次如图 9-2 所示。其中带阴影的类是节点流，其他类是过滤流。

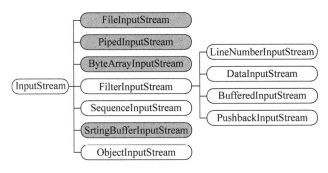

图 9-1　字节输入流类层次

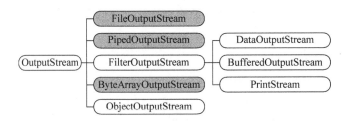

图 9-2　字节输出流类层次

常用的字节流有：用于从文件中读写数据的文件输入/输出流 FileInputStream/FileOutputStream，以及用于从缓冲区流中读写数据的缓冲流 BufferedInputStream/BufferedOutputStream。采用缓冲流可以减少实际上从外部输入设备上读写数据的次数，从而提高效率。

2) 字符流类

Java 语言中，抽象类 java.io.Reader 和 java.io.Writer 是基于字符的数据流的父类。类 Reader 是字符输入流，其中包含了一套所有字符输入流都需要的，可以实现最基本的从字符输入流读取数据的功能。Writer 是字符输出流，包含所有字符输出流都要使用的方法。

字符输入流的类层次如图 9-3 所示。字符输出流的类层次如图 9-4 所示。其中带阴影的类是节点流，其他类是过滤流。

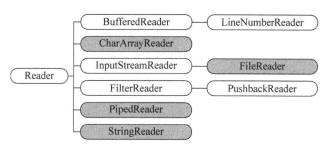

图 9-3 字符输入流类层次

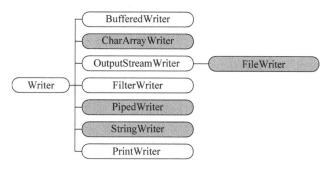

图 9-4 字符输出流类层次

常用的字符流有：文件字符流 FileReader/FileWriter，缓冲字符流 BufferedReader/BufferedWriter。

**3．随机访问文件**

在上面所介绍的数据流能实现对数据的顺序读取和写入，并不能对一个已存在的文件随机的修改。为了实现对文件的随机访问以及对文件在任意位置的修改，Java 语言提供了一个类 java.io.RandomAccessFile。RandomAccessFile 类和 File 类同属于 Java 语言的管理文件系统，其中 RandomAccessFile 类提供了文件的随机读写。这就意味着 RandomAccessFile 可以让用户从文件的不同位置读写不同长度的数据。

**4．对象的序列化**

对象序列化就是将对象的状态转换成字节流，以后可以通过这些值再生成相同状态的对象。在 java.io 包中，接口 Serializable 是实现对象序列化的工具，只有实现了 Serializable 接口的类对象才可以被序列化。对象序列化主要有两方面的应用：一方面是保存文件以及网络中进行对象传送；另一方面是实现分布式应用，可以通过 RMI 远程访问特定对象。

## 9.2 设计 1 文件管理

**1．设计目的**

（1）掌握 Java 的文件管理。

(2) 掌握流式输入输出的基本原理。

(3) 熟悉 FileInputStream、FileOutputStream、FileReader、FileWriter 等常用流的用法。

(4) 理解 BufferedReader、BufferedWriter 等缓冲流的作用。

## 2. 设计要求

请在指定目录下创建一个名为 dir 的子目录，在 dir 目录下创建三个文件：file1.txt、file2.txt、file3.txt，然后列表显示指定目录；重命名 file2.txt 为 file.txt，列表显示 dir 目录；删除文件 file3.txt，再列表显示 dir 目录。

## 3. 设计步骤

### 1) 分析

本设计主要考查对 File 类的使用，在 Java 中，文件和目录都是用 File 对象来表示的，创建和区分方法：先创建一个 File 对象，并指定文件名或目录名，若指定文件名或目录名不存在，则 File 对象的建立并不会创建一个文件或目录，需要用 creatNewFile 方法或 mkdir 方法来分别创建文件或目录。查阅 API 文档中有关 File 类的描述，本设计中还要用到的方法有 isDirectory()、getName()、renameTo()、list()、getPath()、delete() 等。

### 2) 编写程序

程序清单 TestFile.java：

```java
import java.io.*;
public class TestFile{
    public static void main(String []args){
        if(args.length == 0)   {
            args = new String[1];
            args[0] = ".";
        }
        try{
            File curPath = new File(args[0]);
            if(!curPath.isDirectory())
                curPath.mkdir();
            File dirPath = new File(curPath,"dir");
            dirPath.mkdir();
            File f1 = new File(dirPath,"file1.txt");
            f1.createNewFile();
            File f2 = new File(dirPath,"file2.txt");
            f2.createNewFile();
            File f3 = new File(dirPath,"file3.txt");
            f3.createNewFile();
            System.out.println("显示指定目录的内容");
            listDir(curPath);
            File newf2 = new File(dirPath,"file.txt");
            f2.renameTo(newf2);
            System.out.println("文件改名后,显示 dir 子目录的内容");
            listDir(dirPath);
            f3.delete();
            System.out.println("删除文件后,显示 dir 子目录的内容");
```

```
            listDir(dirPath);
        }catch(IOException e){
            System.out.println("无法创建文件");
        }
    }
    static void listDir(File tempPath){
        String[] fileNames = tempPath.list();
        try{
            for(int i = 0;i < fileNames.length;i++){
                File f = new File(tempPath.getPath(),fileNames[i]);
                if(f.isFile())
                    System.out.println(f.getName());
                else{
                    System.out.println(f.getCanonicalPath());
                    listDir(f);
                }
            }
        }catch(IOException e){
            System.err.println("IOException");
        }
    }
}
```

3）编译并运行程序

结果如图 9-5 所示。

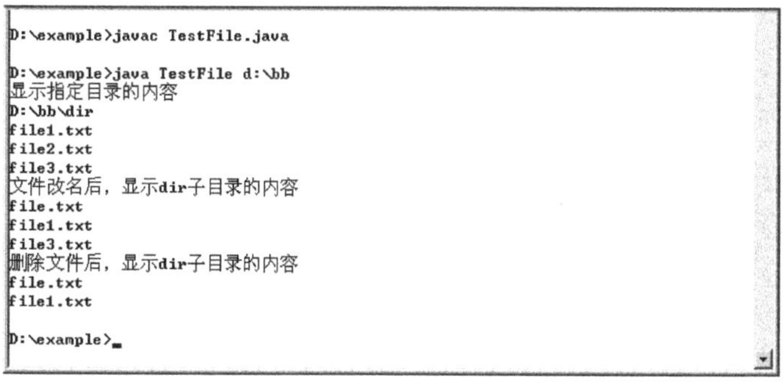

图 9-5　程序 TestFile.java 的编译运行结果

### 4. 拓展设计

（1）请编写程序 TestFileIOStream.java，运用 FileInputStream 流和 FileOutputStream 流，将上述文件 TestFile.java 复制到 d:\bb\dir\file.txt 文件中。

（2）请考虑：Java 语言中，在进行文件复制时，为提高程序的执行性能，可以采取什么方法？请解释原因，并写出相关程序。

### 5. 设计提示

（1）拓展设计(1)中的关键语句：

```java
FileInputStream in = new FileInputStream("d:/example/TestFile.java");
FileOutputStream out = new FileOutputStream("d:/bb/dir/file.txt");
while((b = in.read())!= -1){
    out.write(b);
}
in.close();
out.close();
```

注意程序中文件路径分隔符的书写方式。

(2) Java 语言提供的缓冲流可以用来提高程序的执行性能，BufferedReader 流中提供的方法 readLine()和 BufferedWriter 流中提供的方法 write(String str)都可以减少实际上从外部设备访问数据的次数，从而提高效率。例如，以下程序完成拓展设计(1)中的文件复制：

```java
import java.io.*;
public class TestBufferStream{
    public static void main(String[] args) {
        try {
        BufferedWriter bw = new BufferedWriter(
                new FileWriter("d:\\bb\\dir\\file1.txt"));
        BufferedReader br = new BufferedReader(
                new FileReader("d:\\bb\\dir\\file.txt"));
        String s = null;
        while((s = br.readLine())!= null){
          bw.write(s);
          bw.newLine();
        }
        bw.flush();
        bw.close();
        br.close();
        } catch (IOException e){
            e.printStackTrace();
        }
    }
}
```

## 9.3 设计 2  常用流练习

### 1. 设计目的

(1) 巩固 FileInputStream、FileOutputStream 等节点流的使用方法。

(2) 掌握 DataInputStream、DataOutputStream、ObjectInputStream、ObjectOutputStream 等处理流的使用方法。

(3) 掌握标准输入输出流的使用方法。

### 2. 设计要求

编写程序，实现以下两个功能：

(1) 从键盘输入学生信息,包括学号、姓名和三门课的成绩,将其保存在一个文本文件中。其中文件名和学生人数要求从键盘输入。(要求使用 DataOutputStream 流)

(2) 从以上文本文件读取每个学生的信息,计算每个学生各门课的平均成绩,将其写入文件 average.txt 中(要求使用 BufferedWriter 流实现学生信息的写入)。

**3. 设计步骤**

1) 分析

实现功能(1)时,要从键盘输入信息,只需 System.in 的 read()方法即可。也可以在 System.in 上套接其他处理流,如 InputStreamReader 流和 BufferedReader 流,这样可以更方便地从标准输入流上读取数据。将信息写入文本文件时,要求使用 DataOutputStream 流,这是一个处理流,必须套接在一个节点流 FileOutputStream 上。

实现功能(2)时,要将 DataInputStream 流套接在 FileInputStream 流上以读取各个文件的信息。将平均成绩等信息写入指定文件时,要求用 BufferedWriter 流,则该流必须套接在一个节点流 FileWriter 上。

2) 编写程序

程序清单 Score.java(该程序实现功能(1)):

```java
import java.io.*;
class Score{
    public static void main(String []args){
        int id;
        String name;
        float chinese,math,english;
        BufferedReader br;
        DataOutputStream dos;
        try{
            br = new BufferedReader(new InputStreamReader(System.in));
            System.out.print("请输入文件名:");
            String file = br.readLine();
            dos = new DataOutputStream(new FileOutputStream(file + ".txt"));
            System.out.print("请输入学生人数:");
            int num = Integer.parseInt(br.readLine());
            for(int i = 0;i < num;i++){
                System.out.print("请输入学生学号:");
                id = Integer.parseInt(br.readLine());
                dos.writeInt(id);
                System.out.print("请输入学生姓名:");
                name = br.readLine();
                dos.writeUTF(name);
                System.out.print("请输入语文成绩:");
                chinese = Float.parseFloat(br.readLine());
                dos.writeFloat(chinese);
                System.out.print("请输入数学成绩:");
                math = Float.parseFloat(br.readLine());
                dos.writeFloat(math);
                System.out.print("请输入英语成绩:");
```

```
            english = Float.parseFloat(br.readLine());
            dos.writeFloat(english);
        }
        dos.close();
        br.close();
    }catch(Exception e){e.printStackTrace();}
    }
}
```

程序清单 AveScore.java(该程序实现功能(2))：

```
import java.io.*;
class AveScore {
    int id;
    String name;
    float score;
    void saveAveScore(){
    BufferedReader br;
    DataInputStream dis;
    BufferedWriter bw;
    try{
        br = new BufferedReader(new InputStreamReader(System.in));
        System.out.print("请输入文件名：");
        String file = br.readLine();
        dis = new DataInputStream(new FileInputStream(file+".txt"));
        br.close();
        bw = new BufferedWriter(new FileWriter("average.txt"));
        while(true){
            id = dis.readInt();
            name = dis.readUTF();
            score = (dis.readFloat()+dis.readFloat()+dis.readFloat())/3;
            bw.write(" "+ id + '\t' + name + '\t' + score);
            bw.newLine();
            bw.flush();
        }
    }catch(EOFException e){
        System.out.println("所有数据已读完!");}
     catch(IOException e){
        e.printStackTrace();}
    }
    public static void main(String []args){
        AveScore ts = new AveScore();
        ts.saveAveScore();
    }
}
```

3) 编译并运行程序

程序 Score.java 编译运行结果如图 9-6 所示。

程序 AveScore.java 编译运行结果如图 9-7 所示。

用记事本打开 average.txt 文件，文件内容如图 9-8 所示。

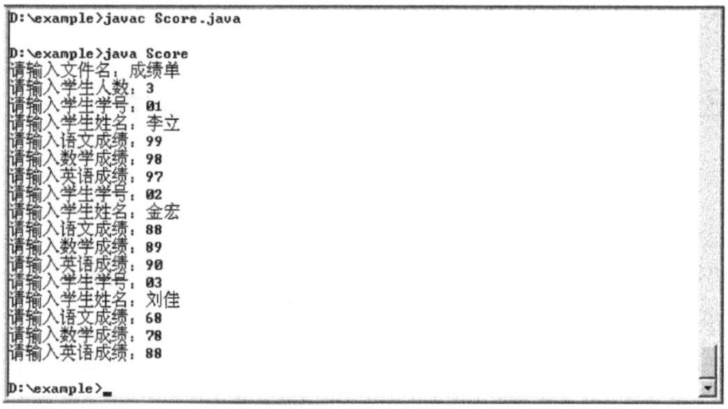

图 9-6  程序 Score.java 的编译运行结果

图 9-7  程序 AveScore.java 的编译运行结果

图 9-8  average.txt 文件内容

### 4. 拓展设计

(1) 请问：将程序 AveScore.java 中的异常捕获语句

catch(EOFException e){System.out.println("所有数据已读完!");}

去掉，程序编译运行将会是什么结果？

(2) 定义学生类，从键盘输入学生的信息，包括学号、姓名、成绩三个属性，创建学生对象，用 ObjectOutputStream 流把学生对象写入文件，再用 ObjectInputStream 流读出学生信息显示在屏幕上。

### 5. 设计提示

(1) 如果到达 InputStream 的末尾之后还继续从中读取数据，就会发生 EOFException 异常。这个异常可以用来检查是否已经到达文件末尾，如程序清单 AveScore.java 所示。

(2) 在拓展设计(2)中，定义学生类时，必须设置 Serializable 接口。此外，可以定义一个数组，用来存放从键盘输入的学生信息构成的多个学生对象，然后将学生对象数组中的元素依次通过 ObjectOutputStream 流中输出到指定文件，关闭该输出流。为实现从刚刚写入信息的文件中读取学生对象，必须创建 FileInputStream 流，再套接 ObjectInputStream 流，依次读出所有对象,并将其属性信息在屏幕输出。

## 9.4 设计 3 RandomAccessFile 类的应用

### 1. 设计目的

(1) 了解和掌握 RandomAccessFile 类的构造方法和常用的方法。
(2) 利用 RandomAccessFile 实现随机访问文件。
(3) 比较用 RandomAccessFile 访问文件与其他字符流和字节流的不同。

### 2. 设计要求

请编写一个程序,从键盘输入联系人姓名,若通讯录中有此人,则输出其联系方式,否则,将该联系人姓名及手机号保存至通讯录。(要求用 RandomAccessFile 类实现对通讯录文件的随机读写)

### 3. 设计步骤

1) 分析

程序中需要从键盘读入数据,可以用标准输入 System.in 的 read()方法,也可以在其上套接其他过滤流,如 InputStreamReader 流以及 BufferedReader 流。

程序要求用 RandomAccessFile 类实现对通讯录文件的随机读写,这样,就需要创建与通讯录文件对象绑定的 RandomAccessFile 类,并指定其读写模式。

2) 编写程序

程序清单 Telephone.java:

```java
import java.io.*;
class Telephone{
    public static void main(String []args){
        String name;
        RandomAccessFile rf;
        BufferedReader br;
        PrintWriter pw;
        try{
            rf = new RandomAccessFile("phone.txt","rw");
            br = new BufferedReader(new InputStreamReader(System.in));
            pw = new PrintWriter(new OutputStreamWriter(System.out));
            System.out.print("请输入姓名: ");
            String s = br.readLine();
            rf.seek(0);
            try{
                while(rf.getFilePointer()<= rf.length()){
                    name = rf.readUTF();
                    if(!s.equals(name)){
                        rf.readUTF();
                    }
```

```
                else {
                    String tel = rf.readUTF();
                    pw.println(tel);
                    pw.flush();
                    rf.close();
                }
            }
        }catch(EOFException e){
            System.out.println("通讯录无此人!");
            rf.writeUTF(s);
            System.out.print("请输入手机号: ");
            s = br.readLine();
            rf.writeUTF(s);
        }
        rf.close();
        br.close();
        pw.close();
    }catch(Exception e){System.out.println("哈哈,找到了!");}
    }
}
```

3) 编译运行程序

若目录 d:\example 中无 phone.txt 文件,则编译并运行程序,将会在此目录下创建文件 phone.txt,相继输入联系人 jack,mary,mary 时,运行结果如图 9-9 所示。

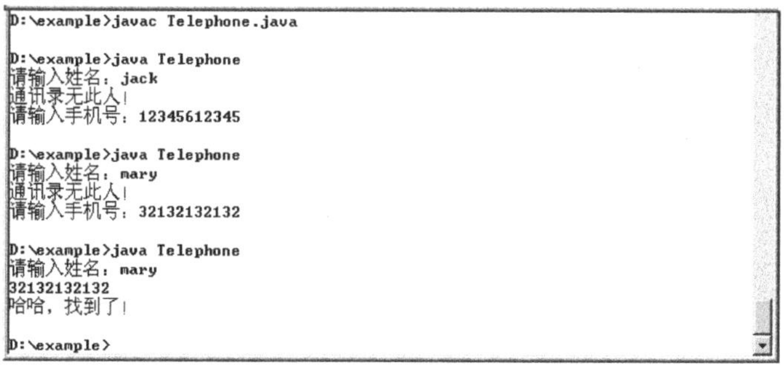

图 9-9　程序 Telephone.java 编译运行结果

### 4. 拓展设计

(1) 请问：当在通讯录中查找有无从键盘输入的联系人时,若想跳过通讯录中每个联系人的联系方式,可以使用 RandomAccessFile 类的 skipBytes()方法,程序应如何修改？

(2) 请将程序中通讯方式(手机号)定义为字节数组,使用 RandomAccessFile 类中的相应 API 实现该程序。

### 5. 设计提示

(1) 当使用 RandomAccessFile 类的 writeUTF()方法写入一个字符串时,在文件的当前指针位置处将写入整个字符串在文本文件中的实际长度(注意不是字符串长度),占两个字节,接着再写入字符串。原程序中,联系方式(手机号)为 String 类型,所以要想在查询联系人时跳过其联系方式这部分内容,可以使用如下语句:

```
rf.skipBytes(rf.readShort());
```

(2) 将字节数组写入文件,可以使用 RandomaccessFile 类的 write(byte[] b)方法,此时需将从键盘输入的字符串利用其 getBytes()方法转换为字节数组,然后再执行写入操作。从文件读出时可以利用 RandomaccessFile 类的 read(byte[] b)方法。为实现本程序中的查找联系人操作,当要跳过联系方式时,可以使用如下语句:

```
rf.skipBytes(b.length);
```

# 第2部分

# Java应用技术

第 10 章　Java 图形用户界面设计
第 11 章　Java 多媒体程序设计
第 12 章　Java 多线程程序设计
第 13 章　综合案例：拼图游戏

# 第10章 Java图形用户界面设计

图形化用户界面GUI(Graphics User Interface)实现应用与用户的交互,是应用程序的重要组成部分。在Java语言中提供了大量的类来实现界面的设计。这些类按照作用域可以分为三个方面:关于GUI组件的类(如按钮、文本框等)、容器处理的类(如窗口、面板等)和辅助的类(如图形、颜色处理、字体处理等)。这些类的详细学习可以查阅Java的GUI API。

GUI编程不仅仅是关于Java的组件在容器中的排列和显示的美观,更重要的是,要使组件能够根据用户的动作产生响应或随着组件的某些状态的变化而变化,即能够处理某些特定的操作,实现某些功能。因此,学习组件除了要了解组件的属性和功能外,一个更重要的方面是学习怎样处理组件上发生的界面事件。

Java语言中处理GUI接口可以依赖于java.awt包和javax.swing包。java.awt包通常被称为AWT(Abstract Windows Toolkit),即抽象窗口工具集,而javax.swing包被称为Swing工具集。

本章利用生活中最常见的两个案例来学习利用Swing构建GUI、GUI事件处理模型、Swing组件及辅助类。

## 10.1 案例:几何图形计算器

### 10.1.1 案例问题描述

由于学习或工作的需要,人们经常要对一些常见的平面图形,如三角形、矩形、正方形、梯形、圆等,计算它们的周长及面积,或者对一些常见的几何体,如圆柱、圆锥、球体等,计算它们的表面积及体积,甚至对一些更复杂图形做一系列的相关计算。本节案例的目的就是设计一个能对这些常见图形进行计算的计算器。希望读者通过本案例的学习,掌握Swing的相关组件及其事件处理、容器及布局管理的知识。

### 10.1.2 案例功能分析及演示

几何图形计算器运行以后,首先显示的是程序的主窗口,见图10-1。程序主窗口显示为两部分,一个是一个组合框对象,一个是一幅图像。组合框的默认选项显示"请单击下拉列表选择",所以当用户单击了下拉列表右端向下的箭头后,可以看到此计算器可以完成哪些图形的计算,见图10-2。

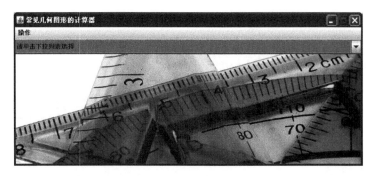

图 10-1　程序的主界面(一)

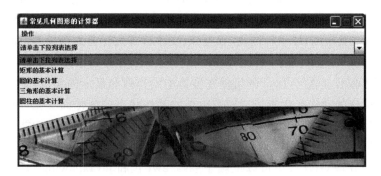

图 10-2　程序的主界面(二)选择图形

假设用户选择了圆形,则程序界面会切换到计算圆形的周长和面积的图形界面,见图 10-3。在这个界面中,用户首先需要输入半径才能得到它的周长和面积。如果不设计图 10-3 所示的数字键盘,也可以帮助用户得到圆的周长和面积,但是考虑到这样做会使用户不停地处于键盘和鼠标的切换当中,用户体验效果差,因此设计了数字键盘,帮助用户一直使用鼠标就可以完成计算。当然在实际录入数字的时候,用户可根据自己的输入习惯,选择键盘录入,或者选择数字键盘上的按钮录入。

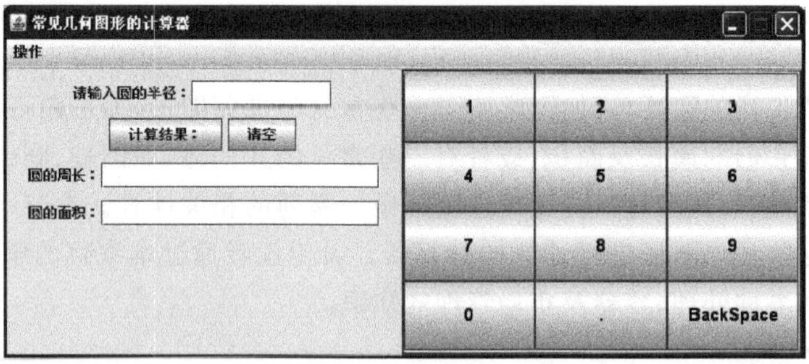

图 10-3　圆的计算器界面(一)

当用户输入圆的半径,单击"计算结果:"按钮后,可以得到圆的周长和面积,见图 10-4。如果求另一个圆的周长和面积,则可以单击"清空"按钮,清空原来的半径,用户重新输入半径即可得到新圆的周长和面积。

在使用数字键盘时,可以看到除了有 0～9 这样的数字,还有"."按钮和退格键

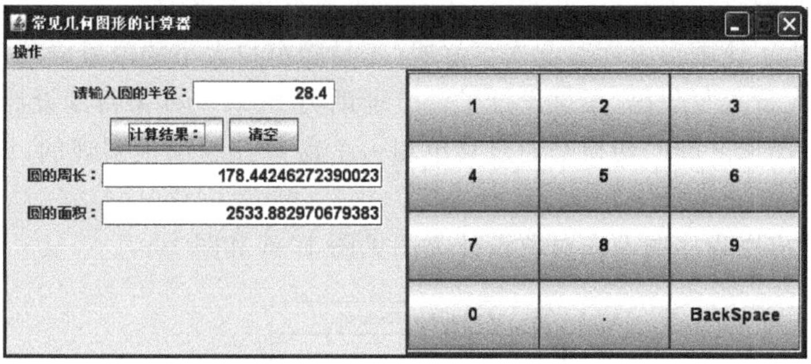

图 10-4　圆的计算器界面(二)

(BackSpace 键),这样可以帮助用户输入浮点数,如果上一个键按错了,还可以按 BackSpace 键删除上一个数。

如果要退出整个应用程序,可以选择"操作"菜单中的"退出"命令;如果想结束圆的计算,进行其他图形的计算,则单击"操作"菜单中的"主界面"命令,这样界面就会恢复到图 10-1 的初始状态,用户可以重新选择新的图形进行计算。见图 10-5。

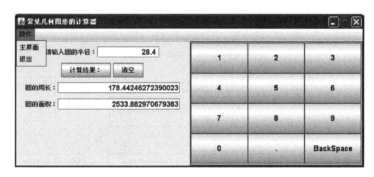

图 10-5　程序"操作"菜单的功能

## 10.1.3　案例总体设计

用 Java 编写一个面向对象的程序,很大程度上是设计类,并编写这些类的定义。因此,通过对该案例功能的分析及演示,下面把问题分解成一组相互协作的对象,然后设计和创建这些对象本身。

为了说明如何处理这样的问题,这里简化了案例,只选择了矩形、圆、三角形、圆柱这 4 种情形来说明对几何图形的基本计算。因此,本案例分别设计了 RectanglePanel 类、CirclePanel 类、TrianglePanel 类、CylinderPanel 类这 4 个类来实现对矩形、圆、三角形、圆柱的基本计算。这 4 个类的父类是抽象类 AbstractPanel 类,设计此类是为了获得子类图形输入参数的当前焦点对象。这样数字软键盘才知道当前要操作的对象是谁,即实现对哪个图形对象的参数录入。

这 4 个类完全可以实现对指定图形的简单计算,只不过图形参数的输入需要通过计算机键盘手动录入。考虑到用户界面设计的友好性,案例设计了一个数字软键盘,这样用户就不必总是在键盘和鼠标之间切换。考虑到这一问题,设计了 KeyJPanel 类来实现数字软键盘的功能,这个软键盘除了有 0~9 数字,还包括小数点和退格键。

除此以外,为了让图形界面看起来更好看(不易随窗口大小的位置或其他因素的变化随意变形)和减少大量重复代码,案例设计了一个辅助类 BoxPanel 来帮助设计。

程序对 4 种图形之间的切换是通过菜单和一个窗口控制面板实现的,详细的设计可以见后文。

图形计算器应用程序的面向对象设计类图见图 10-6 和图 10-7。

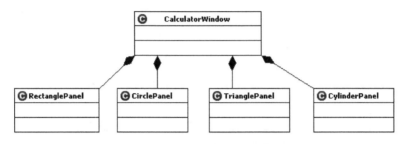

图 10-6　图形计算器面向对象设计图(一)

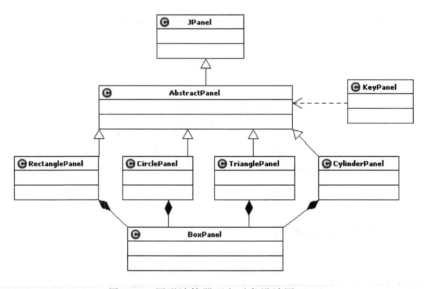

图 10-7　图形计算器面向对象设计图(二)

1. **CalculatorWindow 类**

CalculatorWindow 类实现程序的主界面,它的 UML 图如图 10-8 所示。以下是 UML 图中有关数据和方法的详细说明。

1) 成员变量
- mycard 是 CardLayout 布局管理器。
- controlPanel 是控制面板,负责选择要进行计算的图形。
- pCenter 是一个面板,放置在窗口中间位置,布局为 CardLayout 型,控制面板 controlPanel 和其他 4 类图形面板都在这个容器里。

- chooseList 是个组合框，提供用户进行选择的列表。
- menuBar 是 JMenuBar 类创建的菜单条，可以向 menuBar 中添加菜单。
- menu 是 JMenu 类创建的菜单，这个菜单主要帮助切换回控制面板和退出程序。
- mainWindow 和 exit 是 JMenuItem 类创建的两个菜单项，被添加到 menu 菜单中，这两个菜单项的名称分别是"主界面"和"退出"。

2) 成员方法
- CalculateWindow()是构造方法，负责完成窗口的初始化。
- main(String[])方法是软件运行的入口方法。
- actionPerformed(ActionEvent)方法是 CalculatorWindow 类实现的 ActionListener 接口中的方法，负责执行菜单项发出的有关命令。用户选择菜单中的菜单项可触发 ActionEvnet 事件，导致 actionPerformed(ActionEvent)方法执行相应的操作。
- itemStateChanged(ItemEvent)方法是 CalculatorWindow 类实现的 ItemListener 接口中的方法，负责切换 pCenter 面板中的图形面板。用户选择下拉列表中的某个选项时触发 ItemEvent 事件，导致 itemStateChanged(ItemEvent)方法执行相应的操作。

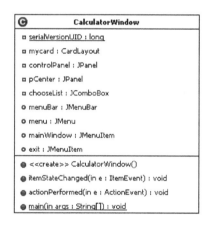

图 10-8　CalculatorWindow 类的 UML 图

### 2. CirclePanel 类

CirclePanel 类的 UML 图如图 10-9 所示。以下是 UML 图中有关数据和方法的详细说明。

1) 成员变量
- resultButton、clearButton 是 JButton 类创建的按钮对象，其上的名字依次为"计算结果："、"清空"。这两个按钮都注册有 ActionEvent 事件监听器。
- leftPanel、rightPanel、buttonPanel 都是 JPanel 对象，是容器。leftPanel 负责显示参数的输入、按钮面板和结果的显示；rightPanel 负责显示数字软键盘；buttonPanel 里有两个按钮，分别是 resultButton 和 clearButton。
- radiusTextField、lengthTextField、areaTextField、inputTextField 是 4 个 JTextField 对象。radiusTextField 文本框接收用户输入的圆的半径，注册 FocusEvent 事件监

听器；lengthTextField 文本框负责显示圆的周长；areTextField 文本框负责显示圆的面积。inputTextField 表明当前面板上哪个参数文本框获得了焦点。
- bpRadius、bpLength、bpArea 是三个 BoxPanel 对象。这三个对象都是一个水平排列组件的 Box 容器。bpRadius 容器负责显示用户输入参数的界面，bpLength 容器负责显示圆的周长界面，bpArea 负责显示圆的面积界面。

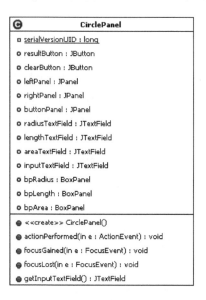

图 10-9　CirclePanel 类的 UML 图

2) 成员方法
- CirclePanel()是构造方法，负责完成圆计算器的窗口初始化。
- actionPerformed(ActionEvent)方法是 CirclePanel 类实现的 ActionListener 接口中的方法。用户单击 resultButton（显示结果）按钮或 clearButton（清空）按钮可触发 ActionEvent 事件，导致 actionPerformed(ActionEvent)方法执行相应的操作。
- focusGained(FocusEvent)和 FocusLost(FocusEvent)方法是 CirclePanel 类实现的 FocusListener 接口中的两个方法。当参数文本框获得输入焦点或失去输入焦点时可触发 FocusEvent 事件，如果是获得焦点，则自动执行 focusGained(FocusEvent)方法；如果是失去焦点，则自动执行 focusLost(FocusEvent)方法中的代码。
- getInputTextField()方法是重写父类 AbstractPanel 中的方法，通过该方法可以知道当前获得输入焦点的是哪个文本框。

3. RectanglePanel 类

RectanglePanel 类的 UML 图如图 10-10 所示。以下是 UML 图中有关数据和方法的详细说明。

1) 成员变量
- resultButton、clearButton、leftPanel、rightPanel、buttonPanel 这 5 个成员变量的含义同上面 CirclePanel 类中对应的成员变量的含义。

- widthTextField、heightTextField、lengthTextField、areaTextField、inputTextField 是 5 个 JTextField 对象。widthTextField 文本框具体负责接收用户输入的表示矩形宽的参数，该文本框注册 FoucsEvent 事件监听器；heightTextField 负责接收用户输入的表示矩形高的参数，该文本框注册 FocusEvent 事件监听器；lengthTextField 和 areaTextField 两个文本框分别用来显示当前矩形周长和面积的计算结果；inputTextField 表示当前获得输入焦点的文本框。
- bpWidth、bpHeight、bpLength、bpArea 是 4 个 BoxPanel 类对象，这 4 个对象是 4 个水平排列组件的 Box 容器。bpWidth 容器负责管理矩形宽的输入，bpHeight 容器负责管理矩形高的输入，bpLength 和 bpArea 分别负责管理矩形的周长和面积的显示。

2）成员方法
- RectanglePanel()是构造方法，负责矩形计算器窗口的初始化。
- actionPerformed(ActionEvent)方法是 RectanglePanel 类实现的 ActionListener 接口中的方法，负责执行用户单击按钮发出的相关命令。具体含义同 CirclePanel 类中的相关说明。
- focusGained(FocusEvent)方法、focusLost(FoucsEvnet)方法是 RectanglePanel 类实现的 FocusListener 接口中的两个方法，根据当前输入参数文本框的焦点状态来决定自动执行哪个方法。具体含义见 CirclePanel 类中的相关说明。
- getInputTextField()方法含义见 CirclePanel 类中相应的说明。

图 10-10　RectanglePanel 类的 UML 图

## 4．TrianglePanel 类

TrianglePanel 类的 UML 图如图 10-11 所示。以下是 UML 图中有关数据和方法的详

细说明。

1) 成员变量
- resultButton、clearButton、leftPanel、rightPanel、buttonPanel 这 5 个成员变量的含义同上面 CirclePanel 类中对应的成员变量的含义。
- sideATextField、sideBTextField、sideCTextField、lengthTextField、areaTextField、inputTextField 是 6 个 JTextField 对象。sideATextField、sideBTextField、sideCTextField 三个文本框分别负责接收用户输入的表示三角形三边的参数,且三个文本框都注册了 FoucsEvent 事件监听器;lengthTextField 和 areaTextField 这两个文本框分别用来显示当前三角形周长和面积的计算结果;inputTextField 表示当前获得输入焦点的文本框。
- bpSideA、bpSideB、bpSideC、bpLength、bpArea 是 5 个 BoxPanel 类对象,这 5 个对象是 5 个水平排列组件的 Box 容器。bpSideA、bpSideB、bpSideC 容器负责管理三角形三边的输入。bpLength 和 bpArea 分别负责管理三角形的周长和面积的显示。

图 10-11　TrianglePanel 类的 UML 图

2) 成员方法
- TrianglePanel()是构造方法,负责三角形计算器窗口的初始化。
- actionPerformed(ActionEvent)方法是 TrianglePanel 类实现的 ActionListener 接口中的方法,负责执行用户单击按钮发出的相关命令。具体含义同 CirclePanel 类中的相关说明。
- focusGained(FocusEvent)方法、focusLost(FoucsEvnet)方法是 TrianglePanel 类实现的 FocusListener 接口中的两个方法,根据当前输入参数文本框的焦点状态来决定自动执行哪个方法。具体含义见 CirclePanel 类中的相关说明。

- getInputTextField()方法得到当前获得焦点的文本框对象。

**5．CylinderPanel 类**

CylinderPanel 类的 UML 图如图 10-12 所示。以下是 UML 图中有关数据和方法的详细说明。

1) 成员变量

- resultButton、clearButton、leftPanel、rightPanel、buttonPanel 这 5 个成员变量的含义同上面 CirclePanel 类中对应的成员变量的含义。
- radiusTextField、heightTextField、surfaceAreaTextField、volumeTextField、inputTextField 是 5 个 JTextField 对象。radiusTextField 和 heightTextField 两个文本框分别负责接收用户输入的表示圆柱底面半径和圆柱高的参数，并且这两个文本框都注册了 FocusEvent 事件监听器；surfaceAreaTextField 和 volumeTextField 两个文本框分别用来显示当前圆柱表面积和体积的计算结果；inputTextField 表示当前获得输入焦点的文本框。
- bpRadius、bpHeight、bpSurfaceArea、bpVolume 是 4 个 BoxPanel 类对象，它们是水平排列组件的 Box 容器。bpRaidus、bpHeight 容器负责管理圆柱底面半径和高的输入。bpSurfaceArea 和 bpVolume 分别负责管理圆柱的表面积和体积的显示。

图 10-12  CylinderPanel 类的 UML 图

2) 成员方法

- CylinderPanel()是构造方法，负责圆柱计算器窗口的初始化。
- actionPerformed(ActionEvent)方法是 CylinderPanel 类实现的 ActionListener 接口中的方法，负责执行用户单击按钮发出的相关命令。具体含义同 CirclePanel 类中的相关说明。

- focus Gained(FocusEvent)方法、focusLost(FoucsEvnet)方法是 CylinderPanel 类实现的 FocusListener 接口中的两个方法,根据当前输入参数文本框的焦点状态来决定自动执行哪个方法。具体含义见 CirclePanel 类中的相关说明。
- getInputTextField()方法得到当前获得焦点的文本框对象。

### 6. AbstractPanel 类

AbstractPanel 类的 UML 图如图 10-13 所示。

这个类中只有一个抽象方法 getInputTextField(),得到面板当前获得焦点的文本框对象。它的子类必须重写这个方法。

图 10-13　AbstractPanel 类的 UML 图

### 7. BoxPanel 类

BoxPanel 类的 UML 图如图 10-14 所示。以下是 UML 图中有关数据和方法的详细说明。

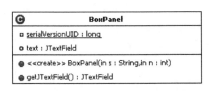

图 10-14　BoxPanel 类的 UML 图

这个类中有两个成员方法,分别是:
- BoxPanel(String,int)是构造方法,初始化容器。BoxPanel 是 JPancl 的子类,它内部有一个水平排列组件的 Box 容器。该 Box 容器中第一个组件是一个标签,第二个组件是一个文本框对象,因此,该构造方法中第一个参数指定标签的显示文本,第二个参数指定文本框对象的列数。
- getJTextField()方法的目的是获得当前面板对象的文本框对象。

### 8. KeyJPanel 类

KeyJPanel 类的 UML 图如图 10-15 所示。以下是 UML 图中有关数据和方法的详细说明。

1) 成员变量
- keyButton 是 JButton 型数组,每个元素是一个 JButton 类创建的"数字键盘"对象,该数组的长度是 12。keyButton 数组中含有的按钮依次为 0,1,2,…,9,.,BackSpace 键。每个数字键盘按钮都注册有 ActionEvent 事件监听器。
- num 是 String 型数组,每个元素是一个字符串,该数组的内容为:"1","2","3",

"4","5","6","7","8","9","0",". ","BackSpace"。
- selectedPanel 表示当前用户选择进行计算的是哪个图形计算器面板。
- inputTextField 表示当前数字软键盘实现对该文本框对象的参数录入工作。

2) 成员方法
- KeyJPanel(AbstractPanel)是构造方法,完成对数字软键盘的初始化工作。参数指定当前数字软键盘实现对哪个图形的计算工作。
- actionPerformed(ActionEvent)方法是 KeyJPanel 类实现的 ActionListener 接口中的方法,负责执行用户单击数字软键盘上按钮发出的相关命令。当用户单击了按钮就触发了 ActionEvent 事件,导致 actionPerformed(ActionEvent)方法执行相应的操作。
- inputNumber(JTextField,JButton)方法负责具体处理当用户单击了按钮后如何实现对文本框的数据录入。分为下面三种情况,当用户单击了数字型的按钮;单击了"."按钮;单击了退格键 BackSpace 时应该如何处理。第一个参数输入数字软键盘操作的 JTextField 对象,第二个参数输入当前用户单击的数字软键盘上的按钮。

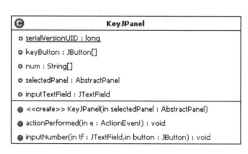

图 10-15  KeyJPanel 类的 UML 图

## 10.1.4 案例代码实现

### 1. 编写程序主窗体(CalculatorWindow.java)

CalculatorWindow.java 文件中包含了程序的入口,是本应用程序的主窗体。通过这个主窗体实现各种图形直接的切换。

CalculatorWindow 类是 javax.swing 包中 JFrame 的一个子类,并实现了 ActionListener 接口和 ItemListener 接口。从图 10-1~图 10-4 可以看到本程序把对矩形、圆、三角形、圆柱的计算都整合到了这个主窗体中,主窗体通过菜单操作和下拉列表的选项实现不同图形之间的切换。下面读者需要了解的是,如何响应用户单击菜单的操作和选择哪种布局管理器可以把这几种功能整合在一起。

1) GUI 中的事件处理

为了使图形界面能够响应用户的操作,必须给各个组件加上事件处理机制。

在事件处理的过程中,主要涉及以下三类对象。
- EventSource(事件源):事件发生的场所,通常就是各个组件,例如按钮、窗口、菜单等。

- Event(事件)：事件封装了 GUI 组件上发生的特定事情(通常就是一次用户操作)。如果程序需要获得 GUI 组件上所发生事件的相关信息,都通过 Event 对象来取得。
- EventListener(事件监听器)：负责监听事件源所发生的事件,并对各种事件做出响应处理。

在本程序中当用户单击某个菜单项,或者选择主窗口下拉列表中的某个选项时,这些动作就会激发一个响应的事件,该事件会由 AWT 封装成一个相应的 Event 对象,该事件就会触发事件源上注册的事件监听器(特殊的 Java 对象),事件监听器调用相应的事件处理器(事件监听器里的实例方法)来做出对用户操作的响应,具体如图 10-16 所示。

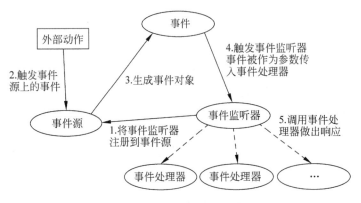

图 10-16　事件处理示意图

具体的事件处理过程见下面的步骤：

(1) 定义监听器类
- 声明监听器类。在负责事件处理的类(监听器类)的声明中指定要实现的监听器接口,例如,本例中要响应用户单击菜单项的行为,就可以这样声明：

    ```
    public class CalculatorWindow implements ActionListener{
        …
    }
    ```

- 实现监听器中的接口。在监听器类中实现监听器接口中的所有方法。例如：

    ```
    public class CalculatorWindow implements ActionListener{
        …
        //ActionListener 接口中只定义了 actionPerformed()一个方法
        public void actionPerformed(ActionEvent e){
            …            //响应某个动作的代码
        }
        …
    }
    ```

(2) 注册监听器

通过调用组件的 addXXXListener()方法,在组件上将监听器类的实例注册为监听器。例如,本例中的菜单项 exit：exit.addActionListener(this);。

同样,本例中对用户选择下拉列表中不同选项时的响应也是这样处理的。当用户选择下拉列表中的选项时,系统自动生成 ItemEvent 事件对象,该事件对象会自动触发注册到事件源上的事件监听器(具体的接口和方法请查阅 API 文档)。

2) 布局管理器

为了使生成的图形用户界面具有良好的平台无关性,Java 语言中提供了布局管理器这个工具来管理组件在容器中的布局,而不使用直接设置组件的位置和大小的方式。

每一个容器组件都有一个默认的布局管理器,也可以通过 setLayout()方法来设置其他布局管理器。一旦确定了布局管理方式,容器组件就可以使用相应的 add()方法加入组件。Java 语言中包含以下布局管理器:FlowLayout、CardLayout、BorderLayout、GridLayout、GridBagLayout、BoxLayout 等。

(1) FlowLayout 管理器

在 FlowLayout 布局管理器中,组件像水流一样向某方向流动(排列),遇到障碍(边界)就折回,重新开始排列。默认情况下,FlowLayout 布局管理器从左到右排列所有组件,遇到边界就会折回下一行重新开始。它是 Panel 的默认布局。

(2) BorderLayout 管理器

BorderLayout 管理器是常见的窗口容器和对话框容器的默认布局,它将整个容器的区域分为东、西、南、北、中 5 个部分,将组件加入容器时,需要指明放置的方向。

(3) GridLayout 管理器

GridLayout 布局管理器将容器分割成纵横线分割的网格,每个网格所占的区域大小相同。当向使用 GridLayout 布局的容器中添加组件时,默认从左到右、从上到下依次添加到每个网格中。使用 GridLayout 布局虽然可以整齐地排列组件,但由于划分的格子大小都一样,无法控制组件占用空间的大小,从而无法获得满意的效果。

(4) GridBagLayout 管理器

GridBagLayout 布局管理器是功能最强大,但也是最复杂的布局管理器,与 GridLayout 布局管理器不同的是:在 GridBagLayout 布局管理器中,一个组件可以跨越一个或多个网格,并可以设置各网格的大小互不相同,从而增加了布局的灵活性。当窗口的大小发生变化时,GridLayout 布局管理器也可以准确地控制窗口各部分的反应。

(5) BoxLayout 管理器

BoxLayout 管理器是 javax.swing 包中的盒子管理器。该管理器允许多个组件按照水平或垂直的方向排列。BoxLayout 管理器通过坐标常量来确定布局的类型。

(6) CardLayout 管理器

CardLayout 是卡片布局管理器。它将加入容器的所有组件看成一叠卡片,每次只有最上面的那个 Component 才可见。就好像一副扑克牌,它们叠在一起,每次只有最上面的一张扑克牌才可见。

(7) null 布局管理器

null 布局管理器是用户使用坐标系统来放置每个组件,即调用组件的 setBounds(int x, int y, int width, int height)方法来设置其放置位置和大小。

从上面的介绍可以看到本程序要想把控制面板、4 种图形的计算面板都整合起来,且同一时间只显示一个面板,CardLayout 布局是一种比较合适的选择。下面介绍使用 CardLayout 的一般步骤(假设有一个容器 con)。

① 创建 CardLayout 对象作为布局,如 CardLayout card=new CardLayout();。

② 容器使用 setLayout() 方法设置布局, 如: con.setLayout(card);。

③ 容器调用 add(String s, Component b) 方法将组件 b 加入到容器, 并给出了显示该组件的代号 s。最先加入 con 的是第一张, 依次排序, 组件的代号是另外给的, 和组件的名字没有必然联系, 不同的组件代号互不相同。

④ 创建的布局 card 用 CardLayout 类提供的 show() 方法, 根据容器 con 和其中的组件代号 s 显示这一组件: card.show(con,s); 也可以按组件加入容器的顺序显示组件, 如 card.first(con); 显示 con 中的第一个组件, 其他方法可查阅 API 文档。

CalculatorWindow 类就是利用上面步骤中的方法完成各个面板的整合和切换的。具体代码可见 CalculatorWinow.java 文件。

3) 代码 (CalculatorWindow.java)

```java
package caida.xinxi.shapeCalculate;

import java.awt.BorderLayout;
import java.awt.CardLayout;
import java.awt.event.ActionEvent;
import java.awt.event.ActionListener;
import java.awt.event.ItemEvent;
import java.awt.event.ItemListener;

import javax.swing.ImageIcon;
import javax.swing.JButton;
import javax.swing.JComboBox;
import javax.swing.JFrame;
import javax.swing.JMenu;
import javax.swing.JMenuBar;
import javax.swing.JMenuItem;
import javax.swing.JPanel;

public class CalculatorWindow extends JFrame implements ActionListener,ItemListener{
    CardLayout mycard;
    JPanel controlPanel,pCenter;
    JComboBox chooseList;
    JMenuBar menuBar;
    JMenu menu;
    JMenuItem mainWindow,exit;

    //CalculatorWindow 类的构造方法, 完成对程序主界面的初始化工作
    public CalculatorWindow(){
        setTitle("常见几何图形的计算器");

        //设置窗口的菜单条、菜单、菜单项, 给菜单项注册 ActionEvent 事件监听器
        menuBar = new JMenuBar();
        menu = new JMenu("操作");
        mainWindow = new JMenuItem("主界面");
        mainWindow.addActionListener(this);
        exit = new JMenuItem("退出");
        exit.addActionListener(this);
```

```java
        menu.add(mainWindow);
        menu.add(exit);
        menuBar.add(menu);
        setJMenuBar(menuBar);

        //在主窗口的中部添加一个面板 pCenter,其布局是 CardLayout 型
        mycard = new CardLayout();
        pCenter = new JPanel();
        pCenter.setLayout(mycard);
        add(pCenter,"Center");

        //设计一个控制面板 controlPanel,可供选择图形进行计算
        controlPanel = new JPanel();
        controlPanel.setLayout(new BorderLayout());
        //创建下拉列表 chooseList,并添加选项,注册 ItemEvent 事件监听器
        chooseList = new JComboBox();
        chooseList.addItem("请单击下拉列表选择");
        chooseList.addItem("矩形的基本计算");
        chooseList.addItem("圆的基本计算");
        chooseList.addItem("三角形的基本计算");
        chooseList.addItem("圆柱的基本计算");
        chooseList.addItemListener(this);
        //创建带图标的按钮
        ImageIcon icon = new ImageIcon("image/MP900438781.JPG");
        JButton imageButton = new JButton(icon);
        //在控制面板 controlPanel 容器内添加下拉列表和按钮
        controlPanel.add(imageButton,"Center");
        controlPanel.add(chooseList,"North");

        //在具有 CardLayout 布局的面板 pCenter 中添加组件
        pCenter.add("0",controlPanel);
        pCenter.add("1",new RectanglePanel());
        pCenter.add("2",new CirclePanel());
        pCenter.add("3",new TrianglePanel());
        pCenter.add("4",new CylinderPanel());

        setBounds(100,100,700,300);
        setResizable(false);
        setDefaultCloseOperation(JFrame.EXIT_ON_CLOSE);
        setVisible(true);
    }

    //当用户选择下拉列表的选项时,触发 ItemEvent 事件,自动执行该方法内的代码
    public void itemStateChanged(ItemEvent e) {
        int index = chooseList.getSelectedIndex();
        String choice = String.valueOf(index);
        mycard.show(pCenter, choice);
    }

    //当用户单击菜单项时,触发 ActionEvent 事件,自动执行该方法内的代码
    public void actionPerformed(ActionEvent e) {
```

```java
        //选择"主界面"命令,界面切换回控制面板 controlPanel 的状态
        if(e.getSource() == mainWindow){
            mycard.first(pCenter);
            chooseList.setSelectedIndex(0);
        }
        //选择"退出"命令,退出程序
        else if(e.getSource() == exit){
            System.exit(0);
        }
    }

    public static void main(String[] args) {
        new CalculatorWindow();
    }
}
```

### 2. BoxPanel.java 和 AbstractPanel.java 的实现

BoxPanel 类是 javax.swing 包中 JPanel 类的一个子类。该面板内有一个水平排列组件的 Box 容器,在该 Box 容器中有两个组件,一个是名为 s 的标签对象,一个是可见列数为 n 的文本框对象。同时设置文本框内文本是右对齐的,对文本框内的字体也进行了相应的设置。设计这样一个类,有助于后面各类图形计算的界面设计,减少了代码的重复。

AbstractPanel 类也是 javax.swing 包中的 JPanel 类的子类。它有一个抽象方法,要求它的子类必须重写(实现)这个方法。该方法要求获取当前图形界面中获得焦点的文本框对象。通过它和它的子类对象可以实现面向对象程序设计的多态特性。

1) 代码(BoxPanel.java)

```java
package caida.xinxi.shapeCalculate;

import java.awt.Font;
import javax.swing.Box;
import javax.swing.JLabel;
import javax.swing.JPanel;
import javax.swing.JTextField;

public class BoxPanel extends JPanel {
    JTextField text;

    //构造方法,初始化面板
    public BoxPanel(String s, int n){
        //创建水平排列组件的 Box 容器对象 box
        Box box = Box.createHorizontalBox();
        box.add(new JLabel(s));
        text = new JTextField(" ",n);
        //设置文本框文本右对齐
        text.setHorizontalAlignment(JTextField.RIGHT);
        //设置文本框内文本的字体格式
        text.setFont(new Font("Arial",Font.BOLD,15));
```

```
            box.add(text);
            add(box);
        }

        //获取当前容器内的文本框对象
        public JTextField getJTextField(){
            return text;
        }
    }
```

2) 代码(AbstractPanel.java)

```
package caida.xinxi.shapeCalculate;

import javax.swing.JPanel;
import javax.swing.JTextField;

public abstract class AbstractPanel extends JPanel{
    public abstract JTextField getInputTextField();
}
```

**3. 矩形计算器的实现(RectanglePanel.java)**

RectanglePanel 类是 AbstractPanel 类的子类,实现了 ActionListener 接口和 FocusListener 接口,创建的对象是 CalculatorWindow 窗口的成员之一。这个面板的运行效果见图 10-17,可见面板界面的左边是矩形的用于参数输入的文本框和计算结果的显示文本框,右边是一个数字软键盘,用户可以通过它来实现对矩形宽和高的录入,也可以用传统的键盘手动录入。

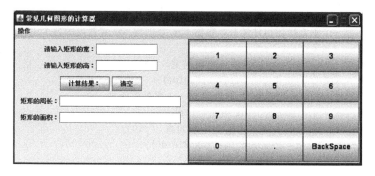

图 10-17  矩形计算器界面

RectanglePanel.java 文件主要处理的工作就是左边的界面如何设计的问题,具体就是采用什么样的布局设计和如何响应用户单击按钮的动作。响应用户的单击行为在上面已经介绍过了,这里不再详细介绍,具体处理过程见代码。至于右边的键盘界面则是 KeyJPanel 类要做的工作。

整个界面采用 GridLayout 布局,添加了两块面板,分别为 leftPanel 和 rightPanel。面板 rightPanel 是一个 KeyJPanel 对象。面板 leftPanel 的布局设计可以有多种处理方法,本程序是这样处理的:面板 leftPanel 内添加一个垂直排列组件的 Box 容器,该容器内依次添加了几个 BoxPanel 对象和一个放置两个按钮的 JPanel 对象。

代码(RectanglePanel.java)：

```java
package caida.xinxi.shapeCalculate;

import java.awt.GridLayout;
import java.awt.event.ActionEvent;
import java.awt.event.ActionListener;
import java.awt.event.FocusEvent;
import java.awt.event.FocusListener;

import javax.swing.Box;
import javax.swing.JButton;
import javax.swing.JOptionPane;
import javax.swing.JPanel;
import javax.swing.JTextField;

public class RectanglePanel extends AbstractPanel implements ActionListener,FocusListener{
    JButton resultButton,clearButton;
    JPanel leftPanel,rightPanel,buttonPanel;
    JTextField widthTextField,heightTextField;
    JTextField lengthTextField,areaTextField,inputTextField;
    BoxPanel bpWidth,bpHeight,bpLength,bpArea;

    //构造方法,初始化矩形计算器界面
    public RectanglePanel(){
        setLayout(new GridLayout(1,2));
        rightPanel = new KeyJPanel(this);
        leftPanel = new JPanel();

        //创建一个垂直排列组件的 Box 容器 box,并创建 box 容器内的各个组件
        Box box = Box.createVerticalBox();
        bpWidth = new BoxPanel("请输入矩形的宽：",10);
        widthTextField = bpWidth.getJTextField();
        //文本框 widthTextField 注册 FocusEvent 事件监听器
        widthTextField.addFocusListener(this);
        bpHeight = new BoxPanel("请输入矩形的高：",10);
        heightTextField = bpHeight.getJTextField();
        heightTextField.addFocusListener(this);
        buttonPanel = new JPanel();
        resultButton = new JButton("计算结果：");
        //按钮注册 ActionEvent 事件监听器
        resultButton.addActionListener(this);
        clearButton = new JButton("清空");
        clearButton.addActionListener(this);
        buttonPanel.add(resultButton);
        buttonPanel.add(clearButton);
        bpLength = new BoxPanel("矩形的周长：",20);
        lengthTextField = bpLength.getJTextField();
        bpArea = new BoxPanel("矩形的面积：",20);
```

```java
        this.areaTextField = this.bpArea.getJTextField();

        //box 容器依次添加 5 个面板
        box.add(bpWidth);
        box.add(bpHeight);
        box.add(buttonPanel);
        box.add(bpLength);
        box.add(bpArea);
        leftPanel.add(box);
        //RectanglePanel 面板添加组件
        add(leftPanel);
        add(rightPanel);
    }

    //当用户单击按钮时,触发 ActionEvent 事件,自动执行该方法内代码
    public void actionPerformed(ActionEvent e) {
        //单击"计算结果："按钮,则计算矩形的周长和面积
        if(e.getSource() == resultButton){
            try {
                double width = Double.parseDouble(widthTextField.getText());
                double height = Double.parseDouble(heightTextField.getText());
                lengthTextField.setText("" + 2 * (width + height));
                areaTextField.setText("" + width * height);
            } catch (NumberFormatException e1) {
                //若参数录入非数字字符,弹出消息对话框警告
                JOptionPane.showMessageDialog(this,"请输入数字：","警告对话框",
JOptionPane.WARNING_MESSAGE);
            }
        }
        //单击"清空"按钮,则清空参数文本框中的内容
        else if(e.getSource() == clearButton) {
            widthTextField.setText(" ");
            heightTextField.setText(" ");
        }
    }

    //FocusListener 接口中的方法,当获得焦点时调用该方法
    public void focusGained(FocusEvent e) {
        inputTextField = (JTextField)e.getSource();
    }
    //FocusListener 接口中的方法,当失去焦点时调用该方法
    public void focusLost(FocusEvent e) {}

    //得到当前获得焦点的文本框对象
    public JTextField getInputTextField(){
        return inputTextField;
    }
}
```

### 4. 圆计算器的实现(CirclePanel.java)

CirclePanel 类是 AbstractPanel 类的子类,实现了 ActionListener 接口和 FocusListener 接

口，创建的对象是 CalculatorWindow 窗口的成员之一。这个面板的运行效果见图 10-3，可见面板界面的左边是圆形的用于参数输入的文本框和计算结果的显示文本框，右边是一个数字软键盘，用户可以通过它来实现对圆半径的录入（通过鼠标），也可以用键盘手动录入。

CirclePanel.java 文件要处理的问题同 RectanglePanel.java，因此这里不再详细介绍，具体见代码。

代码(CirclePanel.java)：

```java
package caida.xinxi.shapeCalculate;

import java.awt.GridLayout;
import java.awt.event.ActionEvent;
import java.awt.event.ActionListener;
import java.awt.event.FocusEvent;
import java.awt.event.FocusListener;

import javax.swing.Box;
import javax.swing.JButton;
import javax.swing.JOptionPane;
import javax.swing.JPanel;
import javax.swing.JTextField;

public class CirclePanel extends AbstractPanel implements ActionListener,FocusListener{
    JButton resultButton,clearButton;
    JPanel leftPanel,rightPanel,buttonPanel;
    JtextField radiusTextField;
    JTextField lengthTextField,areaTextField,inputTextField;
    BoxPanel bpRadius,bpLength,bpArea;

    //构造方法,完成对 CirclePanel 对象的初始化
    public CirclePanel(){
        setLayout(new GridLayout(1,2));
        rightPanel = new KeyJPanel(this);
        leftPanel = new JPanel();

        //创建一个垂直排列组件的 Box 容器 box,并创建容器内要添加的组件
        Box box = Box.createVerticalBox();
        bpRadius = new BoxPanel("请输入圆的半径：",10);
        radiusTextField = bpRadius.getJTextField();
        //radiusTextField 文本框注册 FocusEvent 事件监听器
        radiusTextField.addFocusListener(this);
        buttonPanel = new JPanel();
        resultButton = new JButton("计算结果：");
        //resultButton 按钮注册 ActionEvent 事件监听器
        resultButton.addActionListener(this);
        clearButton = new JButton("清空");
        clearButton.addActionListener(this);
        buttonPanel.add(resultButton);
        buttonPanel.add(clearButton);
        bpLength = new BoxPanel("圆的周长：",20);
```

```java
            lengthTextField = bpLength.getJTextField();
            bpArea = new BoxPanel("圆的面积:",20);
            areaTextField = bpArea.getJTextField();
            //box 容器依次添加 4 个组件
            box.add(bpRadius);
            box.add(buttonPanel);
            box.add(bpLength);
            box.add(bpArea);
            leftPanel.add(box);

            //CirclePanel 对象依次添加组件
            add(leftPanel);
            add(rightPanel);
        }

        //ActionListener 接口中的方法,当用户单击了按钮,就执行该方法
        public void actionPerformed(ActionEvent e) {
            //单击了"计算结果:"按钮,计算并显示圆的周长和面积
            if(e.getSource() == resultButton){
                try {
                    double radius = Double.parseDouble(radiusTextField.getText());
                    lengthTextField.setText("" + 2 * Math.PI * radius);
                    areaTextField.setText("" + Math.PI * radius * radius);
                } catch (NumberFormatException e1) {
                    //若输入参数为非数字的字符,则弹出该警告对话框
                    JOptionPane.showMessageDialog(this,"请输入数字:","警告对话框",
JOptionPane.WARNING_MESSAGE);
                }
            }
            //单击"清空"按钮,则清空半径文本框
            else if(e.getSource() == clearButton) {
                radiusTextField.setText(" ");
            }
        }

        //FocusListener 接口中的方法,当事件源获得焦点时执行该方法
        public void focusGained(FocusEvent e) {
            inputTextField = (JTextField)e.getSource();
        }

        //FocusListener 接口中的方法,当事件源失去焦点时执行该方法
        public void focusLost(FocusEvent e) {}

        //获取当前获得焦点的文本框对象
        public JTextField getInputTextField(){
            return inputTextField;
        }
}
```

## 5. 三角形计算器的实现(TrianglePanel.java)

TrianglePanel 类是 AbstractPanel 类的子类,实现了 ActionListener 接口和 FocusListener

接口，创建的对象是 CalculatorWindow 窗口的成员之一。这个面板的运行效果见图 10-18，可见面板界面的左边是三角形的用于参数输入的文本框和计算结果的显示文本框，右边是一个数字软键盘，用户可以通过它来实现对三角形三边的录入（通过鼠标），也可以用键盘手动录入。

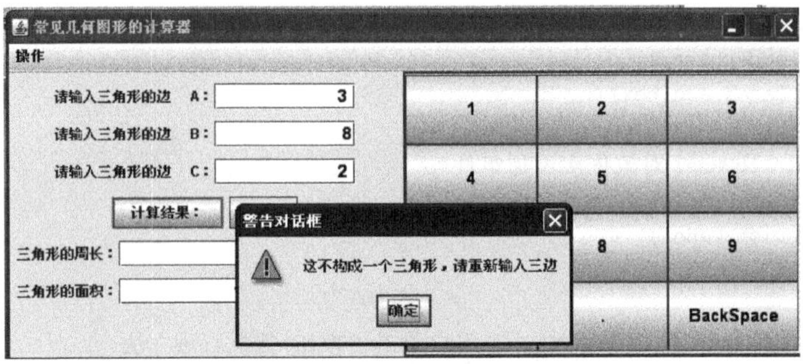

图 10-18　三角形计算器的界面

TrianglePanel.java 文件要处理的问题同 RectanglePanel.java，因此这里不再详细介绍，具体见代码。

代码（TrianglePanel.java）：

```java
package caida.xinxi.shapeCalculate;

import java.awt.GridLayout;
import java.awt.event.ActionEvent;
import java.awt.event.ActionListener;
import java.awt.event.FocusEvent;
import java.awt.event.FocusListener;

import javax.swing.Box;
import javax.swing.JButton;
import javax.swing.JOptionPane;
import javax.swing.JPanel;
import javax.swing.JTextField;

public class TrianglePanel extends AbstractPanel implements ActionListener,FocusListener{
    JButton resultButton,clearButton;
    JPanel leftPanel,rightPanel,buttonPanel;
    JTextField sideATextField,sideBTextField,sideCTextField;
    JTextField lengthTextField,areaTextField,inputTextField;
    BoxPanel bpSideA,bpSideB,bpSideC,bpLength,bpArea;

    //构造方法,负责对三角形面板的初始化
    public TrianglePanel(){
        setLayout(new GridLayout(1,2));
        rightPanel = new KeyJPanel(this);
        leftPanel = new JPanel();

        //创建一个垂直排列组件的 Box 容器 box,并创建要添加的组件
        Box box = Box.createVerticalBox();
        bpSideA = new BoxPanel("请输入三角形的边 A：",10);
```

```java
        sideATextField = bpSideA.getJTextField();
        //sideATextField 文本框注册 FocusEvent 事件监听器
        sideATextField.addFocusListener(this);
        bpSideB = new BoxPanel("请输入三角形的边 B: ",10);
        sideBTextField = bpSideB.getJTextField();
        sideBTextField.addFocusListener(this);
        bpSideC = new BoxPanel("请输入三角形的边 C: ",10);
        sideCTextField = bpSideC.getJTextField();
        sideCTextField.addFocusListener(this);
        buttonPanel = new JPanel();
        resultButton = new JButton("计算结果: ");
        //rusultButton 按钮注册 ActionEvent 事件监听器
        resultButton.addActionListener(this);
        clearButton = new JButton("清空");
        clearButton.addActionListener(this);
        buttonPanel.add(resultButton);
        buttonPanel.add(clearButton);
        bpLength = new BoxPanel("三角形的周长: ",20);
        lengthTextField = bpLength.getJTextField();
        bpArea = new BoxPanel("三角形的面积: ",20);
        areaTextField = bpArea.getJTextField();
        //box 容器依次添加以下 6 个面板对象
        box.add(bpSideA);
        box.add(bpSideB);
        box.add(bpSideC);
        box.add(buttonPanel);
        box.add(bpLength);
        box.add(bpArea);
        leftPanel.add(box);

        //TrianglePanel 对象添加两个面板
        add(leftPanel);
        add(rightPanel);
    }

    //接口 ActionListener 中的方法,当发生了 ActionEvent 事件,就执行该方法
    public void actionPerformed(ActionEvent e) {
        //单击"计算结果: "按钮,计算并显示三角形的周长和面积
        if(e.getSource() == resultButton){
            try {
                double sideA = Double.parseDouble(sideATextField.getText());
                double sideB = Double.parseDouble(sideBTextField.getText());
                double sideC = Double.parseDouble(sideCTextField.getText());
                //若输入的三边满足构成三角形的条件,则计算并显示周长和面积
                if((sideA + sideB > sideC)&&(sideA + sideC > sideB)&&(sideB + sideC > sideA)){
                    double p = (sideA + sideB + sideC)/2.0;
                    lengthTextField.setText("" + 2 * p);
                    double area = Math.sqrt(p * (p - sideA) * (p - sideB) * (p - sideC));
                    areaTextField.setText("" + area);
                }
                //若输入的三边不符合构成三角形的条件,则弹出警告对话框
```

```java
                    else{
                        JOptionPane.showMessageDialog(this,"这不构成一个三角形,请重新输入三
边","警告对话框",JOptionPane.WARNING_MESSAGE);
                        }
                } catch (NumberFormatException e1) {
                    //若输入的三边不是数字,则弹出警告对话框
                        JOptionPane.showMessageDialog(this,"请输入数字: ","警告对话框",
JOptionPane.WARNING_MESSAGE);
                }
            }
            //单击"清空"按钮,则清空表示三角形三边的文本框中的值
            else if(e.getSource() == clearButton) {
                sideATextField.setText(" ");
                sideBTextField.setText(" ");
                sideCTextField.setText(" ");
            }
        }

        //接口 FocusListener 中的方法,当事件源获得焦点时调用此方法
        public void focusGained(FocusEvent e) {
            inputTextField = (JTextField)e.getSource();
        }
        //接口 FocusListener 中的方法,当事件源失去焦点时调用此方法
        public void focusLost(FocusEvent e) {}

        //获取当前获得焦点的文本框对象
        public JTextField getInputTextField(){
            return inputTextField;
        }
}
```

### 6. 圆柱计算器的实现(CylinderPanel.java)

CylinderPanel 类是 AbstractPanel 类的子类,实现了 ActionListener 接口和 FocusListener 接口,创建的对象是 CalculatorWindow 窗口的成员之一。这个面板的运行效果见图 10-19,可见面板界面的左边是圆柱的用于参数输入的文本框和计算结果显示的文本框,右边是一个数字软键盘,用户可以通过它来实现对圆柱底面半径和圆柱高的录入(通过鼠标),也可以用键盘手动录入。

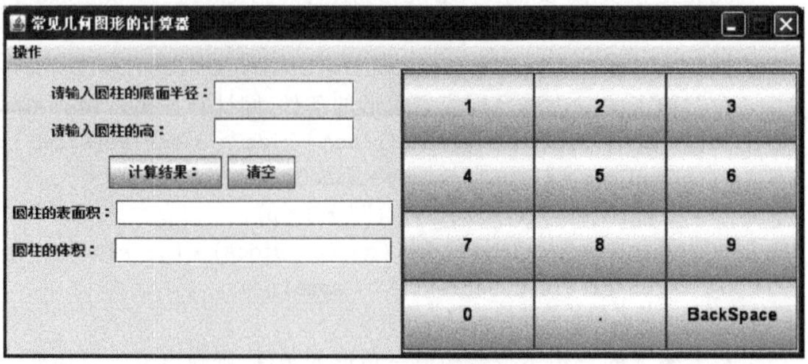

图 10-19 圆柱计算器的界面

CylinderPanel.java 文件要处理的问题同 RectanglePanel.java,因此这里不再详细介绍,具体见代码。

代码(CylinderPanel.java):

```java
package caida.xinxi.shapeCalculate;

import java.awt.GridLayout;
import java.awt.event.ActionEvent;
import java.awt.event.ActionListener;
import java.awt.event.FocusEvent;
import java.awt.event.FocusListener;
import javax.swing.Box;
import javax.swing.JButton;
import javax.swing.JOptionPane;
import javax.swing.JPanel;
import javax.swing.JTextField;

public class CylinderPanel extends AbstractPanel implements ActionListener,FocusListener{
    JButton resultButton,clearButton;
    JPanel leftPanel,rightPanel,buttonPanel;
    JTextField radiusTextField,heightTextField;
    JTextField surfaceAreaTextField,volumeTextField,inputTextField;
    BoxPanel bpRadius,bpHeight,bpSurfaceArea,bpVolume;

    //构造方法,负责 CylinderPanel 对象的初始化
    public CylinderPanel(){
        setLayout(new GridLayout(1,2));
        rightPanel = new KeyJPanel(this);
        leftPanel = new JPanel();

        Box box = Box.createVerticalBox();
        bpRadius = new BoxPanel("请输入圆柱的底面半径: ",10);
        radiusTextField = bpRadius.getJTextField();
        radiusTextField.addFocusListener(this);
        bpHeight = new BoxPanel("请输入圆柱的高:           ",10);
        heightTextField = bpHeight.getJTextField();
        heightTextField.addFocusListener(this);
        buttonPanel = new JPanel();
        resultButton = new JButton("计算结果: ");
        resultButton.addActionListener(this);
        clearButton = new JButton("清空");
        clearButton.addActionListener(this);
        buttonPanel.add(resultButton);
        buttonPanel.add(clearButton);
        bpSurfaceArea = new BoxPanel("圆柱的表面积: ",20);
        this.surfaceAreaTextField = this.bpSurfaceArea.getJTextField();
        bpVolume = new BoxPanel("圆柱的体积:      ",20);
        this.volumeTextField = this.bpVolume.getJTextField();
        box.add(bpRadius);
        box.add(bpHeight);
```

```java
            box.add(buttonPanel);
            box.add(bpSurfaceArea);
            box.add(bpVolume);
            leftPanel.add(box);
            
            add(leftPanel);
            add(rightPanel);
    }
    
    //接口 ActionListener 中的方法,当发生 ActionEvent 事件时,执行此方法
    public void actionPerformed(ActionEvent e) {
        //当单击"计算结果:"按钮时,计算并显示圆柱的表面积和体积
        if(e.getSource() == resultButton){
            try {
                double radius = Double.parseDouble(radiusTextField.getText());
                double height = Double.parseDouble(heightTextField.getText());
                double area = Math.PI * radius * radius;
                surfaceAreaTextField.setText("" + (Math.PI * 2 * radius * height + 2 * area));
                volumeTextField.setText("" + area * height);
            } catch (NumberFormatException e1) {
                //若输入的参数不是数字,则弹出警告对话框
                JOptionPane.showMessageDialog(this,"请输入数字:","警告对话框",
JOptionPane.WARNING_MESSAGE);
            }
        }
        //当单击"清空"按钮时,清空代表底面半径和高的文本框中的值
        else if(e.getSource() == clearButton) {
            radiusTextField.setText(" ");
            heightTextField.setText(" ");
        }
    }
    
    //FocusListener 接口中的方法,当事件源获得焦点时调用此方法
    public void focusGained(FocusEvent e) {
        inputTextField = (JTextField)e.getSource();
    }
    //FocusListener 接口中的方法,当事件源失去焦点时调用此方法
    public void focusLost(FocusEvent e) {}
    
    //获取当前获得焦点的文本框对象
    public JTextField getInputTextField(){
        return inputTextField;
    }
}
```

### 7. 数字软键盘的实现(KeyJPanel.java)

KeyJPanel 类是 javax.swing 包中的 JPanel 类的子类,实现了接口 ActionListener,创建的对象是 RectanglePanel 面板、CirclePanel 面板、TrianglePanel 面板和 CylinderPanel 面板的重要组成部分。

KeyJPanel 类主要处理的问题就是如何布置这 12 个按钮,并且如何响应用户的单击行为,使得用户可以对当前图形中获得焦点的文本框进行操作。这里我们设置了 inputNumber(JTextField,JButton)方法来实现数字键盘对文本框的输入。可以看到 inputNumber 方法的参数指定了要操作的文本框对象和用户当前单击的按钮对象,具体的实现可以见代码。

KeyJPanel 类的布局显然用 GridLayout 布局管理器是最合适的。

代码(KeyJPanel.java):

```java
package caida.xinxi.shapeCalculate;

import java.awt.Color;
import java.awt.Font;
import java.awt.GridLayout;
import java.awt.event.ActionEvent;
import java.awt.event.ActionListener;
import javax.swing.BorderFactory;
import javax.swing.JButton;
import javax.swing.JPanel;
import javax.swing.JTextField;
import javax.swing.border.Border;

public class KeyJPanel extends JPanel implements ActionListener{
    JButton[] keyButton = new JButton[12];
    String[] num = {"1","2","3","4","5","6","7","8","9","0",".","BackSpace"};
    AbstractPanel selectedPanel;
    JTextField inputTextField;

    //构造方法,初始化数字键盘界面,输入参数是某个图形面板对象
    public KeyJPanel(AbstractPanel selectedPanel){
        this.selectedPanel = selectedPanel;
        //设置面板的边框
        Border lb = BorderFactory.createLineBorder(Color.gray, 2);
        setBorder(lb);
        //设置布局是 GridLayout 型
        setLayout(new GridLayout(4,3));

        //创建按钮对象数组,给每个元素注册 ActionEvent 事件监听器
        for(int i = 0;i < 12;i++){
            keyButton[i] = new JButton(num[i]);
            keyButton[i].setFont(new Font("Arial",Font.BOLD,15));
            keyButton[i].setForeground(Color.BLACK);
            keyButton[i].addActionListener(this);
            add(keyButton[i]);
        }
    }

    //接口 ActionListener 中的方法,当单击按钮时,执行此方法
    public void actionPerformed(ActionEvent e) {
        //获取事件源(某个按钮)
        JButton button = (JButton)e.getSource();
        //获取 KeyJPanel 对象所在图形面板上的获得焦点的文本框对象
        inputTextField = selectedPanel.getInputTextField();
        //inputNumber 方法负责 button 对象对 inputTextField 的输入
```

```java
        inputNumber(inputTextField, button);
    }

    //实现数字软键盘对文本框的一次录入工作
    public void inputNumber(JTextField tf,JButton button){
        //获取上一次单击按钮时文本框内的值,作为旧串
        String oldString = tf.getText();
        if(oldString == null){
            tf.setText(" ");
        }
        String subStr = oldString.substring(0, oldString.length() - 1);
        //获得当前按钮的文本信息,作为新串
        String newString = button.getText();
        //若单击了"BackSpace"键,取子串
        if(newString.equals("BackSpace")){
            tf.setText(subStr);
        }
        //若单击了"."按钮,做字符串连接
        else if(newString.equals(".")){
            tf.setText(oldString + ".");
        }
        //若单击了 0~9 中的任意一个按钮,做字符串的连接
        else{
            tf.setText(oldString + newString);
        }
    }
}
```

### 10.1.5　案例练习题目

（1）读者可以看到案例中的控制面板 controlPanel 中添加了一个下拉列表和一个图标按钮,请读者自学 GUI 设计中有关图形的知识,在该面板中把按钮换成一幅图像。

（2）请读者扩充案例的功能,例如增加对梯形的计算或球体的计算,或是其他一些数学问题的求解,掌握 GUI 界面的设计。

（3）请读者按照图 10-20 和图 10-21 所示的运行效果,不看本例的源代码,自己独立编写一个图形计算器。同时也可以重新设计一个更友好的具有自己个性的 GUI 程序。

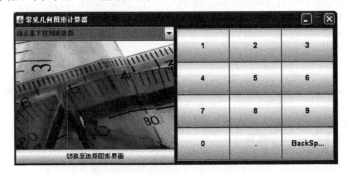

图 10-20　新的图形计算器界面(一)

图 10-21　新的图形计算器界面(二)

## 10.2　案例：饭店点菜

### 10.2.1　案例问题描述

人们在饭店(或学校餐厅、职工食堂)就餐时，经常要根据菜单来点菜，有时候也许拿不定主意吃什么，要根据服务员的推荐来点菜。点菜(就餐)结束后，根据各自的点菜单来结账。这里案例的目的就是设计一个程序来实现这样的饭店点菜流程。希望读者能通过此案例的学习，熟练掌握 Java Swing 的组件、容器、布局管理器、事件处理机制及 Java 的图形处理。

### 10.2.2　案例功能分析及演示

点菜程序运行以后，首先显示的是登录点菜主窗口界面。用户必须输入自己的桌号(包间号)，才能开始具体点菜。如果输入的桌号已有人占用，则弹出对话框提示"此桌已有人，请重新选桌！"。程序运行效果如图 10-22 和图 10-23 所示。

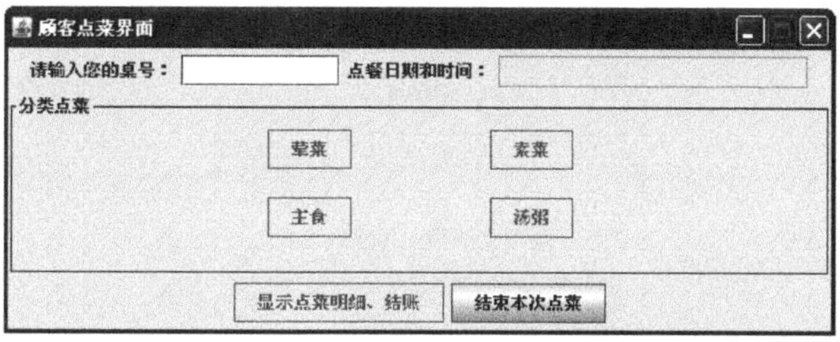

图 10-22　登录点菜主界面(一)

如果用户输入正确的桌号后，会自动显示当前点菜的日期和时间，用户就可以根据荤菜、素菜、主食、汤粥 4 个分类分别进行点菜，如图 10-24 所示。

如果分别单击"荤菜"、"素菜"、"主食"、"汤粥"4 个按钮，会各弹出 4 个分类点菜窗口(说明：这 4 个按钮每次点菜可以选择性地单击)。

例如，图 10-25 只是荤菜点菜界面，其余素菜、主食、汤粥点菜界面也是类似的。每个分

图 10-23　登录点菜主界面(二)

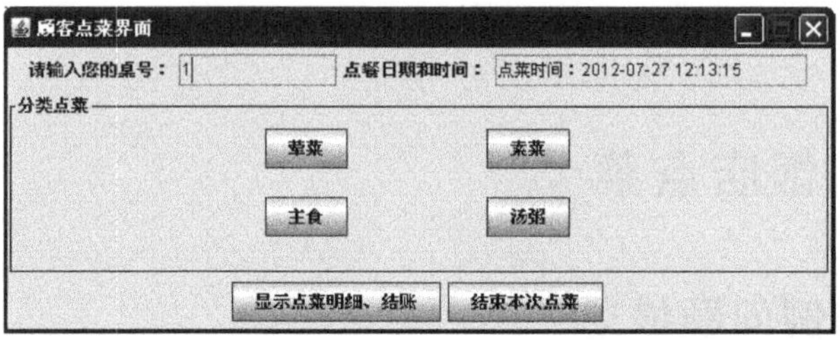

图 10-24　点菜主界面

类点菜界面都有 5 个功能,可以浏览饭店菜单、点菜、随机点菜、撤销点菜、下单。具体操作如下:①用户单击饭店菜单对应的下拉列表,可以分别浏览每个菜的基本情况(如菜名、价格、菜的图像、菜的说明);②如果有中意的菜,则可以单击"点菜"按钮,会把选中的菜添加到"已点"对应的用户下拉列表里;③如果对已点的某个菜不满意,可以选择该菜,再单击"撤销点菜"从用户下拉列表中删除该菜;④如果单击"随机推荐",则程序可以从饭店菜单中随机推荐一种菜,如果用户满意则单击"点菜",不满意则不做操作;⑤如果该界面点菜结束,则必须单击按钮"下单",这样可以保存用户点菜菜单,并关闭当前类别的点菜窗口。

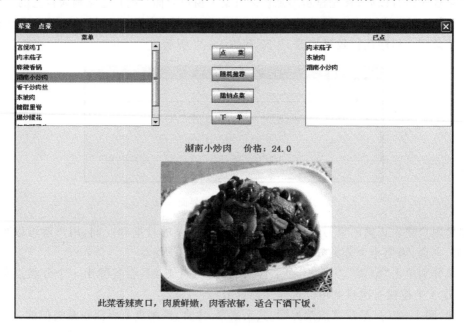

图 10-25　分类点菜界面

如果用户此次就餐点菜结束,则回到的点菜主界面,单击"显示点菜明细、结账",会弹出当前客人的菜单及结账对话框,如图 10-26 所示。

图 10-26 "点菜菜单及结账金额"对话框

若该桌客人整个点菜流程结束,则单击"结束本次点菜",结束点菜工作的最后一步,此点菜程序会切换到图 10-22 显示的登录主界面的初始状态,等待接受另一桌客人继续点菜。

### 10.2.3 案例总体设计

用 Java 编写一个面向对象的程序,很大程度上是设计类并编写这些类的定义。因此,通过对该案例的功能分析及演示,下面把问题分解成一组相互协作的对象,然后设计和创建这些对象本身。

根据演示,读者可以知道程序要处理的最基本对象就是菜及菜单,因此可以设计 Meal 类和 MealMenu 类。程序在点菜过程中是分为荤菜、素菜、主食、汤粥四种类别来进行点菜的,因此这里可以分别建立 MeatMenu 类、VegetarianMenu 类、StapleFoodMenu 类和 SoupAndPorridgeMenu 类。这四个类的父类是 MealMenu 类,这样子类可以继承父类的一些属性和方法,并且可以实现多态。

登录/点菜主界面设计为一个窗口,是 JFrame 的子类,名为 OrderingWindow 类。通过该主界面切换到分类点菜界面和显示点菜明细及结账界面。分类点菜界面用一个对话框来实现,设计为 JDialog 的子类 OrderDishes 类来处理。最后的点菜菜单明细及结账功能用 ShowOrderingRecord 类来实现,是 JDialog 的子类。

整个饭店点菜程序的面向对象设计总体类图如图 10-27 所示。

1. Meal 类

Meal 类的 UML 图如图 10-28 所示。

以下是 UML 图中有关数据和方法的详细说明。

1) 成员变量

name、description、price、image 分别用来表示菜的名字、菜的描述、菜的价格、菜的图像。

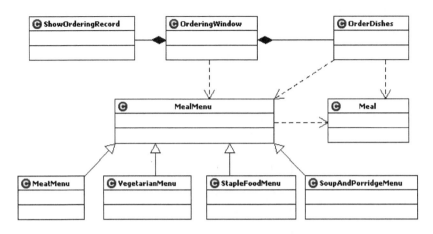

图 10-27　饭店点菜程序的总体设计视图

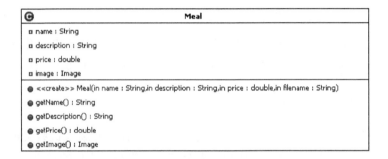

图 10-28　Meal 类的 UML 图

2）成员方法
- Meal(String;String;double;String)是构造方法,创建 Meal 对象时使用该构造方法,实现对 Meal 对象的初始化。
- getName()方法。Meal 对象调用该方法返回它的名字。
- getDescription()方法。Meal 对象调用该方法返回它的描述。
- getPrice()方法。Meal 对象调用该方法返回它的价格。
- getImage()方法。Meal 对象调用该方法返回它的图像的引用。

2．MealMenu 类

MealMenu 类的 UML 类图如图 10-29 所示。设计这样一个类的好处在于：利用继承可以实现代码的"复用"；如果后期需要增加其他新的分类菜单或修改分类菜单,直接创建 MealMenu 的子类即可；而且这里可以利用面向对象的"多态"特性,在运行时灵活地对 OrderDishes 点菜界面进行初始化,提高效率。

以下是 UML 图中有关数据和方法的详细说明。

1）成员变量

菜单对象是菜对象的集合,这里选择 LinkedList 这样的数据结构来处理。MealMenu 是一个 LinkedList<Meal>链表,用来存放饭店菜单的各种菜肴。

```
┌─────────────────────────────────────────────────────────────────┐
│                          MealMenu                               │
├─────────────────────────────────────────────────────────────────┤
│ ○ mealMenu : LinkedList<Meal>                                   │
├─────────────────────────────────────────────────────────────────┤
│ ● <<create>> MealMenu()                                         │
│ ● addItem(in name : String,in description : String,in price : double,in filename : String) : void │
│ ● getMeatMenu() : LinkedList                                    │
└─────────────────────────────────────────────────────────────────┘
```

图 10-29  MealMenu 类的 UML 图

2) 成员方法

- Meal(Strint；String；double；String)是构造方法，创建 Meal 对象时使用该构造方法，实现对 Meal 对象的初始化。
- addItem(String；String；double；String)方法用于往链表 mealMenu 中添加菜元素。
- getMealMenu() 方法。MealMenu 对象调用此方法可以返回菜肴的链表，这个方法也是为后面初始化 OrderDishes 窗口做准备。

3．MeatMenu 类

MeatMenu 类的 UML 图如图 10-30 所示。

这个类只有一个构造方法 MeatMenu()。该方法中实现对饭店荤菜菜单的初始化，在方法内继承父类的 addItem(String；String；double；String)方法，通过它来增加 Meal 对象。同样也继承父类的 mealMenu 链表。

图 10-30  MeatMenu 类的 UML 图

4．VegetarianMenu 类

VegetarianMenu 类的 UML 图如图 10-31 所示。

同样该类也只有一个构造方法 VegetarianMenu()，利用它来实现对素菜菜单的初始化。此类将继承父类 MealMenu 的属性和方法。

图 10-31  VegetarianMenu 类的 UML 图

5．StapleFoodMenu 类

StapleFoodMenu 类的 UML 图如图 10-32 所示。

同样该类也只有一个构造方法 StapleFoodMenu()，利用它来实现对主食菜单的初始化。此类将继承父类 MealMenu 的属性和方法。

图 10-32　StapleFoodMenu 类的 UML 图

#### 6．SoupAndPorridgeMenu 类

SoupAndPorridgeMenu 类的 UML 图如图 10-33 所示。

同样该类也只有一个构造方法 SoupAndPorridgeMenu()，利用它来实现对汤粥菜单的初始化。此类将继承父类 MealMenu 的属性和方法。

图 10-33　SoupAndPorridgeMenu 类的 UML 图

#### 7．OrderingWindow 类

OrderingWindow 类的 UML 图如图 10-34 所示。

图 10-34　OrderingWindow 类的 UML 图

以下是对 UML 图中的有关数据和方法的详细说明。

1）成员变量

- pNorth、pCenter、pSouth 是 JPanel 创建的三个容器，这三个容器分别位于窗口对象的北部、中部和南部。
- meatButton、vegetarianButton、stapleFoodButton、soupAndPorridgeButton、showButton、stopOrderingButton 是 JButton 创建的 6 个按钮，被添加到本窗口中，单击不同的按

钮，分别实现不同的功能。
- idTextField、dateTextField 是 JTextField 类创建的对象。idTextField 用于接收顾客输入的桌号；dateTextField 则在进入点菜界面后，显示最初的点菜日期和时间。
- file 是 File 类创建的文件对象。file 文件对象所引用的文件是记事本文件，在顾客下单时生成，写入客户的点菜明细，当显示用户订餐明细及结账时，又从该记事本文件中依次读出内容。
- idSet 是一个 HashSet<String> 的散列集。用来存放用户输入的桌号。这里我们假定桌号/包间号不能重复，因此采用 Set 集合可以保证集合中没有重复元素。

2）成员方法
- OrderingWindow()是构造方法，负责完成窗口的初始化。
- actionPerformed(ActionEvent)方法是 OrderingWindow 类实现的 ActionListener 接口中的方法，负责执行用户发出的命令。用户在文本框 idTextField 中输入桌号回车后，或者单击了 OrderingWindow 主界面上的 6 个按钮后，都可以触发 ActionEvent 事件，导致 actionPerformed(ActionEvent)方法执行响应的操作。
- main(String[])方法是软件运行的入口方法。

8. OrderDishes 类

OrderDishes 类的 UML 图如图 10-35 所示。

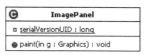

图 10-35　OrderDishes 类的 UML 图

以下是 UML 图中有关数据和方法的详细说明。

1) 成员变量

- mealMenu 是 MealMenu 类创建的一个对象，是 OrderDishes 对象里操作的对象，表示饭店的某类菜单。
- pTop、pBottom、pImage 是 JPanel 类创建的三个容器。pTop 容器区域负责处理点菜的多个功能，pBottom、pImage 容器区域主要用于显示每盘菜的名字、价格、图像和相关说明这些信息。
- menuList、orderList 是 JList 对象，表示两个下拉列表，分别用来显示某类饭店菜单和客户点菜单。
- addButton 负责点菜，randomButton 负责向顾客随机推荐菜，deleteButton 负责撤单，saveButton 按钮负责顾客点餐结束后下单。
- 链表 mealMenus 用来存放某类饭店菜单，向量 mealNames、mealDiscriptions、mealPrice、mealIamges 分别用来存放饭店菜单所有菜的菜名、菜的说明、菜价、菜的图像。
- 向量 orderDishes 则用来存放顾客点的所有菜。
- selectedIndex 的值表示饭店下拉列表中被选择的菜的索引；orderIndex 的值表示顾客已点菜的下拉列表中被选择菜的索引。
- orderMealNames 是 DefaultListModel 对象，正是通过它来实现对客户点菜列表的增加、删除元素的功能。
- file 指向一个具体的 txt 文件，该文件用来保存用户的订餐明细。

2) 成员方法

- OrderDishes()构造方法。
- OrderDishes(MealMenu,String,File)构造方法，主要通过它实现对点菜菜单的初始化。该方法有三个输入参数：mealMenu 表示该窗口要处理的菜单，name 用来设置窗口的标题，file 表示顾客的订餐明细要保存到哪个文件。
- actionPerformed（ActionEvent）方法。该方法是 OrderDishes 类实现的 ActionListener 接口中的方法。当用户单击了 OrderDishes 窗口对象中的按钮时，actionPerformed(ActionEvent)方法将被调用执行，所执行的主要操作就是：点菜、随机推荐菜、撤销点菜、下单四个功能。
- valueChanged（LisSelectionEvent）方法。该方法是 OrderDishes 类实现的 ListSelectionListener 接口中的方法。当用户浏览饭店菜单或者在撤单时都会触发 ListSelectionEvent 事件，valueChanged(ListSelectionEvent)方法被调用执行。
- paint(Graphics)方法用来绘制显示菜的相关信息。

### 9. ShowOrderingRecord 类

ShowOrderingRecord 类的 UML 类如图 10-36 所示。

1) 成员变量

- showArea 区域显示顾客订餐明细及就餐消费总金额。
- file 指向记录顾客订餐明细的文本文件。

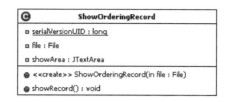

图 10-36　ShowOrderingRecord 类的 UML 图

2）成员方法
- ShowOrderingRecord(File)是构造方法，初始化显示订餐明细窗口。
- showRecord()方法实现从文本文件中读出顾客订餐的名称、价格、计算消费金额，并把它们添加到窗口的 showArea 区域。

### 10.2.4　案例代码实现

**1. 编写饭店点菜主界面（OrderingWindow.java）**

OrderingWindow.java 文件中包含了程序的入口，是本应用程序的主窗体，通过窗口的按钮把其他子窗体联系在一起（见图 10-37）。需要说明的是，本应用程序没有通过数据库技术来实现，完全通过 Java Swing 技术来实现。下面介绍此程序要解决的几个问题。

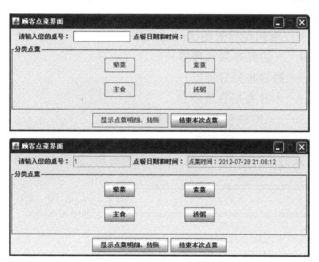

图 10-37　程序主界面

1）实现登录功能

如何保证顾客在点菜前，桌号不能为空且不能重复呢？桌号是用来标识顾客的唯一关键字，它不能有重复值，而且应用程序还要为每桌顾客生成他们自己的菜单并结账。因此 OrderingWindow 类必须有一个成员变量用来表示桌号。常规的方法可以把它保存在一个 String 数组中，但这样做并不能保证数组中元素不重复。由前面基础篇的学习读者可以知道 Set 集合不允许包含相同的元素，如果视图把两个相同的元素加入同一个 Set 集合中，则添加操作失败，add 方法返回 false，且新元素不会被加入。HashSet 是 Set 接口的典型实现,这里选择 HashSet<String>来保存桌号。

具体实现思路是在初始化 OrderingWindow 窗口时,让"荤菜"按钮、"素菜"按钮、"主食"按钮、"汤粥"按钮、"显示点菜明细、结账"按钮设置为不可用状态。然后用如下代码实现:

```
HashSet < String > idSet = new HashSet < String >();
idTextField = new JTextField(10);
if(idTextField!= null){
    if(!idSet.add(idTextField.getText())){
        JOptionPane.showMessageDialog(this, "此桌已有客人,请重新选桌!");
    }
    else {
    //此处代码省略,在实现过程将做详解。功能是激活图 10 – 37 中原来不可用的按钮
    }
}
```

2) 主界面布局设计

Java 语言中,提供了布局管理器这个工具来管理组件在容器中的布局,而不使用直接设置组件位置和大小的方法。布局管理器决定容器的布局策略及容器内组件的排列顺序、组件大小和位置,以及当窗口移动或调整大小后组件如何变换等。每个容器都有一个默认的布局管理器,该布局管理器可通过调用 setLayout()改变。

Java 提供了下列布局管理器,BoxLayout 是 Swing 中新增加的,其他布局管理器都是在 AWT 中进行定义的:

- FlowLayout——流式布局管理器;
- BorderLayout——边界布局管理器;
- GridLayout——网格布局管理器;
- CardLayout——卡片布局管理器;
- GridBagLayout——网格包布局管理器;
- BoxLayout——箱式布局管理器。

根据图 10-37 显示的运行效果图,设计主界面的布局,如图 10-38 的说明。

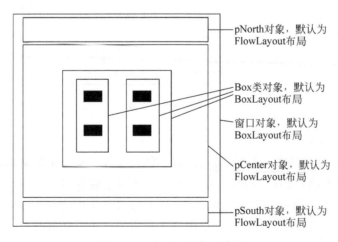

图 10-38　主界面的布局设计

3）显示当前点菜日期和时间

使用 java.util 包中的 Date 类的无参数构造方法创建的对象可以获取本地当前的时间。Date 对象表示时间的默认顺序是星期、月、日、小时、分、秒、年。在应用程序中，我们希望按照我们中国人的习惯来输出时间，这时可以使用 DateFormat 的子类 SimpleDateFormat 来实现日期的格式化。SimpleDateFormat 有一个常用的构造方法：

```
public SimpleDateFormat(String pattern)
```

该构造方法可以用参数 pattern 指定的格式创建一个对象，该对象调用 format(Date date)方法格式化时间对象 date。需要注意的是，pattern 中应当含有一些特殊意义的字符，这些特殊字符被称为元字符，详细情况可查 API。

4）代码（OrderingWindow.java）

```java
package caida.xinxi.Takingorders;

import java.awt.Color;
import java.awt.event.ActionEvent;
import java.awt.event.ActionListener;
import java.io.File;
import java.text.SimpleDateFormat;
import java.util.Date;
import java.util.HashSet;

import javax.swing.Box;
import javax.swing.JButton;
import javax.swing.JFrame;
import javax.swing.JLabel;
import javax.swing.JOptionPane;
import javax.swing.JPanel;
import javax.swing.JTextField;
import javax.swing.border.LineBorder;
import javax.swing.border.TitledBorder;

public class OrderingWindow extends JFrame implements ActionListener{
    private static final long serialVersionUID = 1L;
    JPanel pNorth,pCenter,pSouth;
    JButton meatButton,vegetarianButton;
    JButton stapleFoodButton,soupAndPorridgeButton;
    JButton showButton,stopOrderingButton;
    JTextField idTextField,dateTextField;
    MealMenu mealMenus;
    File file;
    HashSet<String> idSet;

    public OrderingWindow(){
        setTitle("顾客点菜界面");
        idSet = new HashSet<String>();

        //根据图 10-37 设计的布局来添加组件、设置组件的状态及注册监听器
```

```java
            pNorth = new JPanel();
            idTextField = new JTextField(10);
            idTextField.addActionListener(this);
            dateTextField = new JTextField(20);
            //设置 dateTextField 为不可编辑的
            dateTextField.setEditable(false);
            pNorth.add(new JLabel("请输入您的桌号："));
            pNorth.add(idTextField);
            pNorth.add(new JLabel("点餐日期和时间："));
            pNorth.add(dateTextField);

            pCenter = new JPanel();
            //设置 pCenter 的带标题的边框
            pCenter.setBorder(new TitledBorder(new LineBorder(Color.BLUE),"分类点菜",
    TitledBorder.LEFT,TitledBorder.TOP));
            meatButton = new JButton("荤菜");
            meatButton.addActionListener(this);
            meatButton.setEnabled(false);
            vegetarianButton = new JButton("素菜");
            vegetarianButton.addActionListener(this);
            vegetarianButton.setEnabled(false);
            stapleFoodButton = new JButton("主食");
            stapleFoodButton.addActionListener(this);
            stapleFoodButton.setEnabled(false);
            soupAndPorridgeButton = new JButton("汤粥");
            soupAndPorridgeButton.addActionListener(this);
            soupAndPorridgeButton.setEnabled(false);
            Box baseBox = Box.createHorizontalBox();
            Box box1 = Box.createVerticalBox();
            box1.add(meatButton);
            box1.add(Box.createVerticalStrut(20));
            box1.add(stapleFoodButton);
            Box box2 = Box.createVerticalBox();
            box2.add(vegetarianButton);
            box2.add(Box.createVerticalStrut(20));
            box2.add(soupAndPorridgeButton);
            baseBox.add(box1);
            baseBox.add(Box.createHorizontalStrut(100));
            baseBox.add(box2);
            pCenter.add(baseBox);

            pSouth = new JPanel();
            this.showButton = new JButton("显示点菜明细、结账");
            showButton.addActionListener(this);
            showButton.setEnabled(false);
            this.stopOrderingButton = new JButton("结束本次点菜");
            stopOrderingButton.addActionListener(this);
            pSouth.add(showButton);
            pSouth.add(stopOrderingButton);

            add(pNorth,"North");
```

```java
        add(pCenter,"Center");
        add(pSouth,"South");
        
        setBounds(100,100,600,230);
        this.setResizable(false);
        setDefaultCloseOperation(JFrame.EXIT_ON_CLOSE);
        setVisible(true);
    }
    
    //实现接口 ActionListener 中的 actionPerformed(ActionEvent)方法
    public void actionPerformed(ActionEvent e) {
        
        //此事件表明用户在 idTextField 中按了回车键
        if(e.getSource() == this.idTextField){
            // 如果 idTextField 文本框的值不为空
            if(idTextField!= null){
                //如果用户输入的桌号已经有顾客了
                if(!idSet.add(idTextField.getText())){
                    JOptionPane.showMessageDialog(this, "此桌已有客人,请重新选桌!");
                }
                //如果该桌号无人,同意点菜,同时激活相关点菜的各个按钮
                else {
                    idTextField.setEditable(false);
                    this.meatButton.setEnabled(true);
                    this.vegetarianButton.setEnabled(true);
                    this.stapleFoodButton.setEnabled(true);
                    this.soupAndPorridgeButton.setEnabled(true);
                    this.showButton.setEnabled(true);
                    // dateTextField 文本框的值显示为当前格式化了的日期和时间
                    Date nowTime = new Date();
                    SimpleDateFormat matter = new SimpleDateFormat("点菜时间: yyyy - MM - dd HH:mm:ss");
                    String date = matter.format(nowTime);
                    this.dateTextField.setText(date);
                    //如果可以点菜,则生成一个文本文件的引用 file
                    //该文本文件将保存该桌顾客的点菜明细
                    String filename = idTextField.getText() + "号桌点菜清单.txt";
                    file = new File(filename);
                }
            }
        }
        
        //如果用户单击了"点菜明细、结账"按钮,则显示 showOrdering 对话框
        else if(e.getSource() == this.showButton){
            ShowOrderingRecord showOrdering = new ShowOrderingRecord(file);
            showOrdering.setVisible(true);
            //对话框调用 showRecord()方法将显示顾客点菜明细及消费金额
            showOrdering.showRecord();
        }
        //如果用户选择"结束本次点菜"按钮,则将相应组件恢复到初始状态
        else if(e.getSource() == this.stopOrderingButton){
```

```java
                    this.idTextField.setText(null);
                    this.idTextField.setEditable(true);
                    this.dateTextField.setText(null);
                    this.meatButton.setEnabled(false);
                    this.vegetarianButton.setEnabled(false);
                    this.stapleFoodButton.setEnabled(false);
                    this.soupAndPorridgeButton.setEnabled(false);
                    this.showButton.setEnabled(false);
                }
                //如果用户单击了"荤菜"、"素菜"、"主食"、"汤粥"按钮,程序将执行下面的代码
                else{
                    //字符串 menusName 用来设置 OrderDishes 对话框的标题
                    String menusName = null;
                    //如果单击"荤菜"按钮,设置当前饭店菜单为肉菜菜单、当前按钮失效等
                    if(e.getSource() == this.meatButton){
                        mealMenus = new MeatMenu();
                        menusName = "荤菜 点菜";
                        this.meatButton.setEnabled(false);
                    }
                    //如果单击"素菜"按钮,设置当前 mealMenus 为素菜菜单等
                    else if(e.getSource() == this.vegetarianButton){
                        mealMenus = new VegetarianMenu();
                        menusName = "素菜 点菜";
                        this.vegetarianButton.setEnabled(false);
                    }
                    //如果单击"主食"按钮,设置当前 mealMenus 为主食菜单等
                    else if(e.getSource() == this.stapleFoodButton){
                        mealMenus = new StapleFoodMenu();
                        menusName = "主食 点菜";
                        this.stapleFoodButton.setEnabled(false);
                    }
                    //如果单击"汤粥"按钮,设置当前 mealMenus 为汤粥菜单等
                    else if(e.getSource() == this.soupAndPorridgeButton){
                        mealMenus = new SoupAndPorridgeMenu();
                        menusName = "汤粥 点菜";
                        this.soupAndPorridgeButton.setEnabled(false);
                    }
                    //创建各类点菜界面,并显示出来
                    new OrderDishes(mealMenus,menusName,file).setVisible(true);
                }
            }
        }
        //main()方法,程序入口,创建应用程序主界面
        public static void main(String[] args) {
            new OrderingWindow();
        }
    }
```

## 2. 具体点菜界面的实现(OrderDishes.java)

OrderDishes.java 是具体地进行专项分类点菜的对话框窗口。在点菜主界面单击了

"荤菜"、"素菜"、"主食"、"汤粥"会产生 4 个这样相同的对话框,不同的是处理的菜单不同。此窗口要实现的功能有菜单浏览、点菜、随机推荐、撤销点菜、下单 5 个功能(见图 10-25)。这里没有显式地设计点菜数量的组件,此功能本程序是这样处理的:如果用户想点两份或多份菜,可以按两次或多次"点菜"按钮就可实现。下面介绍此任务主要解决的几个问题。

1)JList 类实现菜单的显示

这里需要创建一个显示饭店菜单的列表框和一个显示用户点菜结果的列表框,并把它们添加到 OrderDishes 窗口中。查阅 API 文档,可知 JList 有下面的构造方法:
- public JList()
- public JList(ListModel dataModel)
- public JList(Object[ ]listData)
- public JList(Vector<?>listData)

那么选择哪个显示呢?使用数组 Vector 创建的 JList 不可以直接添加、删除列表项,所以如果想创建列表项可变的 JList 对象,则应该在创建 JList 时显式使用 DefaultListModel 作为构造参数,因为 DefaultListModel 作为 JList 的 model,负责维护 JList 组件的所有列表数据。因此我们可以通过向 DefaultListModel 中添加元素、删除元素来实现向 JList 对象中增加、删除列表项。

从上面的知识介绍可以确定显示用户点菜结果的列表框用第二个构造方法,因为我们需要对这个列表框进行添加菜、删除菜的操作。饭店菜单这里假定是不变的,那么我们可以选择第三个和第四个构造方法。本应用程序中我们把菜名存储在一个 Vector 集合里,所以选择第四个构造方法来创建列表(如果用对象数组来保存数据,那么就可以使用第三个了,对应到本文件,可以创建一个 String 数组)。

如果需要监听列表项选择项的改变,可以通过添加对应的监听器来实现。通常 JList 使用 addListSelectionListener 方法来添加监听器。接口 ListSelectionListener 中只有一个方法:

```
public void valueChanged(ListSelectionEvent e){}
```

因此需要对上面两个 JList 列表框注册该监听器,并且实现接口中的方法。

2)实现菜单浏览功能

菜单浏览功能是指当用户选择菜单列表框中的选项时,可以在窗口中同时刷新显示该选项对应的菜的名字、价格、图像、相关说明这样的信息(见图 10-39),这样有助于顾客选菜。

在图 10-39 中,本案例这样设计:给菜单列表框注册监听器,当选中左边菜单列表框中某个选项时,右边的面板可以同时刷新显示该选项对应的 Meal 对象的信息(名字、价格、图像、说明)。

接下来要解决的问题就是如何显示这些信息?在 Component 类里提供了三个和绘图有关的方法:
- paint(Graphics g):绘制组件的外观。
- update(Graphics g):调用 paint 方法,刷新组件外观。
- repaint():调用 update 方法,刷新组件外观。

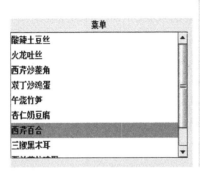

图 10-39　显示菜单浏览功能

上面三个方法的调用关系为：repaint 方法调用 update 方法，update 方法调用 paint 方法。

Container 类中的 update 方法先以组件的背景色填充整个组件区域，然后调用 paint 方法重画组件。

Graphics 是有关抽象的画笔对象，Graphics 可以在组件上绘制丰富多彩的几何图形和位图，Graphics 类提供了大量用于绘制几何图形和位图的方法（见 API 文档）。这里用到的方法有：

- drawString：绘制字符串。
- fillRect：填充一个矩形区域。
- drawImage：绘制位图。
- setColor：设置画笔的颜色，可以使用 RGB、CMYK 等方式设置颜色。
- setFont：设置画笔的字体，需要指定字体名、字体样式、字体大小三个属性。

根据以上知识，案例设计一个 JPanel 的子类 ImagePanel，在该子类中重写 paint 方法。在 paint 方法中，程序在 ImagePanel 对象的指定位置绘制所选选项对应的 meal 的名字、价格、图像、说明这些信息。而当列表框选中选项发生改变时，调用的接口方法中让 ImagePanel 对象调用 repaint 方法即可。

3) 下单功能

下单功能其实就是把顾客所点的菜写入一个文件中，即我们创建的该桌号对应的点菜清单文本文件。每个 OrderDishes 对象窗口都有下单功能，根据应用程序的运行效果，我们看到最多可以有 4 个这样的窗口，分别为"荤菜 点菜"、"素菜 点菜"、"主食 点菜"、"汤粥 点菜"窗口。每个窗口当顾客点了下单，都需要分别将所点的所有荤菜、素菜、主食、汤粥的名字和价格都写到这一个文本文件中。

为了实现这个功能，就需要用到 java.io 包中类。这个包下主要包括输入输出两种 I/O 流，每种输入、输出流又可以分为字节流和字符流两大类。但是这个包下的输入输出流非常多，选择哪个呢？这就要看哪个能实现上面提到的功能。这里选择了 RandomAccessFile 类。

RandomAccessFile 是 Java 输入输出流体系中功能最丰富的文件内容访问类，它提供了众多的方法来访问文件内容，它既可以读取文件内容，也可以向文件输出数据。RandomAccessFile 对象包含了一个记录指针，用以标识当前读写处的位置，当程序新创建一个 RandomAccessFile 对象时,该对象的文件记录指针位于文件头(也就是 0 处),当读写了 n 个字节后,文件记录指针将会向后移动 n 个字节。它可以自由访问文件的任意位置,void seek(long pos)方法将帮助实现此功能。除此以外,它可以读/写各种类型的数据(具体查阅 API 文档)。

4) 窗口的布局设计

OrderDishes 窗口布局设计稍微复杂些,本程序所采用的布局见图 10-40。

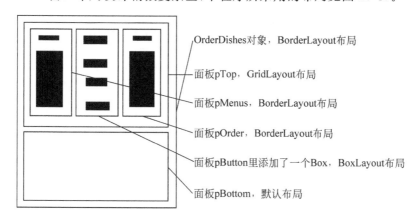

图 10-40　OrderDishes 窗口的布局设计

5) 代码(OrderDishes.java)

```
package caida.xinxi.Takingorders;

import java.awt.BorderLayout;
import java.awt.Color;
import java.awt.Font;
import java.awt.Graphics;
import java.awt.GridLayout;
import java.awt.Image;
import java.awt.event.ActionEvent;
import java.awt.event.ActionListener;
import java.io.File;
import java.io.IOException;
import java.io.RandomAccessFile;
import java.util.Iterator;
import java.util.LinkedList;
import java.util.Random;
import java.util.Vector;

import javax.swing.Box;
import javax.swing.DefaultListModel;
import javax.swing.JButton;
import javax.swing.JDialog;
import javax.swing.JLabel;
import javax.swing.JList;
import javax.swing.JPanel;
import javax.swing.JScrollPane;
```

```java
import javax.swing.ListSelectionModel;
import javax.swing.event.ListSelectionEvent;
import javax.swing.event.ListSelectionListener;

public class OrderDishes extends JDialog implements ActionListener, ListSelectionListener{
    private static final long serialVersionUID = 1L;
    MealMenu mealMenu;
    JPanel pTop,pBottom,pImage;
    private JList menusList,orderList;
    private JButton addButton,randonButton,deleteButton,saveButton;
    private LinkedList<Meal> mealMenus;
    private Vector<String> mealNames,mealDiscriptions;
    private Vector<Image> mealImages;
    private Vector<Double> mealPrice;
    private Vector<Meal> orderDishes;                                    //保存用户点的菜
    private Meal meal;
    private int selectedIndex,orderIndex;
    private DefaultListModel orderMealNames;
    private File file;

    public OrderDishes(){}
    //OrderDishes窗口的构造方法,实现窗口初始化
    public OrderDishes(MealMenu mealMenu,String name,File file){
        this.file = file;
        setTitle(name);
        /**获得mealMenu对象的LinkedList<Meal>类型的成员变量,保存到链表mealMenus中,
        这是OrderDishes对象处理的数据,即顾客要点的菜的集合(饭店菜单)*/
        mealMenus = mealMenu.getMeatMenu();
        /**创建存储菜、菜名、菜价格、菜图像、菜说明的向量对象*/
        orderDishes = new Vector<Meal>();
        mealNames = new Vector<String>();
        mealPrice = new Vector<Double>();
        mealImages = new Vector<Image>();
        mealDiscriptions = new Vector<String>();

        /**遍历mealMenus中的每个Meal元素,并将其相关成员变量作为元素分别添加到向量
        MealNames、mealPrice、mealImage、mealDiscription中*/
        Iterator<Meal> iterator = mealMenus.iterator();
        while(iterator.hasNext()){
            meal = iterator.next();
            mealNames.add(meal.getName());
            mealPrice.add(meal.getPrice());
            mealImages.add(meal.getImage());
            mealDiscriptions.add(meal.getDescription());
        }
        //窗口布局设置为BorderLayout型
        setLayout(new BorderLayout());
        //设置面板pTop的布局和它所添加的各个组件
        pTop = new JPanel();
        pTop.setLayout(new GridLayout(1,3));
```

```java
JPanel pMenus = new JPanel();
pMenus.setLayout(new BorderLayout());
menusList = new JList(mealNames);
//设置显示饭店菜单的列表框默认选择第一选项
menusList.setSelectedIndex(0);
//设置显示饭店菜单的列表框只能单选
menusList.setSelectionMode(ListSelectionModel.SINGLE_SELECTION);
menusList.addListSelectionListener(this);
JScrollPane listScrollPane = new JScrollPane(menusList);
pMenus.add(new JLabel("菜单",JLabel.CENTER),"North");
pMenus.add(listScrollPane,"Center");

JPanel pButton = new JPanel();
Box box = Box.createVerticalBox();
this.addButton = new JButton("点      菜");
this.addButton.addActionListener(this);
this.randonButton = new JButton("随机推荐");
this.randonButton.addActionListener(this);
this.deleteButton = new JButton("撤销点菜");
this.deleteButton.addActionListener(this);
this.saveButton = new JButton("下      单");
this.saveButton.addActionListener(this);
box.add(Box.createVerticalStrut(20));
box.add(addButton);
box.add(Box.createVerticalStrut(15));
box.add(randonButton);
box.add(Box.createVerticalStrut(15));
box.add(deleteButton);
box.add(Box.createVerticalStrut(15));
box.add(saveButton);
pButton.add(box);

JPanel pOrder = new JPanel();
pOrder.setLayout(new BorderLayout());
orderMealNames = new DefaultListModel();
orderList = new JList(orderMealNames);
orderList.addListSelectionListener(this);
JScrollPane listScrollPane2 = new JScrollPane(orderList);
pOrder.add(new JLabel("已点",JLabel.CENTER),"North");
pOrder.add(listScrollPane2,"Center");

pTop.add(pMenus);
pTop.add(pButton);
pTop.add(pOrder);

pBottom = new ImagePanel();
add(pTop,"North");
add(pBottom,"Center");

setBounds(300,10,900,600);
this.setResizable(false);
```

```java
            this.setDefaultCloseOperation(JDialog.DO_NOTHING_ON_CLOSE);
    }

    //实现ActionListener接口中的方法actionPerformed(ActionEvent),处理窗口事件
    public void actionPerformed(ActionEvent e) {
        //单击addButton按钮,实现点菜功能
        if(e.getSource() == addButton){
            Meal meal = mealMenus.get(selectedIndex);        //获取索引处的Meal对象meal
            orderDishes.add(meal);                 //将此meal添加到用户点菜向量orderDishes
            //获取此meal的name,在用户点菜列表框中显示出来
            orderMealNames.addElement(meal.getName());
        }
        //实现撤销点菜功能
        else if(e.getSource() == deleteButton ){
            orderIndex = orderList.getSelectedIndex();       //获取选中索引orderIndex
            orderDishes.remove(orderIndex);         //从orderDishes中删除orderIndex处元素
            orderMealNames.removeElementAt(orderIndex);      //删除列表框中该元素
        }
        //实现随机推荐菜
        else if(e.getSource() == randonButton){
            int size = mealMenus.size();
            Random rand = new Random();
            selectedIndex = rand.nextInt(size);      //产生一个随机数,保存到selectedIndex
            menusList.setSelectedIndex(selectedIndex);       //设置列表框此位置为选中状态
        }
        //实现下单功能
        else if(e.getSource() == saveButton ){
            saveButton.setEnabled(false);
            //将用户点菜向量中所有菜的菜名、价格写入file文件
            try {
                RandomAccessFile out = new RandomAccessFile(file,"rw");
                if(file.exists()){
                    long length = file.length();
                    out.seek(length);
                }
                for(int i = 0;i < this.orderDishes.size();i++){
                    out.writeUTF(orderDishes.get(i).getName());
                    out.writeDouble(orderDishes.get(i).getPrice());
                }
                out.close();
            } catch (IOException e1) {}
            setVisible(false);
        }
    }

    //实现接口ListSelectionListener中方法,响应列表上的事件,实现菜单浏览功能
    public void valueChanged(ListSelectionEvent e) {
        if(e.getSource() == this.menusList){
            selectedIndex = menusList.getSelectedIndex();
            pBottom.repaint();
        }
```

```java
    }
    //内部类,负责绘制菜的相关各类信息
    class ImagePanel extends JPanel{
        private static final long serialVersionUID = 1L;
        //重写 paint 方法
        public void paint(Graphics g){
            //清空 ImagePanel 容器中的内容
            g.setColor(getBackground());
            g.fillRect(0, 0, getWidth(), getHeight());
            //绘制菜的名称、价格、图像、说明
            g.setColor(Color.RED);
            g.setFont(new Font("宋体",Font.BOLD,18));
            g.drawString(mealNames.get(selectedIndex), 350,50 );
            g.drawString("价格: " + mealPrice.get(selectedIndex),470,50);
            g.drawImage(mealImages.get(selectedIndex),300,70,350,260,this);
            g.drawString(mealDiscriptions.get(selectedIndex),170,360);
        }
    }
}
```

### 3. 显示顾客菜单及结账功能(ShowOrderingRecord.java)

ShowOrderingRecord 类是 JDialog 的一个子类,当顾客在主界面单击"显示点菜明细、结账"按钮时,弹出该对话框显示该桌顾客的点菜明细及消费总金额,见图 10-26。

本文件主要处理以下两个问题:

- ShowOrderingRecord 窗口界面的设计及实现。
- 显示该桌点菜清单及消费金额。

ShowOrderingRecord 界面设计了一个 JScrollPane 对象,在该对象中添加了一个 JTextArea 对象,最后将 JScrollPane 对象添加到窗口中。

要显示某桌点菜清单及消费金额(假设该桌号为1),首先要从"1号桌点菜清单.txt"中顺序读出所有菜的名字及价格,并把它添加到 JTextArea 对象中显示出来。对所有菜的价格求和,即可得到消费金额,同样在 JTextArea 对象中显示出来。本程序此处仍然用 RandomAccessFile 类来处理输入流的问题(读者可以试用其他输入流类处理此问题)。

代码(ShowOrderingRecord.java):

```java
package caida.xinxi.Takingorders;

import java.awt.BorderLayout;
import java.awt.Font;
import java.io.File;
import java.io.RandomAccessFile;

import javax.swing.JDialog;
import javax.swing.JScrollPane;
import javax.swing.JTextArea;

public class ShowOrderingRecord extends JDialog {
```

```java
    private static final long serialVersionUID = 1L;
    private File file;
    private JTextArea showArea;

    public ShowOrderingRecord(File file){
        this.file = file;
        showArea = new JTextArea(4,2);
        showArea.setFont(new Font("楷体",Font.BOLD,20));
        add(new JScrollPane(showArea),BorderLayout.CENTER);
        setBounds(200,200,300,400);
    }

    //显示该桌客人点菜清单及消费金额
    public void showRecord(){
        showArea.setText(null);
        showArea.append(" ---- 菜名 ------------ 价格 ---------- ");
        double totalPrice = 0;                              //总消费金额 totalPrice
        try {
            RandomAccessFile in = new RandomAccessFile(file,"r");
            String mealName = null;
            while((mealName = in.readUTF())!= null){
                showArea.append("\n" + mealName);
                double mealPrice = in.readDouble();
                showArea.append("            " + mealPrice);
                totalPrice = totalPrice + mealPrice;
            }
        } catch (Exception e) {}
        showArea.append("\n----------------------- ");
        showArea.append("\n结账:          " + totalPrice);
    }
}
```

**4. Meal 类、MealMenu 类及其子类的实现**

Meal 类、MealMenu 类和其子类 MeatMenu 类、VegetarianMenu 类、StapleFoodMenu 类、SoupAndPorridgeMenu 类的定义可见前面的 UML 图。

1) 代码(Meal.java)

```java
package caida.xinxi.Takingorders;
import java.awt.Image;
import java.awt.Toolkit;

public class Meal {
    private String name;
    private String description;
    private double price;
    private Image image;

    public Meal(String name, String description, double price, String filename) {
        super();
```

```java
            this.name = name;
            this.description = description;
            this.price = price;
            Toolkit toolkit = Toolkit.getDefaultToolkit();
            this.image = toolkit.createImage(filename);
        }

        public String getName() {
            return name;
        }

        public String getDescription() {
            return description;
        }

        public double getPrice() {
            return price;
        }

        public Image getImage() {
            return image;
        }
    }
```

2) 代码(MealMenu.java)

```java
package caida.xinxi.Takingorders;
import java.util.LinkedList;

public class MealMenu {
    LinkedList<Meal> mealMenu;

    public MealMenu(){
        mealMenu = new LinkedList<Meal>();
    }

    public void addItem(String name,String description,double price ,String filename){
        Meal meal = new Meal(name,description,price,filename);
        mealMenu.add(meal);
    }

    public LinkedList<Meal> getMeatMenu() {
        return mealMenu;
    }

}
```

3) 代码(MeatMenu.java)

```java
package caida.xinxi.Takingorders;
public class MeatMenu extends MealMenu{
    public MeatMenu(){
```

```java
        addItem("宫保鸡丁","由鸡丁、干辣椒、花生米等炒制而成；鸡肉鲜嫩配合花生的香脆,广受欢迎。",18.0,"image/宫保鸡丁.jpg");
        addItem("肉末茄子","此菜用料简单,味道鲜美,是道非常受欢迎的家常小菜。",22,"image/肉末茄子.jpg");
        addItem("麻辣香锅","此菜以麻、辣、鲜、香混一锅为特点,香气扑鼻、滋味柔和纯正。",68,"image/麻辣香锅.jpg");
        addItem("湖南小炒肉","此菜香辣爽口,肉质鲜嫩,肉香浓郁,适合下酒下饭。",24,"image/湖南小炒肉.jpg");
        addItem("香干炒肉丝","此菜以香干和鲜肉为主料,具有补肾养血,滋阴润燥的功效。",20,"image/香干炒肉丝.jpg");
        addItem("东坡肉","此菜用猪肉炖制而成,入口肥而不腻,带有酒香,十分美味。",32,"image/东坡肉.jpg");
        addItem("糖醋里脊","由鲜嫩里脊肉制成,肉质滑嫩可口,色泽红亮,酸甜味美,外酥里嫩。",26,"image/糖醋里脊.jpg");
        addItem("爆炒腰花","富含蛋白质、脂肪,另含核黄素、维生素A、硫胺素、钙、铁等成分。",26,"image/爆炒腰花.jpg");
        addItem("红烧狮子头","此菜有肥有瘦的肉红嫩油亮,配上翠绿青菜掩映,色香味俱全。",20,"image/红烧狮子头.jpg");
        addItem("清蒸鲈鱼","以蒸菜为主、口味属于咸鲜。富含多种营养价值,DHA含量高。",58,"image/清蒸鲈鱼.jpg");
    }
}
```

4）代码（VegetarianMenu.java）

```java
package caida.xinxi.Takingorders;
public class VegetarianMenu extends MealMenu{
    public VegetarianMenu(){
        addItem("酸辣土豆丝","食材：土豆,胡萝卜,青椒；功效：改善肠胃、预防高血压。",8.0,"image/酸辣土豆丝.jpg");
        addItem("火龙吐丝","食材：火龙果,青椒,胡萝卜,火腿；功效：美白养颜。",12,"image/火龙吐丝.jpg");
        addItem("西芹炒菱角","食材：鲜菱角,西芹,枸杞；功效：营养丰富且抗癌。",20,"image/西芹炒菱角.jpg");
        addItem("双丁炒鸡蛋","食材：鸡蛋,胡萝卜,青瓜；功效：简单清爽,营养丰富。",10,"image/双丁炒鸡蛋.jpg");
        addItem("午烧竹笋","食材：嫩竹笋,苦瓜,橄榄油；功效：促进肠道蠕动,消除积食,防止便秘。",16,"image/午烧竹笋.jpg");
        addItem("杏仁奶豆腐","食材：琼脂、冷水、纯牛奶、淡奶油、杏仁粉；功效：美容养颜。",14,"image/杏仁奶豆腐.jpg");
        addItem("西芹百合","食材：西芹、新鲜百合、红柿子椒；功效：脆嫩清香、营养丰富。",20,"image/西芹百合.jpg");
        addItem("三椒黑木耳","食材：木耳、黑椒；功效：益气强身、滋肾养胃、活血、降血脂。",16,"image/三椒黑木耳.jpg");
        addItem("西兰花炒鸡蛋","食材：西兰花、鸡蛋；功效：抗癌、养生。",26,"image/西兰花炒鸡蛋.jpg");
        addItem("海带焖黄豆","食材：海带、黄豆；功效：软坚散结、消痰平喘、通行利水、祛脂降压。",12,"image/海带焖黄豆.jpg");
    }
}
```

5）代码(StapleFoodMenu.java)

```java
package caida.xinxi.Takingorders;
public class StapleFoodMenu extends MealMenu{
    public StapleFoodMenu(){
        addItem("扬州炒饭","由虾仁、米饭等多种食材制作,如碎金闪烁,光润油亮,鲜美爽口,有保护肝脏,防治动脉硬化,延缓衰老的功效。",12.0,"image/扬州炒饭.jpg");
        addItem("海鲜面","颜色鲜艳、气味醇香,面筋汤浓,无腥不腻;功效：软坚化痰,利水泄热,降低血压",18,"image/海鲜面.jpg");
        addItem("排骨面","主料是排骨、面和青菜;排骨香酥,面条滑润,具有补肾,益气,润燥之效。",16,"image/排骨面.jpg");
        addItem("小笼包子","皮薄馅大,灌汤流油,软嫩鲜香,洁白光润,提起像灯笼,放下似菊花,具有调理肠胃,开胃消食,安神除烦之效",12,"image/小笼包子.jpg");
        addItem("刀削面","用刀削出的面叶,中厚边薄。棱锋分明,形似柳叶;入口外滑内筋,软而不粘。",8,"image/刀削面.jpg");
        addItem("米饭","性平、味甘;有补中益气、健脾养胃、益精强志、和五脏、通血脉的功效。",6,"image/米饭.jpg");
    }
}
```

6）代码(SoupAndPorridgeMenu.java)

```java
package caida.xinxi.Takingorders;
public class SoupAndPorridgeMenu extends MealMenu{
    public SoupAndPorridgeMenu(){
        addItem("玉米羹","羹汁粉浆稀薄均匀,稠而不粘,肉末鲜嫩,乳香浓郁,略有甜味,味清淡的特色。",15.0,"image/玉米羹.jpg");
        addItem("疙瘩汤","此汤配以鸡蛋、香菇、肉、油菜、西红柿等材料,营养丰富,味道鲜美,利于消化吸收。",12,"image/疙瘩汤.jpg");
        addItem("西红柿蛋汤","此汤将西红柿和鸡蛋的营养价值完美搭配在一起,有健脑益智,美容护肤,减少色斑之效",10,"image/西红柿蛋汤.jpg");
        addItem("酸辣汤","此汤集酸、辣、咸、鲜、香于一体,饭后饮用,有醒酒去腻,助消化作用。",12,"image/酸辣汤.jpg");
        addItem("白米粥","白米粥功效：补虚养身调理,脾调养调理,胃炎调理,便秘调理",10,"image/白米粥.jpg");
        addItem("八宝粥","具有健脾养胃,消滞减肥,益气安神的功效;可作为养生健美之食品。",12,"image/八宝粥.jpg");
    }
}
```

## 10.2.5 案例练习题目

（1）读者自己换一种思路,重新设计 OrderingWindow 界面布局和 OrderDishes 界面布局,并实现它们。比较常用的布局管理方式的不同和使用条件。

（2）本应用程序按照荤菜、素菜、主食、汤粥进行分类点菜,读者可以增加新的分类功能（如冷饮、酒水等）或者完全换一种新的分类（如冷菜、热菜、主食……）。并实现它。

（3）调研实际饭店点菜流程,修改本程序,增加新的功能（如查询）。或者不参考本例,根据你自己的设计,实现一个与案例完全不同的饭店点菜程序。

（4）学习数据库知识,利用数据库技术重新实现此饭店点菜程序。

# 第11章 Java多媒体程序设计

多媒体技术是当前计算机应用中的重要领域。Java语言支持多媒体,它提供了大量的类和接口来实现对文本、图形、图像、动画、声音和视频等多种形式的支持。本章内容将从以下几个方面来学习Java语言对多媒体技术的支持:图形绘制、图像处理、声音播放、用Java实现动画等。

## 11.1 案例:随机绘图与动画

### 11.1.1 案例问题描述

创建一个GUI程序,当用户单击按钮"随机生成图形"时,程序主界面将随机显示若干个图形(图形可能是线、矩形、椭圆)。这些图形的显示位置、显示大小、显示颜色以及是否填充等属性的设置都是随机产生的。用户还可以通过单击"动画开始"让生成的图形以一定的方式在窗口中运动,同时还可以通过按钮"动画结束"、"动画继续"来控制动画播放的停止与继续。

案例设计的目的就是希望读者能了解并掌握图形的绘制以及动画程序的设计。

### 11.1.2 案例功能分析与演示

从案例问题描述中可以看到,本应用程序的功能主要有:
- 随机绘制图形。
- 设计动画,并控制动画播放。

每次运行应用程序,会显示程序主界面的初始状态,见图11-1。整个界面的中心是绘制图形的区域,下边区域是一个控制区域,放置5个按钮,代表5种功能,分别是"随机生成图形"、"动画开始"、"动画结束"、"动画继续"、"退出程序"。

当用户单击"动画开始"播放按钮时,会在中间区域随机绘制图形,结果见图11-2。若再次单击该按钮,就立即刷新界面,随机绘制出新的图形来,见图11-3的运行效果。读者可以观察到,在图11-2和图11-3中,绘制出来的图形是不一样的,它们的形状、出现位置、显示大小、是否填充等都不一样,都是通过应用程序随机生成的。例如,图11-2绘制了7个图形(3个矩形、3个椭圆、1个线段),图11-3绘制了5个图形(3个矩形、1个椭圆、1个线段)。

假设应用程序目前绘制的图形是图11-3所示的效果,如果用户单击"动画开始",那么

图 11-1　绘制图形与动画应用程序主界面初始状态

图 11-2　随机生成图形(一)

界面上的5个图形会分别以一定的运动方式在窗口上运动,形成动画的效果。如果当图形碰到窗口的4个边框时,它们又会以相反方向的运动方式在窗口上运行。如果用户单击"动画结束",则所有图形都停止运动。如果用户单击"动画继续",则图形会在上次停止的位置继续以原来的方式在窗口中运动。动画的运行效果读者可以观察图11-3和图11-4,图11-3是原图,单击了"动画开始"后,5个图形开始移动。图11-4中的第一幅图是图11-3移动了一段时间后,单击"动画结束"后的静止状态;第二幅图是接着单击了"动画继续"按钮后,移动过程中的一幅截图。

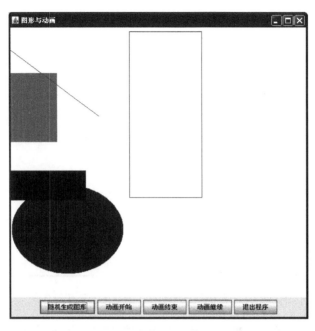

图 11-3　随机生成图形(二)

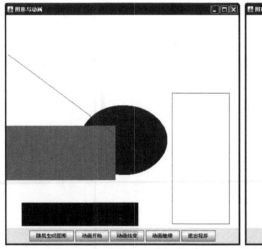 

图 11-4　图形的动画

应用程序主窗口上的"退出程序"按钮的作用是退出应用程序。

### 11.1.3　案例总体设计

用 Java 编写一个面向对象的程序,很大程度上是设计类,并编写这些类的定义。因此,通过对该案例的功能分析及演示,下面把问题分解成一组相互协作的对象,然后设计和创建这些对象本身。

根据上面的运行效果，可以知道案例需要随机绘制图形（假设有线、矩形、椭圆）。读者可以先按下面的思路来处理程序：可以分别创建 MyLine 类、MyRectangle 类和 MyOval 类，利用这三个类来绘制线、矩形和椭圆；DrawShapePanel 类扩展自 JPanel 类，负责创建各种图形，应声明 3 个数组，分别对应每种形状，具体绘制。

如果按照这种思路编写程序后，读者可以观察你写的代码，会注意到各种形状类之间有许多相似性，即有大量的相同代码。因此，这里可以换一个思路，利用继承，将 3 个类中的公共属性"提取"出来，放在一个 Shape 超类中。然后利用该超类变量可以对 3 个具体的形状类进行多态处理。去除冗余代码，使程序更小并且更灵活，也更易于维护。

因此，第一步：可以设计超类 Shape，把表示坐标、颜色等公共的属性放到这里作为成员。由于 Shape 代表任何形状，因此如果不知道具体形状，就不能实现绘制方法。所以 Shape 类应该声明为 abstract，绘制方法定义成一个抽象方法：public abstract void draw(Graphics g)。第二步：设计超类 Shape 的子类 MyLine 类、MyRectangle 类、MyOval 类。子类新增自己的属性。如 MyRectangle 类可以新增一个表示是否填充的 Boolean 型变量，可以实现父类的抽象方法，具体绘制矩形。第三步：设计一个绘制各种图形的面板类 DrawShapPanel。在绘制完图形后，还可以设计动画。第四步：设计一个 GUI 的程序，控制图形绘制与动画功能。此功能由 ShapeAndAnimationWindow 类实现。

整个绘制图形与动画应用程序的面向对象设计总体类图如图 11-5 所示。

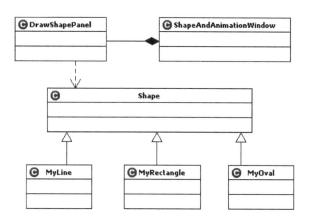

图 11-5　绘制图形与动画程序的总体设计类图

1. Shape 类

已知空间中的两个点可以绘制直线、矩形和椭圆。因此这里可以设计 x1、y1、x2、y2 坐标来绘制直线、矩形和椭圆，椭圆和矩形的 width 和 height 可以通过坐标之间的计算得到。颜色 color 可以随机生成，在具体绘制时初始化。这 7 个变量都设计为 Shape 类的 private 成员。

Shape 类的 UML 图如图 11-6 所示。

以下是 UML 图中有关数据和方法的详细说明。

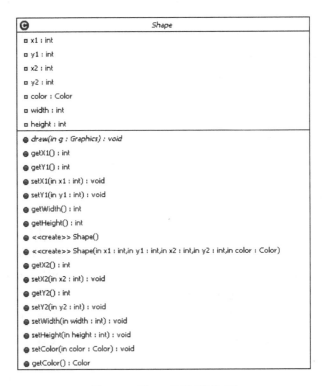

图 11-6　Shape 类的 UML 图

1）成员变量
- x1、y1 是图形的左上角的坐标值；x2、y2 是图形的右下角的坐标值。
- color 是图形的边框颜色或填充颜色。
- width、height 是图形的宽度和高度。

2）成员方法
- Shape()是无参的构造方法。将形状的所有坐标设置为 0，颜色设置为 Color.WHITE。
- Shape(int,int,int,int,Color)是构造方法，用实参提供的值来初始化坐标、颜色和图形的大小。
- draw(Graphics)是一个抽象方法，功能是绘制图形，由子类来具体实现。
- 其余方法是各个坐标和颜色的 set 方法和 get 方法，实现对形状的每个数据成员的设置和检索。

2．MyLine 类

MyLine 类的 UML 图如图 11-7 所示。
以下是 UML 图中的有关数据和方法的详细说明。
该类的成员方法有：
- MyLine()无参构造方法。
- MyLine(int,int,int,int,Color)构造方法。初始化直线的两个坐标和颜色。

- draw(Graphics)方法。在指定位置用指定颜色绘制直线。

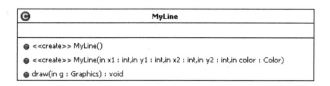

图 11-7　MyLine 类的 UML 图

### 3．MyRectangle 类

MyRectangle 类的 UML 图如图 11-8 所示。

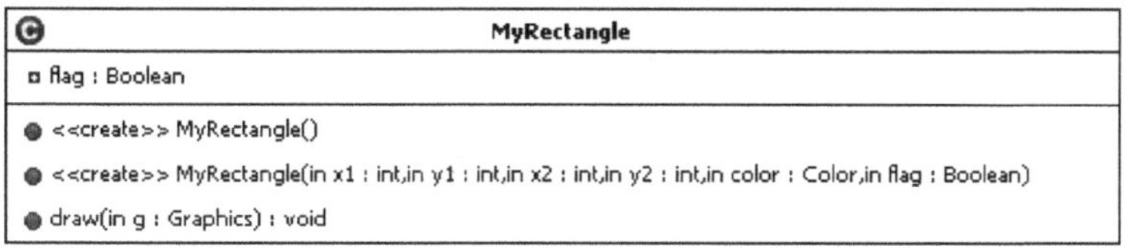

图 11-8　MyRectangle 类的 UML 图

下面是 UML 图中有关数据和方法的说明。
1) 成员变量

flag 表示矩形是否填充。

2) 成员方法

- MyRectangle()构造方法，调用父类的构造方法初始化。
- MyRectangle(int,int,int,int,Color,Boolean)构造方法，初始化矩形的各个成员变量。
- draw(Graphics)方法。绘制指定大小、位置、颜色的矩形。

### 4．MyOval 类

MyOval 类的 UML 图如图 11-9 所示。

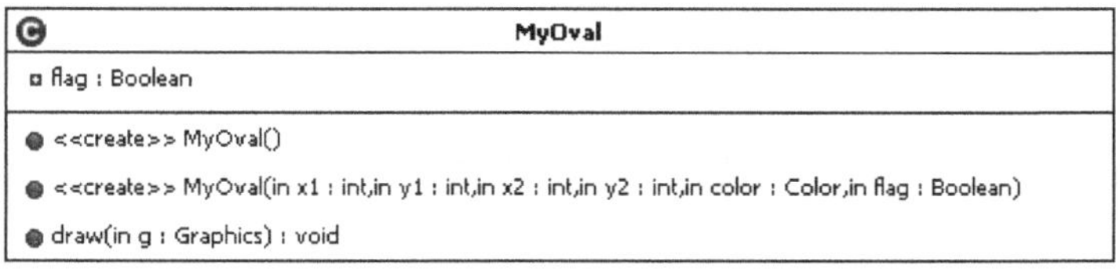

图 11-9　MyOval 类的 UML 图

下面是 UML 图中有关数据和方法的说明。
1) 成员变量

flag 表示椭圆是否填充。true 为填充。

2) 成员方法

- MyOval()构造方法。调用父类构造方法初始化。

- MyOval(int,int,int,int,Color,Boolean)构造方法。用指定的值初始化椭圆的参数。
- draw(Graphics)方法,在指定位置绘制指定大小和颜色的椭圆。

### 5. DrawShapePanel 类

DrawShapePanel 类是 Jpanel 的子类,负责随机绘制各种图形,并实现动画的播放。DrawShapePanel 类的 UML 图如图 11-10 所示。

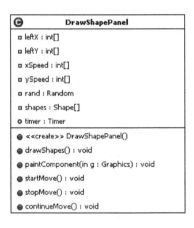

图 11-10  DrawShapePanel 类的 UML 图

下面是 UML 图中有关数据和方法的说明。

1) 成员变量

- leftX、leftY、xSpeed、ySpeed 均是 int 型的一维数组。leftX 用来存放 shapes 数组中所有图形左上角距离容器左端的距离。leftY 用来存放 shapes 数组中所有图形左上角距离容器上端的距离。xSpeed 用来存放 shapes 数组中各个图形的水平移动速度。ySpeed 用来存放 shapes 数组中各个图形的纵向移动速度。
- rand 是 Random 类对象。
- shapes 是 Shape 对象数组。用来存放随机绘制的各种图形。
- timer 是 Timer 类对象,是一个计算器,用来控制动画。

2) 成员方法

- DrawShapePanel()是构造方法。用来初始化各个数组的值。
- drawShapes()方法,随机生成 shapes 数组中的元素,绘制 shapes 数组中的图形。
- paintComponent(Graphics)方法,重写 JComponent 类中的方法,负责绘制 shapes 数组中的各个图形。
- startMove()方法,开始动画播放。
- stopMove()方法,停止动画播放。
- continueMove()方法,负责重新启动动画播放。

### 6. ShapeAndAnimationWindow 类

ShapeAndAnimationWindow 类是 JFrame 类的子类,实现了 ActionListener 接口。该

类对象是一个窗口,中心是一个 DrawShapePanel 面板对象,底部是一组按钮,实现程序的各个功能。ShapeAndAnimationWindow 类的 UML 图如图 11-11 所示。

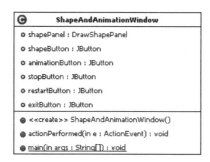

图 11-11　ShapeAndAnimationWindow 类的 UML 图

下面是 UML 图中有关数据和方法的说明。

1) 成员变量
- shapePanel 是 DrawShapePanel 对象,绘制图形的面板。
- shapeButton、animationButton、stopButton、restartButton、exitButton 是 JButton 类对象,shapeButton 按钮:随机绘制图形。animationButton 按钮:开始播放动画。stopButton 按钮:停止播放动画。restartButton 按钮:继续播放动画。exitButton 按钮:退出应用程序。

2) 成员方法
- ShapeAndAnimationWindow()构造方法。负责完成窗口的初始化。
- actionPerformed(ActionEvent)方法,是 ShapeAndAnimationWindow 类实现的接口 ActionListener 中的方法,负责响应用户单击按钮的行为。当用户单击了按钮,产生了 ActionEvent 事件,actionPerformed(ActionEvent)方法被调用执行。
- main(String[])方法。应用程序应用的入口方法。

### 11.1.4　案例代码实现

**1. 编写程序主窗体(DrawAndAnimationWindow.java)**

DrawAndAnimationWindow.java 文件中包含了程序的入口,是本应用程序的主窗体,通过该窗体可以实现图形的绘制与动画。

DrawAndAnimationWindow 类是 javax.swing 包中 JFrame 的一个子类,并实现了 ActionListener 接口。从图 11-1 的运行效果可以看到该窗口的布局设计非常简单,利用默认的 BorderLayout 型即可,在 Center 位置放置绘图面板对象,在 South 位置放置一个容纳系列按钮的面板。要实现程序对用户单击按钮的响应,利用 Java 的事件处理机制就可以(具体见第 10 章)。

代码(DrawAndAnimationWindow.java):

```
package caida.xinxi.ShapeAndAnimation;

import java.awt.BorderLayout;
```

```java
import java.awt.event.ActionEvent;
import java.awt.event.ActionListener;

import javax.swing.JButton;
import javax.swing.JFrame;
import javax.swing.JPanel;

@SuppressWarnings("serial")
public class ShapeAndAnimationWindow extends JFrame implements ActionListener{
    DrawShapePanel shapePanel;
    JButton shapeButton,animationButton;
    JButton stopButton,restartButton,exitButton;

    //构造方法,完成窗口的初始化
    public ShapeAndAnimationWindow(){
        setTitle("图形与动画");

        //创建窗口中的各个组件
        shapePanel = new DrawShapePanel();
        JPanel pSouth = new JPanel();
        shapeButton = new JButton("随机生成图形");
        animationButton = new JButton("动画开始");
        stopButton = new JButton("动画结束");
        restartButton = new JButton("动画继续");
        exitButton = new JButton("退出程序");

        //给下面5个按钮注册ActionEvent事件监听器
        shapeButton.addActionListener(this);
        animationButton.addActionListener(this);
        stopButton.addActionListener(this);
        restartButton.addActionListener(this);
        exitButton.addActionListener(this);

        //将5个按钮顺序添加到面板pSouth中
        pSouth.add(this.shapeButton);
        pSouth.add(this.animationButton);
        pSouth.add(this.stopButton);
        pSouth.add(this.restartButton);
        pSouth.add(this.exitButton);

        //给窗口添加组件
        add(shapePanel,BorderLayout.CENTER);
        add(pSouth,BorderLayout.SOUTH);

        setBounds(100,100,600,600);
        setDefaultCloseOperation(JFrame.EXIT_ON_CLOSE);
        setVisible(true);
    }

    //单击按钮触发ActionEvent事件,执行actionPerformed(ActionEvent)方法
    public void actionPerformed(ActionEvent e) {
```

```java
        //如果单击"随机生成图形"按钮,将在面板随机绘制图形
        if(e.getSource() == this.shapeButton){
            shapePanel.drawShapes();
        }
        //若单击了"动画开始"按钮,则开始播放动画,所有图形开始运动
        else if(e.getSource() == this.animationButton){
            shapePanel.startMove();
        }
        //若单击"动画结束"按钮,将停止播放动画
        else if(e.getSource() == this.stopButton){
            shapePanel.stopMove();
        }
        //若单击"动画继续"按钮,将重启动画
        else if(e.getSource() == this.restartButton){
            shapePanel.continueMove();
        }
        //若单击"退出程序"按钮,将退出应用程序
        else if(e.getSource() == this.exitButton){
            System.exit(0);
        }
    }

    public static void main(String[] args) {
        new ShapeAndAnimationWindow();
    }
}
```

### 2. 图形类及其子类的实现

超类 Shape 中包含各种图形的公共数据,且为了保证正确的封装性,类 Shape 中的所有数据都必须是专有的。这需要声明对应的 set 方法和 get 方法来操作数据。除此以外,类 Shape 还应该声明下列方法:

- 一个无参的构造方法,将形状的所有坐标设置为 0,颜色设置为 Color.WHITE。
- 一个构造方法,用实参提供的值来初始化坐标和颜色、图形大小。
- abstract 方法: public abstract void draw(Graphics g);。它将从程序的 paintComponent 方法中被调用,用来在屏幕上绘制图形。

类 MyLine 应该定义一个无实参构造方法和一个带坐标和颜色的实参构造方法;类 MyRectangle 和类 MyOval 应该定义一个无实参构造方法和一个带有坐标、颜色以及填充模式实参的构造方法。无实参构造函数除了设置默认值以外,还应该将形状设置为不填充的。另外,为了后面控制动画在窗体内的运动,还应该计算图形的宽度和高度。具体实现参见前面的 UML 图和后面的代码。

下面解决如何绘制直线、矩形和椭圆的问题,以及了解 Grahpics 类和 paintComponent 方法。

1) Java 图形处理

要绘制图形,首先需要了解 Java 图形坐标系统,以及可以使用哪些类来绘制图形。

(1) 图形坐标

与大多数其他计算机图形系统所采用的二维坐标系统一样，Java 的坐标原点(0,0)在屏幕的左上角，水平向右为 $x$ 轴的正方向，竖直向下为 $y$ 轴的正方向，每个坐标点的值表示屏幕上的一个像素点的位置，所有的坐标位置都用整数表示。组件都是矩形形状，组件本身有一个默认的坐标系，组件的左上角的坐标值是(0,0)，$x$ 坐标向右增加，$y$ 坐标向下增加。

(2) java.awt.Graphics

Graphics 类是提供设备无关图形界面的一个抽象类，它可以在不同平台的屏幕上显示图形和图像。它是 Java 虚拟机在本地平台上实现的。当使用 paintComponent 方法在一个图形环境 g 中画图时，这个 g 是抽象类 Graphcis 在特定平台上具体子类的一个实例。Grahpics 类封装了平台的细节，使用户能够不考虑特定平台而使画图统一。

通过使用 java.awt.Graphics，可以绘制各种形状的图形，包括线、矩形、多边形、圆弧等。该类提供了非常丰富的绘制图形的方法(读者可查阅 API)，本应用程序用到的方法具体如下。

- public abstract void drawLine(int x1,int y1,int x2, int y2)：绘制从坐标(x1,y1)开始到(x2,y2)为止的直线。
- public void drawRect(int x, int y, int width, int height)：绘制指定矩形的边框。x：要绘制矩形的 x 坐标。y：要绘制矩形的 y 坐标。width：要绘制矩形的宽度。height：要绘制矩形的高度。
- public abstract void fillRect(int x, int y, int width, int height)：填充指定的矩形。矩形的坐标是 x,y；width 和 height 是要填充矩形的宽度和高度。
- public abstract void drawOval(int x, int y, int width, int height)：绘制椭圆的边框。得到一个圆或椭圆，它刚好能放入由 x、y、width 和 height 参数指定的矩形中。
- public abstract void fillOval(int x, int y, int width, int height)：使用当前颜色填充外接指定矩形框的椭圆。x、y 是椭圆左上角的坐标；width、height 是椭圆的宽度和高度。

可以看到 Graphics 类的绘制图形函数有两类，一个仅仅绘制外部线条，而另一个则同时填充内部。

在 Java 的后期版本推出了 Graphics2D 类。这个类比 Graphics 类功能要丰富。本案例仍以常用的 Graphics 类来示范讲解，讲解的实例可以很容易扩充到 Graphics2D 上去。本例练习题就要求用下面 Java2D API 中的方法实现程序。

(3) Java2D API

通过使用 Java2D API，可以轻松地描绘出任意的几何图形、运用不同的填色效果、对图形进行旋转(rotate)、缩放(scale)、扭曲(shear)等。Java2D 为开发人员提供了下列功能：

- 对渲染质量的控制；
- 裁剪、合成、透明度；
- 绘制和填充简单及复杂的形状；
- 图像处理和变换；
- 高级字体处理和字符串格式化。

Java2D 的类位于以下的包中：

- java.awt 包中包含了一些新增的 2D API 类和接口。其中 Graphics2D 继承自 java.

awt.Graphics,是描绘 2D 图形的类。在 Graphics2D 中新增了许多状态属性,例如 Stroke、Paint、Clip、Transform 等。
- java.awt.geom 包含可以勾勒任何形状的 GeneralPath 类。此包中更定义了许多基本几何图形,包括 Arc2D、CubicCurve2D、Line2D 等。
- java.awt.font 中定义了 TextLayout 类,负责构建文本图像、执行适当的文本操作,以及决定文本的适当位置与顺序。
- java.awt.color 定义了类 ColorSpace,提供了转换色盘的各种方法。
- java.awt.print 提供了打印方法,并可以设置打印的属性,如双面列印等。

(4) 颜色

这里可以使用 java.awt 包中的 Color 类来控制颜色。Java 语言中使用的颜色的基本模型非常通用。每种颜色都是红、绿、蓝三种颜色组合而成,称为三原色,每一种原色相关的信息用一个字节来描述,其取值范围为 0~255,其中 0 为最暗,255 为最亮。这种颜色模型称为 RGB 模型。对于已经定义常量的颜色,可以通过"Color.颜色常量名"来调用指定的颜色,例如:要使用白色,可以用 Color.WHITE 来实现使用白色的目的。

Graphics 类记录当前的画笔颜色,可以用方法 getColor()得到,也可以用方法 setColor()来设置。图形和文本的绘制都是以当前的颜色进行的。

2) 代码(Shape.java)

```java
package caida.xinxi.ShapeAndAnimation;

import java.awt.Color;
import java.awt.Graphics;

public abstract class Shape {
    private int x1,y1;
    private int x2,y2;
    private Color color;
    private int width,height;

    //无参的构造方法.初始化坐标为 0,颜色为白色
    public Shape() {
        this.x1 = 0;
        this.y1 = 0;
        this.x2 = 0;
        this.y2 = 0;
        this.setColor(Color.WHITE);
    }

    //构造方法,初始化图形各属性值,(x1,y1)表示图形左上角在面板中的位置
    public Shape(int x1, int y1, int x2, int y2, Color color) {
        super();
        this.x1 = Math.min(x1, x2);
        this.y1 = Math.min(y1, y2);
        this.x2 = Math.max(x1, x2);
        this.y2 = Math.max(y1, y2);
        this.setColor(color);
```

```java
            width = x2 - x1;
            height = y2 - y1;
    }

    //抽象方法,绘制图形
    public abstract void draw(Graphics g);

    //图形类私有成员的 set 方法和 get 方法
    public int getX1() {
        return x1;
    }
    public void setX1(int x1) {
        this.x1 = x1;
    }
    public int getY1() {
        return y1;
    }
    public void setY1(int y1) {
        this.y1 = y1;
    }
    public int getX2() {
        return x2;
    }
    public void setX2(int x2) {
        this.x2 = x2;
    }
    public int getY2() {
        return y2;
    }
    public void setY2(int y2) {
        this.y2 = y2;
    }
    public int getWidth() {
        return width;
    }
    public void setWidth(int width) {
        this.width = width;
    }
    public int getHeight() {
        return height;
    }
    public void setHeight(int height) {
        this.height = height;
    }
    public void setColor(Color color) {
        this.color = color;
    }
    public Color getColor() {
        return color;
    }
}
```

3）代码（MyLine.java）

package caida.xinxi.ShapeAndAnimation;

import java.awt.Color;
import java.awt.Graphics;

```java
public class MyLine extends Shape{

    public MyLine(){
        super();
    }
    //初始化直线的各成员变量的值
    public MyLine(int x1, int y1, int x2, int y2, Color color) {
        super(x1,y1,x2,y2,color);
    }
    //设置画笔颜色为指定的颜色,在指定位置画指定长度的直线
    public void draw(Graphics g) {
        g.setColor(getColor());
        g.drawLine(getX1(), getY1(), getX1() + getWidth(), getY1() + getHeight());
    }
}
```

4）代码（MyRectangle.java）

package caida.xinxi.ShapeAndAnimation;

import java.awt.Color;
import java.awt.Graphics;

```java
public class MyRectangle extends Shape{
    private Boolean flag;
    //无参构造方法
    public MyRectangle() {
        super();
        this.flag = false;
    }
    //构造方法,初始化矩形类的各成员变量值
    public MyRectangle(int x1, int y1, int x2, int y2, Color color, Boolean flag) {
        super(x1,y1,x2,y2,color);
        this.flag = flag;
    }

    //根据指定颜色和是否填充变量的值,绘制指定大小和位置的矩形
    public void draw(Graphics g) {
        g.setColor(getColor());
        //flag 为真,填充矩形
        if(flag){
            g.fillRect(getX1(), getY1(), getX1() + getWidth(), getY1() + getHeight());
        }
        //flag 为假,仅绘制矩形边框
```

```java
        else{
            g.drawRect(getX1(), getY1(), getX1() + getWidth(), getY1() + getHeight());
        }
    }
}
```

5）代码（MyOval.java）

```java
package caida.xinxi.ShapeAndAnimation;

import java.awt.Color;
import java.awt.Graphics;

public class MyOval extends Shape{
    private Boolean flag;
    //无参构造方法
    public MyOval() {
        super();
        this.flag = false;
    }
    //构造方法,初始化椭圆的各个成员属性
    public MyOval(int x1, int y1, int x2, int y2, Color color, Boolean flag) {
        super(x1,y1,x2,y2,color);
        this.flag = flag;
    }

    //根据指定颜色和是否填充变量值,在指定位置绘制指定大小的椭圆
    public void draw(Graphics g) {
        g.setColor(getColor());
        //flag 为真,绘制填充椭圆
        if(flag){
            g.fillOval(getX1(), getY1(), getX1() + getWidth(), getY1() + getHeight());
        }
        //flag 为假,绘制椭圆边框
        else{
            g.drawOval(getX1(), getY1(), getX1() + getWidth(), getY1() + getHeight());
        }
    }
}
```

### 3. 编写绘图面板,实现动画功能

DrawShapePanel 类是 JPanel 类的一个子类。由 DrawShapePanel 类来实现随机绘制若干个图形,并进行动画控制。本文件要实现用多态方式绘图,因此程序中不能有 MyLine、MyRectangle 或 MyOval 变量,只能用 Shape 变量指向 MyLine、MyRectangle 或 MyOval 对象。程序应该生成随机形状,并将它们保存在一个 Shape 类型的数组中。方法 paintComponent 利用这个 Shape 数组,并画出每个形状(即多态地调用各形状的 draw 方法)。

因此,编写此类,需要考虑如何生成随机数(创建包含各种形状的 Shape 数组、决定图形

绘制的位置、大小、颜色和是否填充)、绘图的有关原理和如何实现动画的设计。

1) 生成随机数

下面介绍两个方法来生成随机数。

(1) 利用 random()方法来生成随机数

在 Math 类中，Java 语言提供了一个方法 public static double random()。通过这个方法可以让系统产生随机数。其产生的随机数范围比较小，为大于等于 0 到小于 1 的 double 型随机数。只要对这个方法进行一些灵活的处理，就可以获取任意范围的随机数。例如，如果要生成 1~100 之间的随机整数，可以用下面语句来实现：

```
int i = (int)(Math.random() * 100) + 1;
```

(2) 通过 Random 类来生成随机数

Random 类的对象可以产生随机的 boolean、byte、float、double、int、long 和高斯值。本例就使用 Random 类的对象来产生随机值。

- 新的随机数生成器对象可按如下方式创建：

```
Random  randomNumber = new Random();
```

- 然后就可以使用这个随机数生成器对象随机生成 boolean、byte、float、double、int、long 和高斯值。这里只讨论随机 int 值。关于 Random 类的更多信息可参见 API。

如：

```
randomNumber.nextInt();              //生成一个处于 int 整数取值范围内的随机 int 值
//即生成一个 - 2 147 483 648~ + 2 147 483 647 范围内的随机 int 值
```

再如：

```
randomNumber.nextInt(26);            //生成一个 0~25 之间的随机整数
```

- 为了避免两个 Random 对象产生相同的数字序列，通常推荐使用当前时间来作为 Random 对象的种子，如下代码所示：

```
Random rand = new Random( System.currentTimeMillis() );
```

2) 绘图的有关原理

在 Component 类里提供了以下三个和绘图有关的方法。

- paint(Graphics g)：绘制组件的外观。
- update(Graphics g)：调用 paint 方法，刷新组件外观。
- repaint()：调用 update 方法，刷新组件外观。

上面三个方法的调用关系为：repaint 方法调用 update 方法，update 方法调用 paint 方法。

Container 类中的 update 方法先以组件的背景色填充整个组件区域，然后调用 paint 方法重画组件。普通组件的 update 方法则直接调用 paint 方法。

重写 update 或 paint 方法时，该方法里包含了一个 Graphics 类型的参数，通过该 Graphics 参数就可以实现绘图功能。

paintComponent 方法位于 JComponent 类，该方法与 paint 方法类似，也要求一个

Graphics 类的实例为参数。在 Swing 中通常应该使用 paintComponent 方法，而不是使用 paint 方法绘图，这是因为 JComponent 类的 paint 方法要执行大量的复杂行为，如果重写了该方法，会发生冲突，导致程序不能正常运行。

无论在系统开始执行时，或窗口被最小化后再恢复，以及窗口被覆盖后需要被重新绘制时，paintComponent 方法都会被系统自动调用，所以在程序中不要调用它，否则会与自动化过程冲突。

如果需要重新绘制，就调用 repaint 方法，而不是 paintComponent 方法。repaint 方法将引起系统调用所有组件的 paintComponent 方法，并且使得所有组件的 paintComponent 方法的 Graphics 变量被正确配置。

需要注意的是，在 JPanel 子类的 paintComponent 方法中调用了超类 JComponent 的 paintComponent 以完成组件背景的绘制。所以，在 JPanel 子类中重写 paintComponent 方法时，必须在绘制之前调用 super.paintComponent(g)。这点可以在后面的代码中见到。

3）程序动画的设计

所谓动画，就是隔一定的时间间隔（通常小于 0.1 秒）重新绘制新的图像，两次绘制的图像之间差异较小，肉眼看起来就成了所谓的动画。本程序要使随机生成的多个图形能同时以某种方式各自运动，一旦图形碰到窗口边框，则以相反的方向继续原来的运动。要实现这个功能，可以运用 Java 的多线程实现动画，也可以借助于 Swing 提供的 Timer 类来实现。多线程实现动画的应用在第 12 章多线程程序设计那部分会给读者举例介绍。这里利用 Timer 类介绍动画的制作。

Timer 类是一个定时器，它有一个如下的构造方法。

Timer(int delay, ActionListener listener)：每隔 delay 毫秒，计时器"震铃"一次。计时器发生的震铃事件是 ActionEvent 类型事件。当震铃事件发生时，监听器就会监听到这个事件，就会执行接口 ActionListener 中的方法 actionPerformed。因此，当震铃每隔 dalay 毫秒发生一次时，方法 actionPerformed 就被执行一次。可以看到 delay 就是动画的时间间隔。

计时器创建后，使用 Timer 类的方法 start() 启动计时器；使用 stop() 方法停止计时器；使用 restart() 重新启动计时器。

接下来要解决的就是 actionPerformed 方法中的代码如何编写。actionPerformed 方法要实现的功能其实就是让图形动起来。而要显示一幅动态的图案，在程序设计上要克服以下两个问题：

- 不断地改变图案坐标。

设置一个坐标变化量(dx,dy)，创建图案位置坐标(x+dx,y+dy)，这样图案将随新坐标的改变而移动。

- 当在新位置显示图案时，要清除旧位置上的图案。

清除旧位置上的图案：当图案移往新坐标时，必须清除旧位置上的图案，否则新旧图案重叠，图形会显得混乱而无法辨识。repaint()方法有更新图案与清除旧图案的功能。

因此，actionPerformed 方法中的代码设计为：首先设计图形的新位置坐标，如果位置坐标大于所在窗口的边界，则改变运动方向；接着调用 repaint 方法，那么程序会调用 paintComponent 方法，先清除旧位置上的图形，再在新的坐标位置画图。

4) 代码(DrawShapePanel.java)

```java
package caida.xinxi.ShapeAndAnimation;

import java.awt.Color;
import java.awt.Graphics;
import java.awt.event.ActionEvent;
import java.awt.event.ActionListener;
import java.util.Random;

import javax.swing.JPanel;
import javax.swing.Timer;

@SuppressWarnings("serial")
public class DrawShapePanel extends JPanel{
    private Random rand = new Random(System.currentTimeMillis());
    private Shape[] shapes;
    Timer timer;
    private int leftX[],leftY[];
    private int xSpeed[],ySpeed[];

    //构造方法,实现对成员变量的初始化
    public DrawShapePanel(){
        setBackground(Color.WHITE);
        //产生一个随机整数 size,确定成员变量中各个数组的长度
        int size = 5 + rand.nextInt(6);
        shapes = new Shape[size];
        leftX = new int[size];
        leftY = new int[size];
        xSpeed = new int[size];
        ySpeed = new int[size];

        //初始化图形数组 shapes 中的元素
        for (int i = 0; i < shapes.length; i++) {
            switch (rand.nextInt(3)) {
            case 0:
                shapes[i] = new MyLine();
                break;
            case 1:
                shapes[i] = new MyRectangle();
                break;
            case 2:
                shapes[i] = new MyOval();
                break;
            }
        }

        //分别设置图形数组 shapes 中每个图形的移动速度
        for(int j = 0;j < size;j++){
            xSpeed[j] = 5;
            ySpeed[j] = 5;
```

```java
            }
        }

        //实现随机绘制图形的功能
        public void drawShapes() {
            for (int i = 0; i < shapes.length; i++) {

                //随机生成图形在面板中的左上角坐标和右下角坐标
                int x1 = rand.nextInt(600);
                int y1 = rand.nextInt(500);
                int x2 = rand.nextInt(600);
                int y2 = rand.nextInt(500);
                //随机生成一种颜色
                Color color = new Color(rand.nextInt(256),rand.nextInt(256),rand.nextInt(256));
                //随机生成一个 boolean 值
                boolean flag = rand.nextBoolean();

                //给图形数组 shapes 内随机分配图形对象
                switch (rand.nextInt(3)) {
                case 0:
                    shapes[i] = new MyLine(x1,y1,x2,y2,color);
                    break;
                case 1:
                    shapes[i] = new MyRectangle(x1,y1,x2,y2,color,flag);
                    break;
                case 2:
                    shapes[i] = new MyOval(x1,y1,x2,y2,color,flag);
                    break;
                }
            }
            repaint();

        }

        //重写 paintComponent 方法,实现绘画
        public void paintComponent(Graphics g){
            super.paintComponent(g);
            for(int i = 0;i < shapes.length;i++){
                shapes[i].draw(g);
            }
        }

        //实现动画功能
        public void startMove() {
            ActionListener taskPerformer = new ActionListener(){
                public void actionPerformed(ActionEvent e){
                    for(int i = 0;i < shapes.length;i++){
                        //得到所有图形旧的左上角坐标
                        leftX[i] = shapes[i].getX1();
                        leftY[i] = shapes[i].getY1();
                        // 计算所有图形新的左上角坐标
```

```java
                    leftX[i] = leftX[i] + xSpeed[i];
                    leftY[i] = leftY[i] + ySpeed[i];
                    //设置所有图形的新的左上角坐标
                    shapes[i].setX1(leftX[i]);
                    shapes[i].setY1(leftY[i]);

                    //若图形碰到左或右边界,则改变运动方向
                    if(leftX[i]< 0){
                        xSpeed[i] = 5;
                    }
                    else if((leftX[i] + shapes[i].getWidth())>= getWidth()){
                        xSpeed[i] = -5;
                    }
                    //若图形碰到上或下边界,则改变运动方法
                    if(leftY[i]< 0){
                        ySpeed[i] = 5;
                    }
                    else if((leftY[i] + shapes[i].getHeight())>= getHeight()){
                        ySpeed[i] = -5;
                    }
                }
                repaint();
            }
        };
        timer = new Timer(100,taskPerformer);
        timer.start();
    }

    //停止计时器,即停止动画
    public void stopMove(){
        timer.stop();
    }

    //重新启动计时器,即重新播放动画
    public void continueMove(){
        timer.restart();
    }
}
```

## 11.1.5 案例练习题目

(1)学习Java2D的相关知识,利用Java2D的相关类来实现程序中的绘制图形功能和填充功能等。

(2)利用程序中介绍的Timer类实现动画的方法,改变图形的运动方式(轨迹)。

(3)学习多线程的相关知识,利用Thread类来处理程序中的动画控制。

(4)应用程序中一旦单击"播放动画"按钮,所有图形都开始以某种相同的运动方式同时运动,请读者修改程序,实现下面的功能:当用户单击某个图形时,该图形即开始运动,反之,没有被用户单击的图形则静止不动。

(5) 请读者修改应用程序,仿照 Windows 系统"附件"中的"画图"程序(见图 11-12),增加菜单功能和工具栏按钮,可以实现选择了某类图形按钮,用户就可以利用鼠标拖动在画布上画出该图形。

图 11-12　Windows 中的"画图"程序

(6) 结合本程序中用多态方式绘图的知识,请读者扩充系统,设计和实现更多图形的绘制。

## 11.2　案例:多媒体图片查看器

### 11.2.1　案例问题描述

创建一个程序,实现图片的查看及放映。选择要查看的图片后,在图片显示区以缩略图的方式显示出所选的所有图片。选择"图片显示区"中的某张图片,双击即可打开具体的"图片查看器",在此界面可以实现图片浏览、放映、缩放功能。图片在放映时可以实现背景音乐播放。

案例设计的目的是希望读者能掌握图像加载和显示的方法、Timer 类的使用、音乐的播放等知识。

### 11.2.2　案例功能分析与演示

从案例问题描述中可以看到,本应用程序的功能主要是以多种方式实现图片的查看。

每次运行应用程序,会显示程序主界面的初始状态,见图 11-13。在图 11-13 中,图片显示区初始状态是空白的,用户可以通过"文件"菜单来实现图片的查看。"文件"菜单有"打开"和"退出"两个命令。

用户单击"退出"命令,可以退出本应用程序。用户单击"打开"命令,可以打开文件对话框,选择要显示的图片文件(本应用程序只能查看".jpg"、".jpeg"、".png"、".gif"类型的图片),单击"打开"按钮,则可以在图片显示区内以缩略图的方式查看所选的图片(见图 11-14 和图 11-15)。

图 11-13　媒体图片查看器主界面初始状态

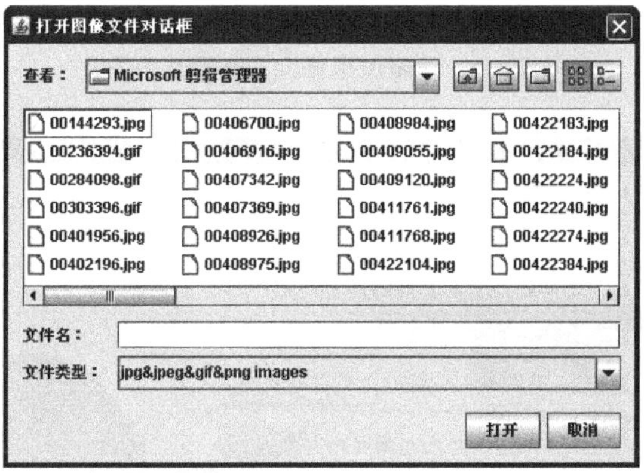

图 11-14　打开文件对话框

在图 11-15 图片显示区内可以看到所选择的所有图片。如果再次单击"文件"菜单中的"打开"命令,则又可以打开文件对话框重新选择图片,图片显示区会根据用户的选择刷新界面,显示新的图片。

当用户双击图片显示区中的某张图片时,会打开"图片查看器"对话框,可以具体查看图片(见图 11-16)。此图片查看器有下面的功能:"上一张"、"下一张"、"开始幻灯片"、"放大"、"缩小"、"退出图片查看器"。

用户可以在图片查看器中通过"上一张"、"下一张"以大图的方式浏览前面选中的所有文件。单击"放大"、"缩小"来实现对当前查看图片的缩放功能。单击"开始幻灯片"可以实现以幻灯片的形式浏览所有选择图片,并伴有背景音乐(见图 11-17)。单击"退出图片查看器"可以退出查看器界面,重新返回图片显示区。如果用户不退出查看器界面,则无法通过文件对话框打开新的图片文件。

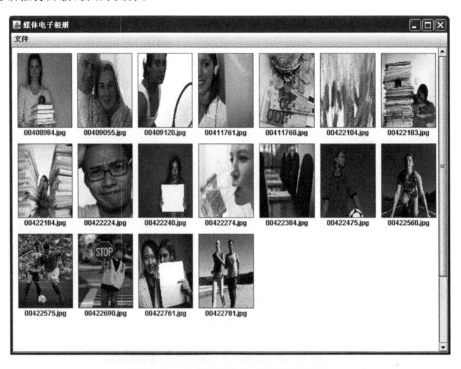

图 11-15　缩略图浏览方式查看选中图片

图 11-16　图片查看器界面

图 11-17 幻灯片方式浏览图片

### 11.2.3 案例总体设计

用 Java 编写一个面向对象的应用程序,很大程度上是设计类,并编写这些类的定义。因此,通过对该案例的功能分析及演示,下面把问题分解成一组相互协作的对象,然后设计和创建这些对象本身。

根据上面的运行效果,可以知道案例主要需要设计三个类:

- 设计一个 JFrame 的子类 PictureWindow,利用 PictureWindow 窗口对象来实现打开文件和退出应用程序的功能,并且可以在该窗口的图片显示区内以缩略图浏览方式查看选择的图片文件。
- 设计一个 JDialog 的子类 PictureViewDialog,利用 PictureViewDialog 对话框对象来实现图片查看器的功能,在此对话框里可以实现浏览"上一张"、"下一张"、"放大"、"缩小"、"开始幻灯片"、"退出图片查看器"的功能。
- 设计一个 JDialog 类的子类 PlayPicture,利用 PlayPicture 对象实现幻灯片放映,放映时伴有背景音乐。

对于 PictureWindow 类对象,它的图片显示区内以缩略图方式浏览选中文件的功能,本应用程序是以下面的思路处理的:图片显示区中每一个图片文件的缩略图是一个 ImagePanel 类对象,这个 ImagePanel 类对象由两个组件组成,中心是一个 ImageButton 类的按钮对象,负责显示图像;底部是一个标签,负责显示图片文件的名字。

整个多媒体图片查看器应用程序的面向对象设计图如图 11-18 所示。

#### 1. PictureWindow 类

PictureWindow 类是 JFrame 的一个子类,负责加载图像,并把图像以缩略图方式显示出来。

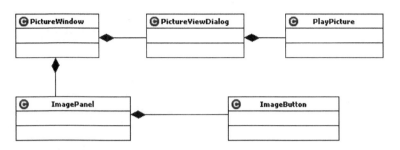

图 11-18　多媒体图片查看器总体设计类图

PictureWindow 类的 UML 图如图 11-19 所示。

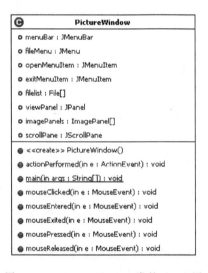

图 11-19　PictureWindow 类的 UML 图

以下是 UML 图中有关数据和方法的详细说明。

1) 成员变量

- menuBar 是 JMenuBar 类创建的菜单条，可以向 menuBar 中添加菜单。
- fileMenu 是 JMenu 类创建的菜单，菜单的名称是"文件"。
- openMenuItem、exitMenuItem 是 JMenuItem 类创建的两个菜单项，被添加到 fileMenu 菜单中。这两个菜单项的名称分别是"打开"和"退出"。通过"打开"命令可以弹出打开文件对话框，来加载图像；通过"退出"命令可以实现退出应用程序。
- filelist 是 File 类对象数组，负责存放打开的图片文件的引用。
- viewPanel 是 JPanel 类对象面板，是图片显示区，负责显示所有图片的缩略图。它被添加到滚动面板 scrollPane 中。
- imagePanels 是 ImagePanel 类对象数组，每个单元都是一个图片的缩略图。
- scrollPane 是 JScrollPane 对象，是一个滚动面板，被添加到 PictureWindow 窗口中，可以滚动查看显示区中的图片。

2) 成员方法

- PictureWindow()是构造方法，负责完成应用程序主窗口的初始化。

- actionPerformed(ActionEvent)方法是 PictureWindow 类实现的 ActionListener 接口中的方法,负责执行菜单项发出的有关命令。用户选择菜单中的菜单项可以触发 ActionEvent 事件,导致 actionPerformed(ActionEvent)方法执行相应的操作。
- main(String[])方法是软件运行的入口方法。
- public void mouseClicked(MouseEvent)方法是 PictureWindow 类实现的 MouseListener 接口中的方法,负责处理在组件上单击或双击触发的鼠标事件。当单击或双击时,监听器自动调用接口中的这个方法对事件做出处理。
- public void mouseEntered(MouseEvent)方法是 PictureWindow 类实现的 MouseListener 接口中的方法,负责处理鼠标指针进入组件触发的鼠标事件,当鼠标指针进入组件上方时,监听器自动调用接口中的这个方法对事件做出处理。
- public void mouseExited(MouseEvent)方法是 PictureWindow 类实现的 MouseListener 接口中的方法,负责处理鼠标指针离开组件触发的鼠标事件,当鼠标指针离开组件时,监听器自动调用接口中的这个方法对事件做出处理。
- public void mousePressed(MouseEvent)方法是 PictureWindow 类实现的 MouseListener 接口中的方法,负责处理在组件上按下鼠标键触发的鼠标事件,当在组件上按下鼠标键时,监听器将自动调用接口中的这个方法对事件做出处理。
- public void mouseReleased(MouseEvent)方法是 PictureWindow 类实现的 MouseListener 接口中的方法,负责处理在组件上释放鼠标键触发的鼠标事件,当在组件上释放鼠标键时,监听器将自动调用接口中的这个方法对事件做出处理。

2. ImageButton 类

ImageButton 类是 JButton 类的子类,负责创建按钮对象,并在按钮对象上绘制图像。ImageButton 类的 UML 图如图 11-20 所示。

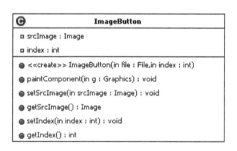

图 11-20 ImageButton 类的 UML 图

以下是 UML 图中有关数据和方法的详细说明。

1) 成员变量
- srcImage 是 Image 对象,是按钮上要绘制的图像。
- index 表示该按钮上所绘图像文件的索引。

2) 成员方法
- ImageButton(File,int)是构造方法,初始化按钮对象,加载 File 指定的图像文件。
- paintComponent(Graphics)方法,绘制按钮上的图像,由 repaint()实现调用。

- setScrImage(Image)方法,设置按钮上加载的图像。
- getSrcImage()方法,获得按钮上加载的图像对象。
- setIndex()方法,设置该按钮上图像文件的索引。
- getIndex()方法,获取按钮上图像文件的索引。

3. ImagePanel 类

ImagePanel 类是 JPanel 类的子类,负责创建图像面板,显示图片的缩略图。ImagePanel 类的 UML 图如图 11-21 所示。

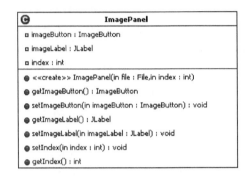

图 11-21 ImagePanel 类的 UML 图

以下是 UML 图中有关数据和方法的详细说明。

1) 成员变量
- imageButton 是 ImageButton 对象按钮,按钮上加载图像。
- imageLabel 是 JLabel 对象,负责显示 imageButton 上加载图像的名字。
- index 是 ImagePanel 图像面板对象显示的图像文件的索引,与 ImagePanel 对象上的 ImageButton 按钮对象上绘制的图像文件的索引相同。

2) 成员方法
- ImagePanel(File,int)构造方法,负责完成面板对象的初始化。
- getImageButton()方法,获取 ImagePanel 对象上的按钮对象。
- setImageButton()方法,设置 ImagePanel 对象上的按钮对象。
- getImageLabel()方法,获取 ImagePanel 对象上的标签对象。
- setImageLabel()方法,设置 ImagePanel 对象上的标签对象。
- setIndex()方法,设置 ImagePanel 对象显示的图像的索引。
- getIndex()方法,获取 ImagePanel 对象显示的图像文件的索引。

4. PictureViewDialog 类

PictureViewDialog 类是 JDialog 类的子类,负责创建图片查看器,实现图片的多种查看方式。

PictureViewDialog 类的 UML 图如图 11-22 所示。

以下是 UML 图中有关数据和方法的详细说明。

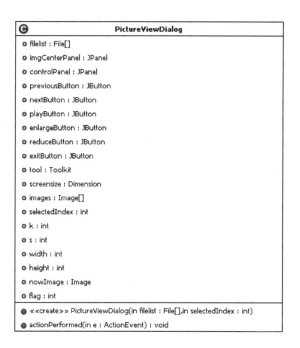

图 11-22 PictureViewDialog 类的 UML 图

1) 成员变量
- filelist 是 File 对象数组,保存所有图像文件的引用。
- imgCenterPanel、controlPanel 是 JPanel 对象。imgCenterPanel 面板负责显示图像。controlPanel 面板内放置控制图片查看的若干个按钮。
- previousButton、nextButton、playButton、enlargeButton、reduceButton、exitButton 是 JButton 对象,每个按钮均注册了 ActionEvent 事件监听器,一旦用户单击相应的按钮,自动响应用户的操作。prieviousButton 按钮负责查看上一张图片;nextButton 负责查看下一张图片;playButton 按钮一旦被用户单击,则开始放映幻灯片;enlargeButton 负责以一定比例放大当前图片;reduceButton 负责以一定比例缩小当前图片;exitButton 负责退出当前图片查看器。
- tool 是 Toolkit 对象。
- screesize 是 Dimension 对象,由它获取当前屏幕的宽和高。
- images 是 Image 对象数组,加载 filelist 中所引用的图像文件。
- selectedIndex 表示用户在 PictureWindow 对象的图片显示区中选择的缩略图的编号。
- k、s 分别用来控制图片缩放的两个整型变量。
- width、height 表示当前图片缩放后的图像的宽和高。
- nowImage 表示图片查看器当前显示的图片。

- flag 是 int 型变量,用来表示图片是否缩放。flag＝0 表示图片没有经过缩放,原图显示;flag＝1 表示图片已缩放,按照缩放大小显示。

2) 成员方法
- PictureViewDialog(File[],int)构造方法,负责图片查看器窗口的初始化。
- actionPerformed(ActionEvent)方式是 PictureViewDialog 类实现的 ActionListener 接口中的方法,负责响应用户单击按钮的行为。一旦用户单击按钮,产生 ActionEvent 事件,监听器对象自动调用 actionPerformed(ActionEvetn)方法响应用户操作。

3) 内部类

PicturePanel 类中的 paintComponent(Graphics)方法,负责绘制图片查看器浏览的图片。

### 5. PlayPicture 类

PlayPicture 类是 JDialog 类的子类,负责放映幻灯片。

PlayPicture 类的 UML 图如图 11-23 所示。

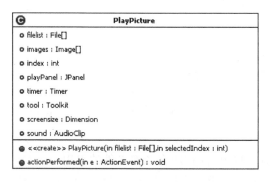

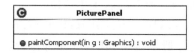

图 11-23 PlayPicture 类的 UML 图

以下是 UML 图中有关数据和方法的详细说明。

1) 成员变量
- filelist 是 File 对象数组,保存所有图像文件的引用。
- images 是 Image 对象数组,保存 PlayPicture 类所放映幻灯片的所有图像。
- index 表示开始放映幻灯片时的第一张图片的索引。
- playPanel 是 JPanel 对象,在该区域放映幻灯片。
- timer 是 Timer 类对象,负责放映幻灯片。
- tool 是 Toolkit 对象。
- screensize 是 Dimension 对象,通过它来获取当前屏幕的宽和高。
- sound 是 AudioClip 对象。

2) 成员方法
- PlayPicture(File[],int)构造方法,负责完成幻灯片的初始化。
- actionPerformed(ActionEvent)方法是 PlayPicture 类实现的 ActionListener 接口中的方法,负责控制幻灯片的播放。timer 对象发生的震铃事件是 ActionEvent 事件。当震铃事件发生时,监听器会监听到这个事件,就会执行接口 ActionListener 中的方法 AcitonPerformed(ActionEvent)。

3) 内部类

PicturePanel 类是 PlayPicture 类的内部类,其 paintComponent(Graphics)方法负责图像的具体显示工作。

### 11.2.4 案例代码实现

#### 1. 编写应用程序主界面(PictureWindow.java)

PictureWindow.java 文件包含了程序的入口,是本应用程序的主窗体,它把其他的 PictureViewDialog、PlayPicture、ImagePanel 和 ImageButton 类对象都整合到了一起。通过该界面,用户可以选择图像文件,并把它们在图片显示区中以缩略图的方式显示出来。当用户选择某一图像缩略图并双击它可以打开图片查看器,通过图片查看器可以放映幻灯片。

PictureWindow 类是 javax.swing 包中 JFrame 的一个子类,并实现了 ActonListener 接口和 MouseListener 接口。从前面的运行效果图里可以看到该窗口的布局设计非常简单,窗口默认布局是 BorderLayout 型,在窗口中添加一个 JScrollPane 面板对象,在面板对象中再添加一个 JPanel 对象即可。该 JPanel 对象是 FlowLayout 型的,这样就可以保证随着窗口大小的变化,容器内的组件的位置也随着变化。该 FlowLayout 型布局管理器是一个左对齐的布局对象,组件之间的水平和垂直间隙是 10 个像素。要实现用户单击菜单项命令的响应和用户双击 ImagePanel 对象的响应,利用 Java 的事件处理机制就可以实现。具体有关布局管理器的知识和 Java 的事件处理机制的相关内容可参看第 10 章 Java 图形用户界面设计部分的介绍。

本程序还需要打开一个文件对话框选择图片文件,这要用到 JFileChooser 类。下面主要介绍 JFileChooser 类,该类与 AWT 的 FileDialog 类似,用于选择文件(图 11-14 所示即 JFileChooser 示例)。

1) javax.swing.JFileChooser 类

JFileChooser 的功能与 AWT 中的 FileDialog 基本相似,也是用于生成"打开文件"、"保存文件"的对话框,与 FileDialog 不同的是,JFileChooser 无须依赖于本地平台的 GUI,它由 100% 纯 Java 实现,在所有平台上具有完全相同的行为,并可以在所有平台上具有相同的外观风格。

JFileChooser 并不是 JDialog 的子类,所以不能使用 setVisible(true)来显示该文件对话框,而是调用 showXxxDialog 方法来显示文件对话框。JFileChooser API 支持打开和保存两种功能的文件选择器,用户还可以对文件选择器进行定制。

使用 JFileChooser 来建立文件对话框并允许用户选择文件的步骤如下:

(1) 采用构造方法创建一个 JFileChooser 对象,该 JFileChooser 对象无须指定 parent

组件。创建 JFileChooser 对象时可以指定初始化路径,如下面代码:

```
//以当前路径创建文件选择器
JFileChooser chooser = new JFileChooser(".");
```

(2) 调用 JFileChooser 一系列可选的方法对 JFileChooser 执行初始化操作(可查 JFileChooser API)。它常用的方法如下。

- setSelectedFile/setSelectedFiles:指定该文件选择器默认选择的文件(也可以默认选择多个文件)。
- setMultiSelectionEnabled(boolean b):默认情况下,该文件选择器只能选择一个文件,通过调用该方法可以设置允许选择多个文件(设置参数为 true 即可)。
- setFileSelectionMode(int mode):默认情况下,该文件选择器只能选择文件,通过调用该方法可以设置允许选择文件、路径、文件与路径,对应的设置参数值为 JFileChooser. FILES_ONLY、JFileChooser. DIRECTORIES_ONLY、JFileChooser. FILES_AND_DIRECTORIES。

(3) 如果需要让文件对话框实现文件过滤功能,则需要结合 FileFilter 类来进行文件过滤。JFileChooser 提供了以下两个方法来安装文件过滤器。

- addChoosableFileFilter(FileFilter filter):添加文件过滤器。
- setFileFilter(FileFilter filter):设置文件过滤器。一旦调用该方法将导致该文件对话框只有一个文件过滤器。

(4) 如果需要改变文件对话框中文件的视图外观,可以结合 FileView 类来改变对话框中文件的视图外观。

(5) 调用 showXxxDialog 方法来打开文件对话框,通常有下面三个方法。

- int showDialog(Component parent, String approveButtonText):弹出文件对话框,该对话框的标题、"同意"按钮的文本由 approveButtonText 来指定。
- int showOpenDialog(Component parent):弹出文件对话框,该对话框具有默认标题,"同意"按钮的文本是"打开"。
- int showSaveDialog(Component parent):弹出文件对话框,该对话框具有默认标题,"同意"按钮的文本是"保存"。

当用户单击"同意"、"取消"按钮,或直接关闭文件对话框时才可以关闭该文件对话框,关闭该对话框时返回一个 int 类型的值,分别是 JFileChooser. APPROVE_OPTION、JFileChooser. CANCEL_OPTION、JFileChooser. ERROR_OPTION。

(6) 获取用户选择文件或文件集,JFileChoose 提供了如下两个方法来获取用户选择的文件。

- File getSelectedFile():返回用户选择的文件。
- File[] getSelectedFiles():返回用户选择的多个文件。

PictureWindow.java 文件即按上面的步骤创建了一个"打开文件"的对话框,并利用 getSelectedFiles()方法获取用户选择的多个文件,把它保存到了一个 File[]数组中,这个数组在后面的多个源文件中都要用到。

2) 代码(PictureWindow.java)

```java
package caida.xinxi.Album;

import java.awt.Color;
import java.awt.Dimension;
import java.awt.FlowLayout;
import java.awt.event.ActionEvent;
import java.awt.event.ActionListener;
import java.awt.event.MouseEvent;
import java.awt.event.MouseListener;
import java.io.File;

import javax.swing.JFileChooser;
import javax.swing.JFrame;
import javax.swing.JMenu;
import javax.swing.JMenuBar;
import javax.swing.JMenuItem;
import javax.swing.JPanel;
import javax.swing.JScrollPane;
import javax.swing.filechooser.FileNameExtensionFilter;

@SuppressWarnings("serial")
public class PictureWindow extends JFrame implements ActionListener, MouseListener{
    JMenuBar menuBar;
    JMenu fileMenu;
    JMenuItem openMenuItem,exitMenuItem;
    File[] filelist;
    JPanel viewPanel;
    ImagePanel[] imagePanels;
    JScrollPane scrollPane;

    //构造方法,完成程序主界面的初始化
    public PictureWindow(){
        setTitle("媒体电子相册");

        //创建菜单条、菜单、菜单项,并给菜单项注册 ActionEvent 事件监听器
        menuBar = new JMenuBar();
        fileMenu = new JMenu("文件");
        openMenuItem = new JMenuItem("打开");
        openMenuItem.addActionListener(this);
        exitMenuItem = new JMenuItem("退出");
        exitMenuItem.addActionListener(this);
        fileMenu.add(openMenuItem);
        fileMenu.add(exitMenuItem);
        menuBar.add(fileMenu);
        this.setJMenuBar(menuBar);

        //创建滚动面板 scrollPane
        scrollPane = new JScrollPane();
        //设置滚动面板 scroolPane 水平滚动条需要时可见
```

```java
        scrollPane.setHorizontalScrollBarPolicy(JScrollPane.HORIZONTAL_SCROLLBAR_AS_NEEDED);
        //设置滚动面板 scrollPane 垂直滚动条永远可见
        scrollPane.setVerticalScrollBarPolicy(JScrollPane.VERTICAL_SCROLLBAR_ALWAYS);
        viewPanel = new JPanel();
        //设置面板 viewPanel 布局是 FlowLayout 型,上下左右间隔是 10 个像素
        viewPanel.setLayout(new FlowLayout(FlowLayout.LEFT,10,10));
        viewPanel.setBackground(Color.WHITE);
        viewPanel.setPreferredSize(new Dimension(600,700));
        //将面板 viewPanel 添加到滚动面板 scrollPane 中
        scrollPane.setViewportView(viewPanel);
        //将滚动面板 scrollPane 添加到主窗口中
        add(scrollPane,"Center");

        //设置窗口的一些属性
        setBounds(100,100,800,600);
        setVisible(true);
        setDefaultCloseOperation(JFrame.EXIT_ON_CLOSE);
        validate();

    }

    //接口 ActionListener 中的方法,当单击菜单项时,监听器自动调用该方法
    public void actionPerformed(ActionEvent e) {

        // 如果单击"打开"菜单项,则打开文件对话框,选择图片文件
        if(e.getSource() == this.openMenuItem){
            //创建文件选择器 chooser
            JFileChooser chooser = new JFileChooser();
            //设置该文件选择器允许选择多个文件
            chooser.setMultiSelectionEnabled(true);
            //图片过滤器
            FileNameExtensionFilter filter = new FileNameExtensionFilter("jpg&jpeg&gif&png images","jpg","gif","png","jpeg");
            //设置文件过滤器
            chooser.setFileFilter(filter);
            //设置该文件选择器只能选择文件
            chooser.setFileSelectionMode(JFileChooser.FILES_ONLY);
            //设置文件对话框的标题
            chooser.setDialogTitle("打开图像文件对话框");
            //弹出打开文件对话框
            int result = chooser.showOpenDialog(this);
            //如果用户在文件对话框中单击了"打开"按钮时,执行下面的操作
            if(result == JFileChooser.APPROVE_OPTION){
                //得到目录下用户所选择的所有图片文件,保存到 filelist 中
                filelist = chooser.getSelectedFiles();

                imagePanels = new ImagePanel[filelist.length];
                //每次在刷新图片显示区以前,要把以前的所有图像面板对象移走
                viewPanel.removeAll();
                viewPanel.validate();
```

```java
            //给图片显示区中添加新的图像面板
            for(int i = 0;i < filelist.length;i++)
            {
                //创建图像面板对象数组
                imagePanels[i] = new ImagePanel(filelist[i],i);
                //给图像面板中的 ImageButton 按钮注册 MouseEvent 监听器
                imagePanels[i].getImageButton().addMouseListener(this);
                //给显示面板 viewPanel 中添加图像面板
                viewPanel.add(imagePanels[i]);
                viewPanel.validate();
            }
        }
    }

    //如果单击"退出"菜单项,执行下面的代码,退出应用程序
    else if(e.getSource() == this.exitMenuItem){
        System.exit(0);
    }
}

public static void main(String[] args) {
    new PictureWindow();
}

//MouseListener 接口中的方法,当用户单击或双击鼠标时,监听器调用该方法
public void mouseClicked(MouseEvent e) {
    //如果用户双击或连接图像面板时,打开图片查看器
    if(e.getClickCount()> 1){
        //获得事件源
        ImageButton imgButton = (ImageButton)e.getSource();
        //获得事件源的索引值
        int index = imgButton.getIndex();
        //打开图片查看器
        new PictureViewDialog(filelist,index).setVisible(true);
    }
}
// MouseListener 接口中的方法,当鼠标指针进入图像面板上时,自动执行该方法
public void mouseEntered(MouseEvent e) {
}

// MouseListener 接口中的方法,当鼠标指针离开图像面板时,自动执行该方法
public void mouseExited(MouseEvent e) {
}

// MouseListener 接口中的方法,当在图像面板上按下鼠标键时,自动执行该方法
public void mousePressed(MouseEvent e) {
}

// MouseListener 接口中的方法,当在图像面板上释放鼠标键时,自动执行该方法
public void mouseReleased(MouseEvent e) {
}
}
```

### 2. ImageButton.java 和 ImagePanel.java 的实现

本案例设计 ImageButton 类和 ImagePanel 类的主要目的是为了实现图片以缩略图的浏览方式显示图片。在应用程序主窗口的图片显示区中每一个文件的缩略图用一个 ImagePanel 对象实现,该 ImagePanel 对象中的显示图像的部分其实是一个按钮组件。此按钮组件是 ImageButton 类的对象。要实现这部分功能需要考虑如何设计 ImageButton 类和 ImagePanel 类的构造方法、图形处理、如何加载和显示图像。

1) 设计 ImageButton 和 ImagePanel 构造方法

考虑到当用户在主窗口图片显示区中用鼠标双击选择了一个图片后,会以该图片为当前图片打开图片查看器,以及在放映幻灯片时也是以当前默认图片为开始循环放映的,因此需要设计一个 int 型变量 index 来标识这些不同的图片文件。所以在设计 ImageButton 类和 ImagePanel 类的构造方法时,分别加了两个参数 ImageButton(File file,int index)和 ImagePanel(File file,int index)。第一个参数 file 表明按钮对象和图像面板所加载的图片文件的引用,第二个参数则是这个图片文件的索引值。

2) 图形处理

Component 类是许多 Java 组件类的父类,它具有三种图形绘制的基本方法:paint()方法、update()方法和 repaint()方法。对于 Component 的子类而言,也继承了这三种方法,可以实现图形的处理。当程序运行时,repiant()方法调用 update()方法,用 update()方法尽可能清除 Component 上次绘制的背景,然后 update()方法直接调用 paint()方法。repaint()方法是程序员用于迫使一个 paint 操作发生的方法。

同样,javax.swing.JComponent 类是 Componnet 的子类,也是许多 Swing 包的父类,这就意味着,Swing 包中的许多组件类也具有 paint()、update()和 repaint()这三种方法。JComponent 组件本身有一个方法 paintComponent(Graphics g),可以用来绘制 JComponent 组件的内容。同样,程序员不能直接调用 paintComponent()方法,也不得不通过调用 repaint()方法来实现间接调用 paintComponent()方法,从而达到对 JComponent 组件内容的绘制。

3) 图像的装载和显示

Java 语言支持多媒体,它提供了大量的类和接口来实现对文本、图形、图像、动画、声音和视频等多种形式的支持。本应用程序要实现对现有图像的处理,而对图像处理,最基本的就是对图像的装载和显示。

在 java.applet、java.awt、java.awt.image 包中包含了许多支持图像的类和方法。在程序中图像由一个 java.Image 类的对象来表示。Image 类是抽象类,不能直接创建对象。为了实现用 Image 获得对象的处理,就需要对指定的图像进行加载,然后再对加载的图像进行相应的处理。

(1) 在 Applet 中加载和显示图像

在 Applet 中加载图像使用 Applet 类提供的 getImage()方法,获得包含该图像的一个 Image 类对象。然后利用 java.applet.Applet 的 paint()方法来显示图像。

Applet 类的 getImage()方法的定义如下。

- public Image getImage(URL url):url 是包含图像文件名的绝对 URL。

- public Image getImage(URL url, String name)：url 是图像文件所在目录的 URL，当 Applet 与图像文件在一个目录下时，可以使用 getCodeBase()方法获取该 URL；而当图像文件与 Applet 嵌入的 HTML 文件在同一个目录下时，可以使用 getDocumentBase()方法获得该 URL。参数 name 是相对于 URL 参数指定目录的文件名。

（2）在 Application 中加载和显示图像

在 Application 中同样可以加载和显示图像。利用 java.awt.Toolkit 类的 getDefaultToolkit()方法获得默认的工具包 Toolkit 引用；然后利用 Toolkit 引用的 getImage()方法装载图像。在组件的 paint()方法或 paintComponent()方法中，用 Graphics 的 drawImage()显示图像。

下面是 Application 中加载图像的方法。

- public abstract Image getImage(String filename)；加载图像，参数是 GIF、PNG、JPG 文件的名称。
- public abstract Image getImage(URL url)；加载图像，用于浏览器。

上面两种通过 getImage()方法取得的是 java.awt.Image 类型的对象，也可以使用 javax.imageio.ImageIO 类的 read()取得一个图像，返回的是 BufferedImage 对象。BufferedImage 是 Image 的子类，它描述了具有可访问图像数据缓冲区的 Image。

在 Application 中显示图像使用的是 Graphics 类中的 drawImage()方法。Graphics 类定义了如下 4 种 drawImage()方法，它们都返回一个布尔值。如果图像完全加载并且显示出来，返回 true，否则返回 false。

- boolean drawImage(Image img, int x, int y, ImageObserver observer)
- boolean drawImage(Image img, int x, int y, int width, int height, ImageObserver observer)
- boolean drawImage(Image img, int x, int y, Color bgcolor, ImageObserver observer)
- boolean drawImage(Image img, int x, int y, int width, int height, Color bgcolor, ImageObserver observer)

Graphics 类的 drawImage()方法的参数含义如下。

- Image img：要绘制的图像对象。
- int x，int y：图像的左上角坐标，以像素为单位。
- int width，int height：图像的宽度和高度，以像素为单位。
- Color bgcolor：图像的背景色，当图像有透明色时使用。
- ImageObserver observer：实现了 ImageObserver 接口类的对象。这使得该对象成为要显示图像的观察者。这样关于图像的信息一更新，就将通知该对象。一般用 this 作为参数的值。这是因为 Component 类实现了 ImageObserver 接口，它在 imageUpdate()方法的实现中调用了 repaint()方法，使得图像能够在图像加载的同时逐步刷新显示。

4）代码(ImageButton.java)

```
package caida.xinxi.Album;
```

```java
import java.awt.Graphics;
import java.awt.Image;
import java.io.File;

import javax.imageio.ImageIO;
import javax.swing.JButton;
import javax.swing.JOptionPane;

@SuppressWarnings("serial")
public class ImageButton extends JButton {
    private Image srcImage;
    private int index;

    //构造方法,完成 ImageButton 按钮的初始化
    //file 表示按钮上显示的图像文件的引用,index 表示该按钮的编号
    public ImageButton(File file,int index){
        this.setIndex(index);
        try {
            //读取 file 文件指向的原始图像
            setSrcImage(ImageIO.read(file));
            repaint();
        } catch (Exception e) {
            //如果图像加载出现问题,弹出警告窗口
            JOptionPane.showMessageDialog(this,"该图像不能正常加载!","警告对话框",JOptionPane.WARNING_MESSAGE);
        }
    }

    //重写 JComponent 的 paintComponent 方法,实现绘画
    public void paintComponent(Graphics g){
        super.paintComponent(g);
        //将 srcImage 绘制到按钮上
        g.drawImage(srcImage,0,0,100,150,this);
    }

    public void setSrcImage(Image srcImage) {
        this.srcImage = srcImage;
    }
    public Image getSrcImage() {
        return srcImage;
    }

    public void setIndex(int index) {
        this.index = index;
    }
    public int getIndex() {
        return index;
    }

}
```

5) 代码(ImagePanel.java)

```java
package caida.xinxi.Album;

import java.awt.BorderLayout;
import java.awt.Color;
import java.awt.Dimension;
import java.io.File;

import javax.swing.JLabel;
import javax.swing.JPanel;

@SuppressWarnings("serial")
public class ImagePanel extends JPanel {
    private ImageButton imageButton;
    private JLabel imageLabel;
    private int index;

    //构造方法,实现图像面板的初始化
    public ImagePanel(File file,int index){
        this.setIndex(index);
        //设置面板的布局为 BorderLayout 型
        setLayout(new BorderLayout());
        //创建图像按钮
        imageButton = new ImageButton(file,index);
        // 创建标签,标签文本是图像文件的名字,对齐方式为居中
        imageLabel = new JLabel(file.getName(),JLabel.CENTER);
        // 给面板添加组件
        add(imageButton,"Center");
        add(imageLabel,"South");
        //设置面板的最佳尺寸大小
        this.setPreferredSize(new Dimension(100,150));
        //设置面板的背景色为白色
        setBackground(Color.WHITE);
    }

    public ImageButton getImageButton() {
        return imageButton;
    }
    public void setImageButton(ImageButton imageButton) {
        this.imageButton = imageButton;
    }

    public JLabel getImageLabel() {
        return imageLabel;
    }
    public void setImageLabel(JLabel imageLabel) {
        this.imageLabel = imageLabel;
    }

    public void setIndex(int index) {
```

```java
        this.index = index;
    }
    public int getIndex() {
        return index;
    }

}
```

### 3. 实现图片查看器(PictureViewDialog.java)

PictureViewDialog 类是 JDialog 类的一个子类，实现了接口 ActionListener，通过该对话框窗口实现图片查看器功能。因此本类的设计主要处理的问题就是如何实现浏览上一张图片、浏览下一张图片、如何放映幻灯片、如何缩放图片和如何退出图片查看器的问题。

PictureViewDialog 类设计了一个 Image 对象数组 images 来加载实参 filelist 传递进来的所有图像；一个 int 型变量 selectedIndex，其初始值是实参 selectedIndex 传递进来的值，表示当前处理的图像的索引；一个 Image 类的对象 nowImage 来表示当前处理(或查看)的图像，其初始值为 images[selectedIndex]。那么要实现浏览上一张图片和下一张图片，只需要改变 selectedIndex 索引的值，在屏幕上重绘 nowImage 即可。要实现缩放功能，这里设计了两个控制缩放变量 k、s，按相同比例修改了图像的宽和高，在屏幕上重绘修改了宽和高的 nowImage 对象(见图 11-24 图片查看器的缩放功能)。有关图像的加载与显示及图像的绘制知识上面详细介绍过，这里不再重复。

代码(PictureViewDialog.java)：

```java
package caida.xinxi.Album;

import java.awt.BorderLayout;
import java.awt.Dimension;
import java.awt.Font;
import java.awt.Graphics;
import java.awt.Image;
import java.awt.Toolkit;
import java.awt.event.ActionEvent;
import java.awt.event.ActionListener;
import java.io.File;
import java.io.IOException;

import javax.imageio.ImageIO;
import javax.swing.JButton;
import javax.swing.JDialog;
import javax.swing.JPanel;

@SuppressWarnings("serial")
public class PictureViewDialog extends JDialog implements ActionListener {
    File[] filelist;
    JPanel imgCenterPanel,controlPanel;
    JButton previousButton,nextButton,playButton,enlargeButton,reduceButton,exitButton;
    Toolkit tool = Toolkit.getDefaultToolkit();
```

```java
Dimension screensize = tool.getScreenSize();
Image[] images;
//k,s 控制缩放程度的变量;width,height 为图片缩放后的大小
int selectedIndex,k = 0,s = 0,width = 0,height = 0;
Image nowImage;
//flag = 0 时表示图片以原图显示;flag = 1 时表示图片以缩放格式显示
int flag = 0;

//构造方法,实现图片查看器窗口的初始化
//filelist 代表主窗口图像显示区中的所有文件,selectedIndex 为被选中的图片文件
public PictureViewDialog(File[] filelist,int selectedIndex) {
    this.filelist = filelist;
    this.selectedIndex = selectedIndex;
    setTitle("图片查看器");
    setLayout(new BorderLayout());

    //创建图像数组,并赋值
    images = new Image[filelist.length];
    for(int i = 0;i < filelist.length;i++){
        try {
            //加载图像
            images[i] = ImageIO.read(filelist[i]);
        } catch (IOException e) {
            e.printStackTrace();
        }
    }
    //nowImage 表示当前正处理的图像
    nowImage = images[selectedIndex];

    //给对话框窗口添加组件
    imgCenterPanel = new PicturePanel();
    imgCenterPanel.repaint();
    controlPanel = new JPanel();
    add(imgCenterPanel,BorderLayout.CENTER);
    add(controlPanel,BorderLayout.SOUTH);

    //分别创建按钮,注册 ActionEvent 事件监听器,并添加到 controlPanel 面板中
    this.previousButton = new JButton("上一张");
    this.previousButton.addActionListener(this);
    //设置按钮的字体格式
    this.previousButton .setFont(new Font("宋体",Font.BOLD,20));
    this.nextButton = new JButton("下一张");
    this.nextButton.addActionListener(this);
    this.nextButton .setFont(new Font("宋体",Font.BOLD,20));
    this.playButton = new JButton("开始幻灯片");
    this.playButton.addActionListener(this);
    this.playButton .setFont(new Font("宋体",Font.BOLD,20));
    this.enlargeButton = new JButton("放大");
    this.enlargeButton .addActionListener(this);
    this.enlargeButton .setFont(new Font("宋体",Font.BOLD,20));
    this.reduceButton = new JButton("缩小");
```

```java
        this.reduceButton .addActionListener(this);
        this.reduceButton    .setFont(new Font("宋体",Font.BOLD,20));
        this.exitButton = new JButton("退出图片查看器");
        this.exitButton .addActionListener(this);
        this.exitButton    .setFont(new Font("宋体",Font.BOLD,20));
        controlPanel.add(this.previousButton);
        controlPanel.add(this.nextButton );
        controlPanel.add(this.playButton );
        controlPanel.add(this.enlargeButton );
        controlPanel.add(this.reduceButton );
        controlPanel.add(this.exitButton );

        //设置对话框窗口的基本属性
        setBounds(0,0,screensize.width,screensize.height - 30);
        //设置对话框窗口是有模式的
        //只有退出了图片查看器,才能激活应用程序主窗口
        this.setModal(true);
        validate();
    }

    //ActionListener 接口中的方法,一旦用户单击按钮,监听器自动调用该方法
    public void actionPerformed(ActionEvent e) {

        //如果单击"上一张",则图片查看器显示上一张图片
        //改变图片索引值、nowImage值,并重绘屏幕
        if(e.getSource() == this.previousButton ){
            this.selectedIndex = this.selectedIndex - 1;
            if(this.selectedIndex == - 1){
                this.selectedIndex = this.filelist.length - 1;
            }
            nowImage = images[this.selectedIndex ];
            flag = 0;
            imgCenterPanel.repaint();
        }

        //如果单击"下一张",则图片查看器显示下一张图片
        if(e.getSource() == this.nextButton ){
            this.selectedIndex = this.selectedIndex + 1;
            if(this.selectedIndex >= this.filelist .length ){
                this.selectedIndex = 0;
            }
            nowImage = images[this.selectedIndex ];
            flag = 0;
            imgCenterPanel.repaint();
        }

        //如果单击"开始幻灯片",则弹出窗口,循环放映图片
        if(e.getSource() == this.playButton   ){
            new PlayPicture(filelist,selectedIndex).setVisible(true);
        }
```

```java
//如果单击"放大"按钮,则以一定比例放大当前图片的宽和高,并重绘屏幕
if(e.getSource() == this.enlargeButton   ){
    k++;
    width = nowImage.getWidth(this) + k * 15 - s * 15;
    height = nowImage.getHeight(this) + k * 15 - s * 15;
    flag = 1;
    imgCenterPanel.repaint();
}

//如果单击"缩小"按钮,则以一定比例缩小当前图片的宽和高,并重绘屏幕
if(e.getSource() == this.reduceButton ){
    s++;
    width = nowImage.getWidth(this) + k * 15 - s * 15;
    height = nowImage.getHeight(this) + k * 15 - s * 15;
    flag = 1;
    imgCenterPanel.repaint();
}

//如果单击"退出图片查看器",则关闭当前图片查看器
if(e.getSource() == this.exitButton ){
    setVisible(false);
}
}

//内部类,相当于画布,实现在画布上绘制图像
public class PicturePanel extends JPanel{
    //重写 paintComponent()方法
    public void paintComponent(Graphics g){
        super.paintComponent(g);
        //获取画布的宽和高
        int panelWidth = this.getWidth();
        int panelHeight = this.getHeight();
        
        //如果 flag = 0,则在画布中心以原图显示图像
        if(flag == 0){
            g.drawImage (nowImage, (panelWidth - nowImage.getWidth(this))/2, (panelHeight - nowImage.getHeight(this))/2, this);
        }
        
        //如果要放大或缩小图片,则在画布中心绘制缩放后的图像
        else if(flag == 1){
            g.drawImage(nowImage, (panelWidth - width)/2, (panelHeight - height)/2, width, height, this);
        }
    }
}
}
```

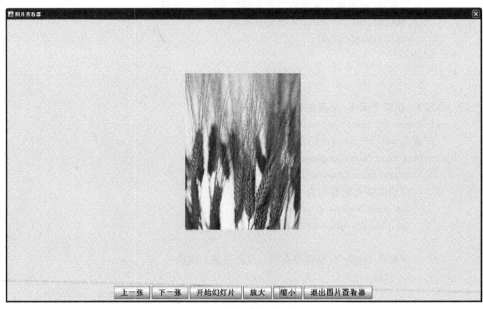

图 11-24　图片查看器缩放功能

### 4. 实现放映幻灯片（PlayPicture.java）

PlayPicture 类是 JDialog 的一个子类,实现了接口 ActionListener。该类可以实现放映幻灯片的功能。首先设置放映幻灯片的背景的一些属性：让背景屏幕大小为当前屏幕的大小；去掉背景屏幕的标题栏（即不显示标题、最大化、最小化、关闭按钮）；设置背景色为黑色；设置背景音乐；接下来实现播放幻灯片。要播放幻灯片,其实就是实现动画,每隔一定时间间隔在屏幕上重绘图片。这里利用 Timer 类来实现动画。最后要停止放映幻灯片。这里设置当用户按 Esc 键时,停止放映。只需要处理键盘事件就可以了。

以上内容的设计用到的大部分相关知识,本书前面内容都有所介绍,下面主要介绍背景音乐的实现。

1)实现播放音乐

在 java.applet 包中的 Applet 类和 AudioClip 接口提供了演播声音的基本支持。用 Java 可以编写播放.au、.aiff、.wav、.midi、.rfm 格式的音频。要实现在 Application 中播放声音,主要是使用 Applet 类中定义的一个静态方法:

```
public static AudioClip newAudioClip(URL url)
```

在 Application 中使用上述静态方法从指定的 URL 获得一个 AudioClip 的对象,该对象中包含要播放的声音文件,然后通过该对象调用 AudioClip 类的 play()、loop()和 stop()播放声音文件。

AudioClip 接口中定义的方法如下。
- public void play():开始播放声音文件,这个方法每次被调用时,都是对声音文件从头播放。
- public void loop():开始声音文件的循环播放。
- public void stop():停止播放声音文件。

例如:假设音频文件"beijing.wav"位于应用程序的当前目录中,有关播放音乐的步骤如下。

(1)创建 File 对象(File 类属于 java.io 包):

```
File musicFile = new File("beijing.wav");
```

(2)获取 URI 对象(URI 类属于 java.net 包):

```
URI uri = musicFile.toURI();
```

(3)获取 URL 对象(URL 类属于 java.net 包):

```
URL url = uri.toURL();
```

(4)创建音频对象(AudioClip 和 Applet 类属于 java.applet 包):

```
AudioClip clip = Applet.newAudioClip(url);
```

(5)播放、循环与停止:

```
clip.play();
clip.loop();
clip.stop();
```

2)代码(PlayPicture.java)

```
package caida.xinxi.Album;

import java.applet.Applet;
import java.applet.AudioClip;
import java.awt.Color;
import java.awt.Dimension;
```

```java
import java.awt.Graphics;
import java.awt.Image;
import java.awt.Toolkit;
import java.awt.event.ActionEvent;
import java.awt.event.ActionListener;
import java.awt.event.KeyAdapter;
import java.awt.event.KeyEvent;
import java.io.File;
import java.io.IOException;
import java.net.MalformedURLException;
import java.net.URI;
import java.net.URL;

import javax.imageio.ImageIO;
import javax.swing.JDialog;
import javax.swing.JPanel;
import javax.swing.Timer;

@SuppressWarnings("serial")
public class PlayPicture extends JDialog implements ActionListener {
    File[] filelist;
    Image[] images;
    int index;
    JPanel playPanel;
    Timer timer;
    Toolkit tool = Toolkit.getDefaultToolkit();
    Dimension screensize = tool.getScreenSize();
    AudioClip sound;

    public PlayPicture(File[] filelist, int selectedIndex ){
        this.filelist = filelist;
        index = selectedIndex;

        //创建 Image 数组,保存 filelist 中所引用的所有图像
        images = new Image[filelist.length];
        for(int i = 0;i < filelist.length ;i++){
            try {
                images[i] = ImageIO.read(filelist[i]);
            } catch (IOException e) {
                e.printStackTrace();
            }
        }

        //给窗口添加放映面板,面板背景色设置为黑色
        playPanel = new PicturePanel();
        playPanel.setBackground(Color.BLACK);
        add(playPanel);

        //设置退出放映模式的方法;按 Esc 键退出幻灯片放映模式
        addKeyListener(new KeyAdapter(){
            public void keyPressed(KeyEvent e){
```

```java
            //如果按了Esc键,则关闭窗口,停止播放音乐
            if(e.getKeyCode() == KeyEvent.VK_ESCAPE){
                setVisible(false);
                sound.stop();
                dispose();
            }
        }
    });

    // 设置窗口是有模式的,这要在放映模式下,不允许进行其他操作
    this.setModal(true);
    //设置放映窗口为: 不显示标题栏的对话框
    this.setUndecorated(true);
    //设置放映窗口大小为当前屏幕的宽和高
    setBounds(0,0,screensize.width ,screensize.height );

    //设置循环播放背景音乐
    File file = new File("WAV/beijing.wav");
    URI uri = file.toURI();
    URL url;
    try {
        url = uri.toURL();
        sound = Applet.newAudioClip(url);
        sound.loop();
    } catch (MalformedURLException e1) {
        e1.printStackTrace();
    }

    //设置计时器 timer,控制幻灯片的放映,每隔3秒放映一张图片
    timer = new Timer(3000,this);
    timer.start();

}

//计时器启动后,每隔3秒后,监听器自动调用该方法,重绘屏幕
public void actionPerformed(ActionEvent e) {
    playPanel.repaint();
}

//内部类,相当于画布,实现在画布上绘制图像
class PicturePanel extends JPanel{
    //重写 paintComponent()方法
    public void paintComponent(Graphics g){
        super.paintComponent(g);

        //获取当前要放映的图像的引用
        Image nowImage = images[index];
        //获取当前图像的宽和高
        int picWidth = nowImage.getWidth(this);
        int picHeight = nowImage.getHeight(this);
        //在画布中心位置绘制当前图像
```

```
                g.drawImage(nowImage,(this.getWidth() - picWidth)/2,(this.getHeight() -
picHeight)/2,this);
                //改变索引的值,切换图片
                index = (index + 1) % filelist.length ;
            }

        }
    }
```

## 11.2.5 案例练习题目

(1) 请读者换一种思路,重新设计实现以缩略图浏览方式显示图片的功能。

(2) 请读者查阅 API 文档,重新实现图片的缩放功能。

(3) 请读者使用 Thread 类来实现放映幻灯片的功能。并且设置组件,可由用户来控制播放幻灯片的放映速度。

(4) 请读者查阅相关 API,学习新的多媒体技术,设计实现多媒体音乐播放器,在播放背景音乐时,用户可以自由选择音乐文件,并控制音乐的播放。

(5) 请读者在图片查看器中增加新的功能,如顺时针旋转、逆时针旋转、删除图像、保存图像、打印图像等功能。

# 第12章 Java多线程程序设计

以往开发的程序大都是单线程的,即一个程序只有一条从头到尾的执行线索。然而现实世界中的很多过程都具有多条线索同时动作的特性。多线程就是指一个程序中可以包含两个或两个以上同时并发运行的程序单元,每个程序单元称为一个线程。Java 为多线程编程提供了语言级的支持。本节将介绍线程及其创建方法、线程的基本控制、线程同步、线程实现动画等内容,多线程和网络编程方面的内容在后面章节中将会介绍。

## 12.1 案例:两按钮反向运动——使用 Thread 子类

### 12.1.1 案例问题描述

创建一个 GUI 程序,当单击"开始"按钮,界面中的两个按钮开始朝相反方向运动,单击"停止"按钮,两按钮同时停止运动。要求通过 Thread 类的子类来实现多线程,再现两按钮反向运动的情景。

希望通过此案例,能让读者理解线程的概念;了解线程的状态和生命周期,学习继承 Thread 类来创建线程。

### 12.1.2 案例功能分析与演示

程序的运行效果如图 12-1 和图 12-2 所示。

图 12-1 两按钮初始状态

每次运行程序,会显示图 12-1 所示的初始状态,两个可运动的按钮分别位于窗口的左边界和右边界,一旦用户单击"开始/继续"按钮,一个按钮开始从左向右水平运动,一个按钮开始从右向左水平运动;如果单击"停止"按钮,两个运动按钮同时停止水平运动,静止不动;如果再次单击"开始/继续"按钮,两个按钮会从停止状态又开始反向运动;如果单击

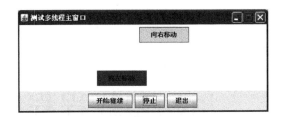

图 12-2 两按钮反向运动过程截图

"退出"按钮,可退出程序。图 12-2 是两个按钮反向运动过程中的一个截图。

## 12.1.3 案例总体设计

本例要求通过继承 Thread 类来实现多线程的创建。在编写 Thread 类的子类时,需要重载父类的 run()方法,其目的是规定线程的具体操作,否则线程就什么也不做。

因此根据问题,可以设计三个独立的线程:一个线程用于 GUI,一个线程控制按钮从左向右水平移动,一个线程控制按钮从右向左水平移动。本例分别用 MoveButtonWindow 类、MoveToRight 类和 MoveToLeft 类实现它们。

### 1. MoveButtonWindow 类

MoveButtonWindow 类是 JFrame 的一个子类,这里设计两个可运动的按钮,并初始化它们的状态,再设计三个按钮来控制按钮的运动。

MoveButtonWindow 类的 UML 图如图 12-3 所示。

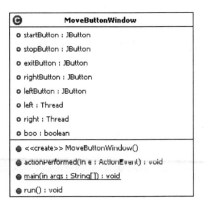

图 12-3 MoveButtonWindow 类的 UML 图

以下是 UML 图中有关数据和方法的详细说明。

1) 成员变量

- startButton、stopButton、exitButton 是 JButton 对象,用来控制两个按钮的运动。startButton 按钮负责启动线程,两个按钮开始反向水平运动;stopButton 按钮负责终止线程,两个按钮停止运动;exitButton 按钮负责退出程序。
- rightButton、leftButton 是 JButton 对象,rightButton 按钮将会从左向右水平运动;leftButton 按钮将会从右向左水平运动。

- right 是 MoveToRight 对象，是控制按钮从左向右运动的线程。
- left 是 MoveToLeft 对象，是控制按钮从右向左运动的线程。

2）成员方法
- MoveButtonWindow()构造方法，完成窗口的初始化。
- actionPerformed(ActionEvent)方法，是 MoveButtonWindow 类实现的 ActionListener 接口中的方法，一旦用户单击控制按钮，发生 ActionEvent 事件，监听器自动调用该方法响应用户的操作。
- main(String[])是程序的入口方法。

2. MoveToRight 类

该类控制按钮从左向右移动。
MoveToRight 类的 UML 图如图 12-4 所示。

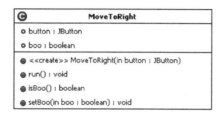

图 12-4　MoveToRight 类的 UML 图

以下是 UML 图中有关数据和方法的详细说明。

1）成员变量
- button 是 JButton 对象，MoveToRight 线程负责移动的按钮对象。
- boo 是 boolean 变量，负责标识线程是否进入死亡状态。

2）成员方法
- MoveToRight(JButton)构造方法。
- run()方法，重写父类中的 run()方法，规定线程的具体动作。
- isBoo()方法，获取线程的状态。
- setBoo(boolean)方法，设置线程的状态。

3. MoveToLeft 类

该类控制按钮从右向左运动。
MoveToLeft 类的 UML 图如图 12-5 所示。

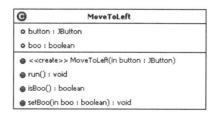

图 12-5　MoveToLeft 类的 UML 图

以下是 UML 图中有关数据和方法的详细说明。

1) 成员变量
- button 是 JButton 对象，MoveToLeft 线程负责移动的按钮。
- boo 是 boolean 型变量，标识 MoveToLeft 线程是否进入死亡状态。

2) 成员方法
- MoveToLeft(JButton)构造方法。
- run()方法，重写父类中的 run()方法，定义线程体。
- isBoo()方法，获取线程的状态。
- setBoo(boolean)方法，设置线程的状态。

### 12.1.4 案例代码实现

**1. 编写应用程序主界面（MoveButtonWindow 类）**

MoveButtonWindow 类是 javax.swing 包中 JFrame 的一个子类，并实现了 ActionListener 接口，是本应用程序的主窗体。

首先这里需要进行 GUI 设计。MoveButtonWindow 窗口的布局是 BorderLayout 型，在 Center 位置和 South 位置分别添加两个 JPanel 类面板，在 Center 位置面板中添加两个可运动的按钮，分别定义它们的大小、位置、背景色；在 South 位置面板中依次添加三个控制按钮，利用 Java 事件处理机制控制 Center 位置按钮的移动。

接着进行线程类的设计。在实现按钮运动的功能时，本例用到了线程，涉及线程的状态变迁。下面介绍线程、线程的状态和生命周期。

1) 程序、进程、线程、多线程、主线程

程序是一段静态的代码，它是应用软件执行的蓝本。

进程是程序的一次动态执行过程，它对应了从代码加载、执行至执行完毕的一个完整过程，这个过程也是进程本身从产生、发展至消亡的过程。

线程是比进程更小的执行单位，可以独立、并发执行，一个进程在其执行过程中，可以产生多个线程，形成多条执行线索，每条线索，即每个线程也有它自身的产生、发展和消亡的过程，也是一个动态的概念。

多线程指程序中同时存在多个执行体，它们按照自己的执行线路并发工作，独立完成各自的功能，互不干扰。使用多线程技术可以加快程序的响应时间，提高计算机资源的利用率和整个应用系统的性能。

每个 Java 程序都有一个默认的主线程。Java 应用程序总是从主类的 main 方法开始执行。当 JVM 加载代码，发现 main 方法之后，就会启动一个线程，这个线程称作"主线程"，该线程负责执行 main 方法。在 main 方法的执行中再创建的线程就称作程序中的其他线程。如果 main 方法中没有创建其他的线程，那么当 main 方法执行完最后一个语句，即 main 方法返回时，JVM 就会结束 Java 应用程序。如果 mian 方法中又创建了其他线程，那么 JVM 就要在主程序和其他线程之间轮流切换，保证每个线程都有机会使用 CPU 资源，main 方法即使执行完最后的语句，JVM 也不会结束程序，JVM 一直要等到程序中的所有线程都结束之后，才结束 Java 应用程序。

2)线程状态与生命周期

一个线程从它的创建到运行结束的过程,称为线程的生命周期。在这个生命周期里,线程有着不同的状态,并在这些状态间进行转换。线程的生命周期主要分为:新建状态、可运行状态、运行状态、阻塞状态、终止状态,如图12-6所示。

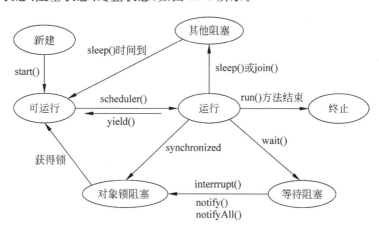

图12-6　线程的生命周期

(1) 新建状态(new)

当程序使用new关键字创建了一个线程后,该线程就处于新建状态,此时它和其他Java对象一样,仅仅由Java虚拟机为其分配了内存,并初始化了其成员变量的值,此时的线程对象没有表现出任何线程的动态特征,程序也不会执行线程的线程体。

(2) 就绪状态(Runnable)

新建的线程调用start()方法之后,该线程处于就绪状态。start()方法使系统为线程分配必要的资源,将线程中虚拟的CPU置为Runnable状态,并将线程交给系统调度。处于这个状态中的线程并没有开始运行,它只是表示该线程可以运行了。至于该线程何时开始运行,取决于Java虚拟机里线程调度器的调度。

(3) 运行状态(Running)和阻塞状态(Blocked)

如果处于就绪状态的线程获得了CPU,开始执行run()方法的线程体,则该线程处于运行状态,如果计算机只有一个CPU,则在任何时刻只有一条线程处于运行状态。如果在一个多处理器的机器上,将会有多个线程并行执行;但当线程数大于处理器数时,依然会有多条线程在同一个CPU上轮换的现象。此时线程状态的变迁有如下三种情况:如果线程正常执行结束或应用程序停止运行,线程将进入终止状态;如果当前线程执行了yield()方法,或者当前线程因调度策略由系统控制进入可运行状态;线程使用CPU资源期间,执行某个操作进入阻塞状态。

- 线程调用了sleep()方法或join()方法,进入阻塞状态。当调用sleep方法的线程经过了指定时间或调用join()方法的线程结束时,解除阻塞。
- 线程调用wait()方法时,由运行状态进入阻塞状态。当线程正等待某个通知时,其他线程发出了一个通知,解除阻塞状态。
- 如果线程中使用synchronized来请求对象的锁未获得时,进入阻塞状态。该状态下的线程成功地获得了对象锁后,将进入就绪状态。

- 如果线程中有输入输出操作，也将进入阻塞状态，待输入输出操作结束后，线程进入就绪状态。

当前正在执行的线程被阻塞之后，其他线程就可以获得执行的机会了。被阻塞的线程会在合适的时候重新进入就绪状态，注意不是运行状态。也就是说，被阻塞线程的阻塞解除后，必须重新等待线程调度器再次调度它。

(4) 终止状态(Dead)

线程会以以下三种方式之一结束，结束后就处于死亡状态：
- run()方法执行完成，线程正常结束。
- 线程抛出一个未捕获的 Exception 或 Error。
- 直接调用该线程的 stop()方法来结束线程——该方法容易导致死锁，通常不推荐使用。

为了测试某条线程是否已经死亡，可以调用线程对象的 isAlive()方法，当线程处于就绪、运行、阻塞三种状态时，该方法将返回 true；当线程处于新建、死亡两种状态时，该方法返回 false。

不要试图对一个已经死亡的线程调用 start()方法使它重新启动，该线程将不可再次作为线程执行。同样对新建状态的线程两次调用 start()方法也是错误的。

3) 代码(MoveButtonWindow 类)

```java
//MoveButtonWindow.java
package caida.xinxi.TestThread1;

import java.awt.BorderLayout;
import java.awt.Color;
import java.awt.event.ActionEvent;
import java.awt.event.ActionListener;

import javax.swing.JButton;
import javax.swing.JFrame;
import javax.swing.JPanel;

@SuppressWarnings("serial")
public class MoveButtonWindow extends JFrame implements ActionListener{
    JButton startButton,stopButton,exitButton;
    JButton rightButton,leftButton;
    MoveToRight right;
    MoveToLeft left;

    //构造方法,完成窗口的初始化
    public MoveButtonWindow(){
        //设置窗口的标题
        setTitle("测试多线程主窗口");

        //创建面板,并添加到窗口中
        JPanel centerPanel = new JPanel();
        JPanel controlPanel = new JPanel();
        add(centerPanel,BorderLayout.CENTER );
```

```java
        add(controlPanel,BorderLayout.SOUTH );

        //创建按钮,注册 ActionEvent 事件监听器,并添加到控制面板 controlPanel 中
        startButton = new JButton("开始/继续");
        startButton.addActionListener(this);
        stopButton = new JButton("停止");
        stopButton.addActionListener(this);
        exitButton = new JButton("退出");
        exitButton.addActionListener(this);
        controlPanel.add(this.startButton );
        controlPanel.add(this.stopButton );
        controlPanel.add(this.exitButton );

        //定义面板 centerPanel 的布局,并添加按钮组件
        //分别创建两个可移动按钮,并定义其大小、位置、背景色
        centerPanel.setLayout(null);
        centerPanel.setBackground(Color.white );
        rightButton = new JButton("向右移动");
        rightButton.setBackground(Color.yellow );
        rightButton.setBounds(0, 5, 100, 30);
        leftButton = new JButton("向左移动");
        leftButton.setBackground(Color.red );
        leftButton.setBounds(395,90,100,30);
        centerPanel.add(rightButton);
        centerPanel.add(leftButton);

        //新建控制按钮移动的线程 right 和 left,处于新建状态
        right = new MoveToRight(rightButton);
        left = new MoveToLeft(leftButton);

        //设置窗口的一些属性
        setBounds(100,100,500,200);
        this.setDefaultCloseOperation(JFrame.EXIT_ON_CLOSE );
        this.setResizable(false);
        setVisible(true);
        validate();

    }

    //接口 ActionLiestener 中的方法,响应用户单击按钮的操作
    public void actionPerformed(ActionEvent e) {

        //如果单击"开始/继续"按钮,则启动线程
        if(e.getSource() == this.startButton ){
            //如果线程 right 死亡了,则重新给线程分配实体
            if(!right.isAlive()){
                right = new MoveToRight(rightButton);
            }
            //如果线程 left 死亡了,则重新给线程分配实体
            if(!left.isAlive()){
                left = new MoveToLeft(leftButton);
```

```java
            }
            
            //启动 right 线程和 left 线程
            right.start();
            left.start();
        }
        
        //如果单击"停止"按钮,结束线程
        else if(e.getSource() == this.stopButton){
            right.setBoo(false);
            left.setBoo(false);
        }
        
        //如果单击"退出"按钮,退出应用程序
        else if(e.getSource() == this.exitButton){
            right.setBoo(false);
            left.setBoo(false);
            System.exit(0);
        }
    }
    
    //应用程序的入口方法
    public static void main(String[] args) {
        new MoveButtonWindow();
    }
}
```

### 2. 实现两个按钮的反向运动(MoveToRight 类和 MoveToLeft 类)

MoveToRight 类和 MoveToLeft 类是 Thread 类的子类。这里是根据案例的要求设计这样两个线程类来实现按钮朝不同方向运动的。那么如何使用 Thread 的子类来创建线程呢?

1) 继承 Thread 类创建线程类

在 java.lang 包中,Thread 类的声明如下:

```
public class Thread extends Object implements Runnable
```

因此,Thread 类本身实现了 Runnable 接口,在 Thread 类的定义中可以发现 run()方法,通过继承 Thread 类创建线程的步骤如下:

(1) 定义 Thread 类的子类,并重写该类的 run()方法,该 run()方法的方法体就代表了线程需要完成的任务。因此,经常把 run()方法称为线程体。

(2) 创建 Thread 子类的实例,即创建了线程对象。

(3) 用线程对象的 start()方法来启动该线程。

本应用程序显式地创建并启动了 MoveToRight 和 MoveToLeft 两个线程类对象,但实际上程序至少有三条线程:程序显式创建的两个子线程和主线程。main 方法的方法体就是主线程的线程执行体。

除此以外,下面代码用到了线程的一个方法。

static void sleep(long millis)：让当前正在执行的线程暂停 millis 毫秒,并进入阻塞状态,该方法受到系统计时器和线程调度器的精度和准确度的影响。

当当前线程调用 sleep 方法进入阻塞状态后,在其 sleep 时间段内,该线程不会获得执行的机会,即使系统中没有其他可运行的线程,处于 sleep 中的线程也不会运行,因此 sleep 方法常用来暂停程序的执行。本程序中按钮两次移动之间的时间间隔为 0.2 秒。

2）代码（MoveToRight 类）

```java
//MoveButtonWindow.java

class MoveToRight extends Thread{
    JButton button;
    boolean boo = true;

    //构造方法,完成初始化
    MoveToRight(JButton button){
        this.button = button;
    }

    //重写 run()方法,定义线程体
    public void run(){
        while(true){
            int x = button.getBounds().x;
            x = x + 5;
            if(x > 400){
                x = 5;
            }
            button.setLocation(x, 5);
            try {
                //调用 sleep 方法让当前线程暂停 0.2 秒
                Thread.sleep(200);
            } catch (InterruptedException e) {
                e.printStackTrace();
            }
            //结束线程体的执行
            if(!boo){
                return;
            }
        }
    }
    public boolean isBoo() {
        return boo;
    }
    public void sctBoo(boolean boo) {
        this.boo = boo;
    }
}
```

3）代码（MoveToRight 类）

//MoveButtonWindow.java

```java
class MoveToLeft extends Thread{
    JButton button;
    boolean boo = true;
    
    //构造方法,完成初始化
    MoveToLeft(JButton button){
        this.button = button;
    }
    
    //重写 run()方法,定义线程体
    public void run(){
        while(true){
            int x = button.getBounds().x;
            x = x - 5;
            if(x < - 20){
                x = 395;
            }
            button.setLocation(x, 90);
            try {
                //当前线程调用 sleep 方法暂停 0.2 秒
                Thread.sleep(200);
            } catch (InterruptedException e) {
                e.printStackTrace();
            }
            //结束当前线程
            if(!boo){
                return;
            }
        }
    }
    public boolean isBoo() {
        return boo;
    }
    public void setBoo(boolean boo) {
        this.boo = boo;
    }
}
```

### 12.1.5　案例练习题目

（1）为该应用程序增加一个功能,用户可以控制按钮移动的速度(让用户可以控制按钮两次移动之间的时间间隔)。

（2）读者可以看到,本程序中有一些代码是重复的,请读者重新编写本案例,减少代码的重复率,仍用 Thread 的子类来实现线程。

（3）请设计一个采用继承 Thread 类方式的多线程程序,用两个小球模拟物理中的自由落体运动和平抛运动。

（4）请设计一个采用继承 Thread 类方式的多线程程序,用两个小球模拟正弦函数和余弦函数曲线运动。

（5）请设计一个采用继承 Thread 类方式的多线程程序，实现猜数字游戏。一个线程负责随机给出一个 1～100 之间的正整数，另一个线程负责猜数字。

## 12.2 案例：两按钮反向运动——使用 Runnable 接口

### 12.2.1 案例问题描述

创建一个 GUI 程序，当单击"开始"按钮时，界面中的两个按钮开始朝相反方向水平运动，单击"停止"按钮，两按钮同时停止运动。要求使用 Runnable 接口实现多线程，再现两按钮反向运动情景。

希望通过此案例，能让读者掌握使用 Runnable 接口实现多线程，并比较两种创建线程方法的优缺点。

### 12.2.2 案例功能分析与演示

程序的运行效果如图 12-7 和图 12-8 所示。

图 12-7　案例运行结果初始状态

图 12-8　案例运行过程截图

每次运行程序，会显示如图 12-7 所示的初始状态，两个可运动的按钮分别位于窗口的左边界和右边界，一旦用户单击"开始/继续"按钮，一个按钮开始从左向右水平运动，一个按钮开始从右向左水平运动；如果单击"停止"按钮，两个运动按钮同时停止水平运动，静止不动；如果再次单击"开始/继续"按钮，两个按钮会从停止状态又开始反向运动；如果单击"退出"按钮，可退出程序。图 12-8 是两个按钮反向运动过程中的一个截图。

### 12.2.3 案例总体设计

本例要求使用 Runnable 接口来实现多线程的创建。根据问题的描述和程序运行效果的展示,可以按照下面三种思路来设计。

(1) 设计三个类,一个类负责实现 GUI 界面设计;另两个类实现接口 Runnable,并实现接口中的方法 run(),其中一个控制其中一个按钮向左运动,一个控制另一个按钮向右运动。

(2) 设计一个 JFrame 的子类,如 MoveButtonWindow 类来实现 GUI 功能;再设计一个类,如 MoveButton 类,让这个类实现接口 Runnable,在这个类里来实现接口中的方法 run(),由这个类的对象作为目标对象创建线程控制按钮的运动。

(3) 设计一个类,如 MoveButtonWindow 类,由这个类负责 GUI 界面,并且实现接口 Runnable,实现接口中的方法 run()来控制按钮的运动。

本例采用第(3)种思路来设计。图 12-9 是 MoveButtonWindow 类的 UML 图。

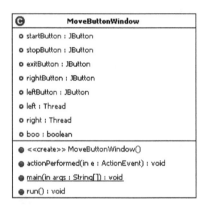

图 12-9　MoveButtonWindow 类的 UML 图

以下是 UML 图中有关数据和方法的详细说明。

### 1. 成员变量

- startButton、stopButton、exitButton 是 JButton 类对象,都注册了 ActionEvent 事件监听器,负责控制按钮运动和退出程序。startButton 按钮负责控制两个按钮反向运动;stopButton 按钮负责中止运动;exitButton 按钮实现退出应用程序。
- rightButton、leftButton 是 JButton 类对象。rightButton 按钮将从左向右运动;leftButton 按钮将从右向左运动。
- left、right 分别是 Thread 类对象,两个线程分别控制按钮反向水平运动。
- boo 是 boolean 型变量,用来标识线程什么时候结束。

### 2. 成员方法

- MoveButtonWindow()构造方法,完成窗口的初始化。
- actionPerformed（ActionEvent）方法,是 MoveButtonWindow 类实现的接口 ActionListener 中的方法,一旦用户单击了控制按钮,产生 ActionEvent 对象,将触发监听器对象调用 actionPerformed(ActionEvent)方法,对用户单击按钮的行为做出响应。

- main(String[])是应用程序的入口方法。
- run()方法,是 MoveButtonWindow 类实现 Runnable 接口中的方法,定义线程的执行体。当线程调用 start()方法后,一旦轮到它来享用 CPU 资源,MoveButtonWindow 类对象就会自动调用接口中的 run()方法,实现按钮的反向水平运动。

## 12.2.4 案例代码实现

MoveButtonWindow 类是 javax.swing 包中 JFrame 的一个子类,实现了 ActionListener 接口和 Runnable 接口,负责 GUI 界面的实现和两个按钮的反向水平运动。

首先需要进行 GUI 设计。MoveButtonWindow 窗口的布局是 BorderLayout 型,在"Center"位置和"South"位置分别添加一个 JPanel 类面板,在"Center"位置面板中添加两个可运动的按钮,分别定义它们的大小、位置、背景色;在"South"位置面板中依次添加三个控制按钮,并利用 Java 事件处理机制,控制 Center 位置按钮的移动。

接着让 MoveButtonWindow 类实现接口 Runnable 中的 run()方法,定义线程的线程执行体,分别控制两个按钮的不同运动。下面介绍如何通过实现 Runnable 接口来创建线程。

### 1. 实现接口创建线程类

使用 Thread 类创建线程对象时,通常使用的构造方法是(其他的构造方法请查阅 API):

```
Thread ( Runnable target )
```

该构造方法中的参数是一个 Runnable 类型的接口。那如何实现 Runnable 接口来创建线程类呢?

下面是实现 Runnable 接口来创建并启动多条线程的步骤:

(1) 定义 Runnable 接口的实现类,并重写该接口的 run()方法,该 run()方法的方法体同样是该线程的线程执行体。

(2) 创建 Runnable 实现类的实例,并以此实例作为 Thread 的 target 来创建 Thread 对象,该对象才是真正的线程对象。代码如下:

```
//创建 Runnable 实现类 TargetObject 的对象 st
TargetObject st = new TargetObject();
//以 st 作为目标对象,创建 Thread 对象,即线程对象
new Thread(st);
```

也可以在创建 Thread 对象时为该 Thread 对象指定一个名字,代码如下所示:

```
//创建 Thread 对象时指定 target 和新线程的名字
new Thread(st,"新线程 1");
```

(3) 调用线程对象的 start()方法来启动线程。

本应用程序显式地创建并调用 start()方法,启动两个线程(一个控制其中一个按钮朝左运动;另一个控制另一个按钮朝右运动),还有一个主线程。MoveButtonWindow 类对象作为目标对象,创建了两个线程。在使用实现 Runnable 接口时要获得当前线程对象必须使

用 Thread.currentThread()方法。currentThread()是 Thread 类的静态方法,该方法总是返回当前正在执行的线程对象。

### 2．代码(MoveButtonWindow.java)

```java
package caida.xinxi.TestThread2;

import java.awt.BorderLayout;
import java.awt.Color;
import java.awt.event.ActionEvent;
import java.awt.event.ActionListener;

import javax.swing.JButton;
import javax.swing.JFrame;
import javax.swing.JPanel;

@SuppressWarnings("serial")
public class MoveButtonWindow extends JFrame implements ActionListener,Runnable{
    JButton startButton,stopButton,exitButton;
    JButton rightButton,leftButton;
    Thread left,right;
    boolean boo = true;

    //构造方法,完成窗口的初始化工作
    public MoveButtonWindow(){
        setTitle("测试多线程主窗口");

        //创建面板对象,并添加到窗口中
        JPanel centerPanel = new JPanel();
        JPanel controlPanel = new JPanel();
        add(centerPanel,BorderLayout.CENTER );
        add(controlPanel,BorderLayout.SOUTH );

        //分别创建控制按钮,注册 ActionEvent 事件监听器,添加到 controlPanel 中
        startButton = new JButton("开始/继续");
        startButton.addActionListener(this);
        stopButton = new JButton("停止");
        stopButton.addActionListener(this);
        exitButton = new JButton("退出");
        exitButton.addActionListener(this);
        controlPanel.add(this.startButton );
        controlPanel.add(this.stopButton );
        controlPanel.add(this.exitButton );

        //定义 centerPanel 面板布局为空,设置背景色为白色,并添加可移动按钮
        //创建可移动按钮,分别设置按钮背景色、按钮大小及位置
        centerPanel.setLayout(null);
        centerPanel.setBackground(Color.white );
        rightButton = new JButton("向右移动");
        rightButton.setBackground(Color.yellow );
```

```java
            rightButton.setBounds(0, 5, 100, 30);
            leftButton = new JButton("向左移动");
            leftButton.setBackground(Color.red );
            leftButton.setBounds(395,90,100,30);
            centerPanel.add(rightButton);
            centerPanel.add(leftButton);

            //创建线程right、left,当前窗口对象为target,线程处于新建状态
            right = new Thread(this);
            left = new Thread(this);

            //设置窗口的一些属性
            setBounds(100,100,500,200);
            this.setDefaultCloseOperation(JFrame.EXIT_ON_CLOSE );
            this.setResizable(false);
            setVisible(true);
            validate();
    }

    //实现ActionListener接口中的方法,当用户单击控制按钮时,自动调用该方法
    public void actionPerformed(ActionEvent e) {

        //如果单击"开始/继续"按钮,启动线程,两按钮开始反向水平运动
        if(e.getSource() == this.startButton ){
            //如果right线程死亡,则重新给right分配实体
            if(!right.isAlive()){
                right = new Thread(this);
            }
            //如果left线程死亡,则重新给left分配实体
            if(!left.isAlive()){
                left = new Thread(this);
            }
            //设置标志变量为true,并启动线程
            boo = true;
            right.start();
            left.start();
        }

        //如果单击"停止"按钮,设置标志变量为false,则结束线程
        else if(e.getSource() == this.stopButton){
            boo = false;
        }

        //如果单击"退出"按钮,则退出应用程序
        else if(e.getSource() == this.exitButton){
            boo = false;
            System.exit(0);
        }
    }
    //main方法,程序的入口
    public static void main(String[] args) {
```

```java
        new MoveButtonWindow();
    }

    //接口 Runnable 中的方法,实现 run()方法,定义线程执行体
    public void run() {
        while(true){
            //如果当前线程是 right,则控制按钮水平向右移动
            if(Thread.currentThread() == right){
                int x = this.rightButton.getBounds().x;
                x = x + 5;
                if(x > 400){
                    x = 5;
                }
                rightButton.setLocation(x, 5);
                try {
                    //当前线程调用 sleep 方法,暂停 0.2 秒
                    Thread.sleep(200);
                } catch (InterruptedException e) {
                    e.printStackTrace();
                }
            }

            //如果当前线程为 left,则控制线程水平向左运动
            else if(Thread.currentThread() == left){
                int x = leftButton.getBounds().x;
                x = x - 5;
                if(x < - 20){
                    x = 395;
                }
                leftButton.setLocation(x, 90);
                try {
                    //当前线程调用 sleep 方法,暂停 0.2 秒
                    Thread.sleep(200);
                } catch (InterruptedException e) {
                    e.printStackTrace();
                }
            }

            //如果 boo 为 false,则结束线程,线程进入死亡状态
            if(!boo){
                return;
            }
        }
    }
}
```

### 3. 两种方式所创建线程的对比

通过继承 Thread 类或实现 Runnable 接口都可以实现多线程,但两种方式存在一定的

差别,相比之下,两种方式的主要差别如下。

1) 采用继承 Thread 类方法的特点

劣势:因为线程已经继承了 Thread 类,所以不能再继承其他父类。

优势:编写简单,如果需要访问当前线程,直接使用 this 即可获得当前线程,并可以在 run()方法中直接调用线程的其他方法。

2) 实现 Runnable 接口的特点

劣势:编程稍微复杂,如果需要访问当前线程,需要使用 Thread.currentThread() 方法。

优势:实现了 Runnable 接口的类,还可以用 extends 继承其他的类;符合面向对象设计的思想,在这种方式下,可以多个线程共享同一个 target 对象,所以非常适合多个线程来处理同一份资源的情况,从而可以将 CPU、代码和数据分开,形成清晰的模型。

因此,提倡采用实现 Runnable 接口的方式。但在具体应用中,可以根据具体情况确定采用哪种方法。

### 12.2.5 案例练习题目

(1) 为该应用程序增加一个功能,用户可以控制按钮移动的速度(让用户可以控制按钮两次移动之间的时间间隔)。

(2) 请设计一个使用 Ruannable 接口实现多线程的程序,实现让两个人其中一个人报 1~100 的平方,另一个人报 1~100 的立方。

(3) 请设计一个使用 Ruannable 接口实现多线程的程序,用两个小球模拟物理中的自由落体运动和平抛运动。

(4) 请设计一个使用 Ruannable 接口实现多线程的程序,用两个小球模拟正弦函数和余弦函数曲线进行运动。

(5) 请设计一个使用 Ruannable 接口实现多线程的程序,实现猜数字游戏。一个线程负责随机给出一个 1~100 之间的正整数,另一个线程负责猜数字。

## 12.3 案例:使用 Thread 类实现图像动画

### 12.3.1 案例问题描述

设计一个 GUI 程序,在程序主窗口界面初始有一个由几幅图像形成的动画。用户可以通过按方向键(←、→、↑、↓)来控制图像的运动方向。例如,如果用户按向下箭头,图像将会沿竖直方向运动,如果运动到窗口下边界,图像将会改变运动方向,沿相反的方向向上运动,如果运动到窗口的上边界,图像将会改变运动方向,沿相反的方向向下运动,就这样图像在竖直方向不停运动。除此以外,程序还有控制动画播放、控制背景音乐播放、设置窗口背景色等功能。

前面章节的案例中是使用 Timer 类来实现动画的,这里希望通过此案例,能让读者了解动画制作的基本原理和过程,并掌握使用 Thread 进行动画的制作。

## 12.3.2 案例功能分析与演示

每次执行应用程序,都会显示一个主窗口,在该窗口的固定位置显示一个初始的动画效果,如图 12-10 所示。这个动画是通过 10 幅图像的轮显实现的,图 12-10 中的 4 个图是这个动画过程中截取的 4 个状态,第 1 张图是动画的初始状态,第 4 张图是动画的最终状态,中间两张是动画过程中某两个活动状态的截图。

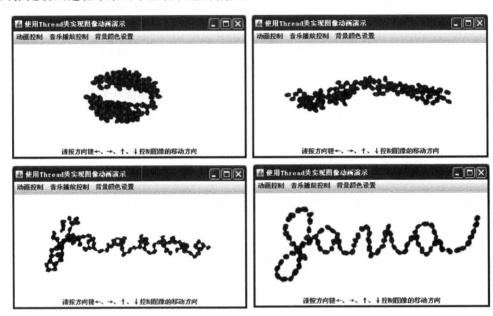

图 12-10 应用程序主窗口初始动画效果截图

如果用户按下方向键,多幅图像将根据方向键所指的方向以动画的形式在主窗口中循环运动。假设用户按下向下方向键,图像将按图 12-11 所示的效果不停地在主窗口中运动。用户还可以通过菜单"动画控制"来控制动画的播放,如图 12-12 所示。

图 12-11 按下向下方向键的动画效果

如果按下图 12-12 中"动画控制"菜单中的"中止动画"命令,图 12-11 所示的动画将被中止,在主窗口中静止不动;如果选择"继续动画",则会从中止的位置继续以中止以前的运动方向循环运动;如果选择"退出",则会退出应用程序。

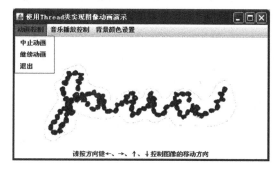

图 12-12　动画控制菜单

除此以外,应用程序还有两个功能:控制动画背景音乐的播放和动画背景色的设置,如图 12-13 所示。从图 12-13 可以看到应用程序通过菜单命令来实现这两个功能。"音乐播放控制"菜单有"播放"、"循环"播放、"停止"播放三个菜单项。"背景颜色设置"菜单通过"颜色选择"命令,可以打开颜色选择器来设置动画背景色,如图 12-14 所示,用户选择背景色为"黄色"。

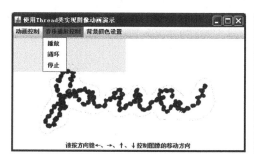 

图 12-13　背景音乐播放及背景色选择

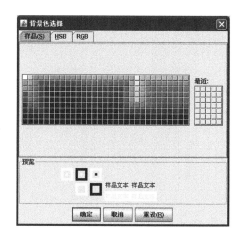 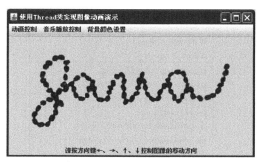

图 12-14　设置动画背景色

## 12.3.3 案例总体设计

本例要求使用 Thread 来实现动画,根据前面内容的介绍,我们知道有两种创建线程的方式,本案例采用实现 Runnable 接口的方式来创建线程,由这个线程来实现上面运行效果所演示的动画效果。

本案例设计一个类 AnimationAndThread,这个类是 JFrame 类的子类,创建应用程序主窗口,并让 AnimationAndThread 类实现接口 ActionListener、接口 KeyListener 和接口 Runnable,通过重写不同接口中的方法实现应用程序的功能:

- 重写 ActionListener 接口中的方法 actionPerformed(ActionEvnt),负责响应用户单击菜单项的行为,即负责实现动画控制、背景音乐控制、背景色设置和退出应用程序的功能。
- 重写 KeyListener 接口中的方法 keyPressed(KeyEvent),负责响应用户按下→、←、↑、↓这几个方向键的行为,即负责对图像运动的方向进行控制。
- 重写 Runnable 接口中的方法 run(),负责定义线程的执行体,即具体设计并实现动画的过程。

图 12-15 是 AnimationAndThread 类的 UML 图。

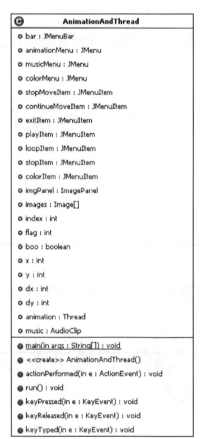

图 12-15 AnimationAndThread 类的 UML 图

以下是 UML 图中有关数据和方法的详细说明。

1. 成员变量

- bar 是 JMenuBar 类创建的菜单条,可以向 bar 中添加菜单。
- animationMenu、musicMenu、colorMenu 是 JMenu 类创建的三个菜单,这三个菜单的名称分别是"动画控制"、"音乐播放控制"、"背景颜色设置"。
- stopMoveItem、continueMoveItem、exitItem 是 JMenuItem 类创建的三个菜单,被添加到 animationMenu 菜单中。这三个菜单项的名称分别是"终止动画"、"继续动画"、"退出"。
- playItem、loopItem、stopItem 是 JMenuItem 类创建的三个菜单项,被添加到 musicMenu 菜单中。这三个菜单项的名字分别是"播放"、"循环"、"停止"。
- colorItem 是 JMenuItem 类创建的菜单项,被添加到 colorMenu 菜单中,其名称是"颜色选择"。
- imgPanel 是 JPanel 类对象,是一个图像面板,负责显示图像。
- images 是一个 Image 数组,保存一系列图像。
- index、flag 是 int 型变量,二者一起来实现对 images 数组下标的变换,从而实现图像的轮显。
- boo 是 boolean 型变量,用来控制是否结束线程。
- x、y 是 int 型变量,表示图像在程序主窗口中的坐标位置。
- dx、dy 是 int 型变量,表示图像在主窗口中运动时的坐标偏移量。
- animation 是一个线程,用来实现动画。
- music 是 AudioClip 对象,实现背景音乐的播放。

2. 成员方法

- main(String[])是应用程序的入口方法。
- AnimationAndThread()是构造方法,实现程序主窗口的初始化。
- actionPerformed(ActionEvent)方法是 AnimationAndThread 类实现的 ActionListener 接口中的方法,负责执行菜单项发出的有关命令。用户选择菜单中的菜单项可触发 ActionEvent 事件,导致 actionPerformed(ActionEvent)方法执行相应的操作。
- run()方法是 AnimationAndThread 类实现的 Runnable 接口中的方法,用来定义 animation 线程的线程执行体,负责实现图像动画。
- keyPressed(KeyEvent)方法是 AnimationAndThread 类实现的 KeyListener 接口中的方法,当用户按下某个键时触发 KeyEvent 事件,监听器自动调用该方法执行相应操作。
- keyReleased(KeyEvent)方法同样是 KeyListener 接口中的方法,当用户松开某个键时触发 KeyEvent 事件,监听器自动执行该方法中的代码。
- keyTyped(KeyEvent)同样是 KeyListener 接口中的方法,当用户单击某个键时触发 KeyEvent 事件,监听器自动执行该方法中的代码。

### 3. 内部类

ImagePanel 是 JPanel 类对象，负责显示 images 数组中的图像。

## 12.3.4 案例代码实现

AnimationAndThread 类是 javax.swing 包中 JFrame 类的子类，实现了 ActionListener 接口、KeyListener 接口和 Runnable 接口，负责 GUI 界面的实现和动画的制作。

首先需要进行 GUI 设计。AnimationAndThread 窗口对象的布局是 BorderLayout 型，在"Center"位置添加一个图像面板对象 imgPanel，它的初始背景为白色；在"South"位置添加一个标签对象 messageLabel，由它来提示用户如何来对动画进行方向控制。

接下来的工作即创建控制动画的线程，由它来实现动画的具体制作过程。实现接口 KeyListener 中的 KeyPressed(KeyEvent)方法，来控制动画的运动方向。由"动画控制"菜单实现对动画的播放。

最后通过菜单"音乐播放控制"和"背景颜色设置"实现动画过程中背景音乐的控制和背景色的选择。背景音乐控制的代码用到了多媒体内容中提到的 AudioClip 类，颜色设置用到了 JColorChooser 类。

以上内容所有知识点在前面的章节案例中都介绍过，这里不再赘述，下面具体看代码（AnimationAndThread.java）：

```java
package caida.xinxi.TestThread3;

import java.applet.Applet;
import java.applet.AudioClip;
import java.awt.BorderLayout;
import java.awt.Color;
import java.awt.Graphics;
import java.awt.Image;
import java.awt.Toolkit;
import java.awt.event.ActionEvent;
import java.awt.event.ActionListener;
import java.awt.event.KeyEvent;
import java.awt.event.KeyListener;
import java.io.File;
import java.net.MalformedURLException;
import java.net.URL;

import javax.swing.JColorChooser;
import javax.swing.JFrame;
import javax.swing.JLabel;
import javax.swing.JMenu;
import javax.swing.JMenuBar;
import javax.swing.JMenuItem;
import javax.swing.JPanel;

@SuppressWarnings("serial")
```

```java
public class AnimationAndThread extends JFrame implements ActionListener, Runnable, KeyListener{
    JMenuBar bar;
    JMenu animationMenu,musicMenu,colorMenu;
    JMenuItem stopMoveItem,continueMoveItem,exitItem,playItem,loopItem,stopItem,colorItem;
    ImagePanel imgPanel;
    Image images[];
    int index = 0,flag;
    boolean boo = true;
    int x = 20,y = 20,dx,dy;
    Thread animation;
    AudioClip music;

    //程序的入口方法
    public static void main(String[] args) {
        new AnimationAndThread();
    }

    //构造方法,完成窗口的初始化
    public AnimationAndThread(){
        setTitle("使用 Thread 类实现图像动画演示");

        //创建图像数组,并加载图像
        Toolkit tool = Toolkit.getDefaultToolkit();
        images = new Image[10];
        for(int i = 0;i < 10;i++){
            images[i] = tool.getImage("image/T" + i + ".gif");
        }

        //分别创建菜单条、菜单、菜单项,并给菜单项注册 ActionEvent 事件监听器
        bar = new JMenuBar();
        setJMenuBar(bar);
        animationMenu = new JMenu("动画控制");
        musicMenu = new JMenu("音乐播放控制");
        colorMenu = new JMenu("背景颜色设置");
        bar.add(animationMenu);
        bar.add(musicMenu);
        bar.add(colorMenu);
        stopMoveItem = new JMenuItem("终止动画");
        stopMoveItem.addActionListener(this);
        stopMoveItem.setBackground(Color.white);
        continueMoveItem = new JMenuItem("继续动画");
        this.continueMoveItem.addActionListener(this);
        this.continueMoveItem.setBackground(Color.white);
        this.exitItem = new JMenuItem("退出");
        this.exitItem.addActionListener(this);
        this.exitItem.setBackground(Color.white);
        this.animationMenu.add(this.stopMoveItem);
        this.animationMenu.add(this.continueMoveItem);
        this.animationMenu.add(this.exitItem);
        this.playItem = new JMenuItem("播放");
```

```java
            this.playItem.addActionListener(this);
            this.playItem.setBackground(Color.white);
            this.loopItem = new JMenuItem("循环");
            this.loopItem.addActionListener(this);
            this.loopItem.setBackground(Color.white);
            this.stopItem = new JMenuItem("停止");
            this.stopItem.addActionListener(this);
            this.stopItem.setBackground(Color.white);
            this.musicMenu.add(this.playItem);
            this.musicMenu.add(this.loopItem);
            this.musicMenu.add(this.stopItem);
            this.colorItem = new JMenuItem("颜色选择");
            this.colorItem.setBackground(Color.white);
            this.colorItem.addActionListener(this);
            this.colorMenu.add(colorItem);

            //创建图像面板和标签对象,并把它们添加到窗口中
            //设置窗口背景色为白色
            JLabel messageLabel = new JLabel("请按方向键←、→、↑、↓控制图像的移动方向",
JLabel.CENTER);
            imgPanel = new ImagePanel();
            add(imgPanel,BorderLayout.CENTER);
            add(messageLabel,BorderLayout.SOUTH);
            this.getContentPane().setBackground(Color.white);

            //设置窗口的一些常见属性
            addKeyListener(this);
            setBounds(100,100,800,600);
            setVisible(true);
            this.setDefaultCloseOperation(JFrame.EXIT_ON_CLOSE);

            //新建线程,并启动线程
            animation = new Thread(this);
            animation.start();

            //创建音频对象music,实现背景音乐的播放
            File musicFile = new File("wav/Jingle Bell Rock.wav");
            URL url;
            try {
                url = musicFile.toURI().toURL();
                music = Applet.newAudioClip(url);
            } catch (MalformedURLException e) {
                e.printStackTrace();
            }
        }

        //接口AcitonListener中的方法,负责响应用户单击菜单项的行为
        public void actionPerformed(ActionEvent e) {
            //如果单击"终止动画",则结束动画线程
            if(e.getSource() == this.stopMoveItem){
                boo = false;
```

```java
        }
        //如果单击"继续动画",则重新给 animation 线程分配实体,重新启动线程
        else if(e.getSource() == this.continueMoveItem){
            //如果 animation 线程死亡,则执行下面的代码
            if(!animation.isAlive()){
                animation = new Thread(this);
                boo = true;
                animation.start();
            }
        }
        //如果单击"退出"菜单项,则退出应用程序
        else if(e.getSource() == this.exitItem ){
            boo = false;
            System.exit(0);
        }
        //如果单击"播放"菜单项,则播放背景音乐,播放一次
        else if(e.getSource() == this.playItem ){
            music.play();
        }
        //如果单击"循环"菜单项,则循环播放背景音乐
        else if(e.getSource() == this.loopItem ){
            music.loop();
        }
        //如果单击"停止"菜单项,则停止播放背景音乐
        else if(e.getSource() == this.stopItem ){
            music.stop();
        }
        //如果单击"选择颜色"菜单项,则弹出颜色选择器,并根据用户选择的颜色设置窗口的背
        //景色
        else if(e.getSource() == this.colorItem){
            Color newColor = JColorChooser.showDialog(this, "背景色选择", this.getBackground());
            if(newColor!= null){
                this.getContentPane().setBackground(newColor);
            }
        }
    }

    // 重写 run()方法,定义线程体,实现动画的制作
    public void run() {
        while(true){
            //设置当前显示图像在主窗口中的坐标
            x = x + dx;
            y = y + dy;

            //改变图像数组下标,并重绘图像
            flag = index % 10;
            repaint();
            index = index + 1;

            //如果图像运动到窗口的左边界或右边界,则让图像向右或向左水平运动
```

```java
            if(x < 0){
                dx = 10;
            }
            else if((x + images[flag].getWidth(this)) > getWidth()){
                dx = -10;
            }
            //如果图像运动到窗口的上边界或下边界,则让图像向下或向上垂直运动
            if(y < 0){
                dy = 10;
            }
            else if((y + images[flag].getHeight(this)) > getHeight()){
                dy = -10;
            }
            //设置动画的每次间隔时间为 0.3 秒
            try {
                Thread.sleep(300);
            } catch (InterruptedException e) {
                e.printStackTrace();
            }

            //如果 boo 变量值为 false,则结束动画线程
            if(!boo){
                return;
            }
        }
    }

    //定义内部类——图像面板,显示(绘制)图像
    public class ImagePanel extends JPanel{
        public void paintComponent(Graphics g){
            super.paintComponents(g);
            g.drawImage(images[flag],x,y,this);
        }
    }

    //当用户按下键盘上的某个键时,监听器就自动调用 keyPressed 方法
    //设置图像运动的方式,即图像每次坐标变换的偏移量及运动方向
    public void keyPressed(KeyEvent e) {
        //KeyEvent 类的 public int getKeyCode()方法,判断哪个键被按下、敲击或释放
        //如果用户按下向上方向键,则垂直向上运动
        if(e.getKeyCode() == KeyEvent.VK_UP){
            dx = 0;
            dy = -10;
        }
        //如果用户按下向下方向键,则垂直向下运动
        else if(e.getKeyCode() == KeyEvent.VK_DOWN ){
            dx = 0;
            dy = 10;
        }
        //如果用户按下向左方向键,则水平向左运动
        else if(e.getKeyCode() == KeyEvent.VK_LEFT ){
```

```
                dx = -10;
                dy = 0;
            }
            //如果用户按下向右方向键,则水平向右运动
            else if(e.getKeyCode() == KeyEvent.VK_RIGHT){
                dx = 10;
                dy = 0;
            }
        }
        //当用户释放键盘上的某个键时,监听器自动调用 keyReleased 方法
        public void keyReleased(KeyEvent e) {}
        //当键被按下又释放时,keyTyped 方法被调用
        public void keyTyped(KeyEvent e) {}
    }
```

### 12.3.5 案例练习题目

(1) 请读者自行修改 AnimationAndThread.java,使得图像可以以其他方式(改变运动轨迹)在主窗口中运动,同时考虑 x、y 的坐标如何改动,即 dx、dy 如何设置。

(2) 请给菜单"音乐播放控制"中添加"打开"菜单项,一旦用户单击该命令,可以弹出打开文件对话框,用户可以自己灵活选择要播放的背景音乐。(建议:使用 JFileChooser 类)

(3) 请使用 Thread 类实现文字在窗口来回滚动显示。

(4) 请使用 Thread 类实现小球在主窗口中运动的动画效果。

(5) 请使用 Thread 类实现幻灯片放映功能。

(6) 请使用 Thread 的子类实现本案例的功能。

## 12.4 案例:线程同步——模拟跑步接力

### 12.4.1 案例问题描述

设计一个 GUI 程序,模拟 4×100 米跑步接力。在应用程序中,除了主线程外,还有 4 个线程:first、second、third、fourth。first 负责模拟一个红色按钮从坐标(0,60)运动到(100,60);second 负责模拟一个绿色按钮从坐标(100,60)运动到(200,60);third 负责模拟一个蓝色按钮从坐标(200,60)运动到(300,60);fourth 负责模拟一个黄色按钮从坐标(300,60)运动到(400,60)。用户按下字母键"S",则开始 4×100 米接力;如果用户按下字母键"Q",则退出应用程序。

通过此案例,希望读者能学习并掌握有关线程同步的知识。

### 12.4.2 案例功能分析与演示

应用程序执行效果如图 12-16 所示。

从图 12-16 可以看到,窗口中依次排列着 4 个按钮(红、绿、蓝、黄),一旦用户按下 S 键,

就开始模拟 4×100 米的跑步接力,红色按钮先跑,跑到 100 米处,绿色按钮接着跑 100 米,再接下来是蓝色按钮跑,最后是黄色按钮跑完最后 100 米。这 4 个按钮会一直演示这个 4×100 米的跑步接力过程。如果用户按下"Q"键,则会退出该应用程序。

图 12-16　线程同步(跑步接力)

### 12.4.3　案例总体设计

从问题描述和运行效果可以看到,需要设计一个 GUI 类(RelayRaceWindow 类)。该类是 JFrame 类的子类,实现了接口 KeyListener 和接口 Runnable,通过重写接口中的方法来实现不同的功能:

- 重写 KeyListener 接口中的方法 KeyPressed(KeyEvent),负责响应用户按下 S 键和按下 Q 键时的行为。当用户按下 S 键时,开始启动 4 个线程,即接力演示开始。当用户按下 Q 键时,则退出应用系统。
- 重写 Runnable 接口中的方法 run(),负责定义线程的线程执行体,即通过控制 4 个按钮来模拟 4×100 米跑步接力。

在重写 run()时,需要注意下面的问题:

在包含多个线程的应用程序中,线程间有时会共享存储空间。当两个或多个线程同时访问同一个共享资源时,必然会出现冲突问题。如一个线程可能尝试从一个文件中读取数据,而另一个线程则尝试在同一个文件中修改数据。在这种情况下,数据可能会变得不一致。这时需要做的是让一个线程彻底完成其任务后,再允许下一个线程执行,必须保证一个共享资源一次只能被一个线程使用,实现此目的的过程就是同步。只有同步还是不够的,线程之间还必须有能够告之对方已经完成工作等的消息,这就是多线程间的协作,也是多线程完成任务时必须考虑的问题。

在本例中,由 4 个线程分别模拟 4 个按钮共同跑完全程,这里需要设计一个 int 型变量 distance 表示全程,其初始值为 0,跑到终点值为 400。这个 distance 变量 4 个线程都要使用到它,这里必须考虑线程同步的问题。在跑步接力的过程中必须保证在同一时刻只有一个队员(按钮)在跑步,其他队员(按钮)必须处于等待状态。当这个队员跑完了,它要通知其他队员准备跑,它自己放弃 CPU 资源的使用,处于等待状态。本案例设计了一个同步方法 move(JButton)来实现这样的同步控制。另外,在设计 move(JButton)方法时,需要注意的是必须让按钮按顺序地接力,即第一个跑完了,接下来是第二个跑,再是第三个,第四个,而不会出现无序的乱跑。

图 12-17 是 RelayRaceWindow 类的 UML 图。

以下是 UML 图中有关数据和方法的详细说明。

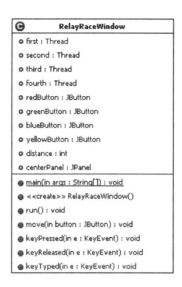

图 12-17　RelayRaceWindow 类的 UML 图

### 1. 成员变量

- first、second、third、fourth 是 Thread 对象，这 4 个线程分别控制着 redButton 按钮、greenButton 按钮、blueButton 按钮和 yellowButton 按钮的运动。
- redButton、greenButton、blueButton、yellowButton 是 JButton 类对象，分别模拟 4×100 米接力的 4 个成员。
- distance 是 int 型变量，用来表示按钮（即队员）已经跑了多远。其初始值为 0，终值为 400。
- centerPanel 是 JPanel 类对象，4 个按钮对象被添加到这个面板中。

### 2. 成员方法

- main(String[])方法，应用程序的入口方法。
- RelayRaceWindow()构造方法，负责初始化窗口。
- run()方法，Runnable 接口中的方法，用来定义线程的线程执行体，这里用来分别定义 4 个线程在 4×100 米中要完成的任务。
- move(JButton)方法，负责让参数指定的 JButton 类对象移动。
- keyPressed(KeyEvent)方法是 RelayRaceWindow 类实现的 KeyListener 接口中的方法，负责响应用户的按键行为。当用户按下某个键时，触发 KeyEvent 事件，监听器自动调用该方法响应用户操作。这里通过用户按键来控制是否开始接力以及是否退出应用程序。
- keyReleased(KeyEvent)方法，同样是 KeyListener 接口中的方法，当用户释放某键时，自动执行该方法。
- keyTyped(KeyEvent)方法，同样是 KeyListener 接口中的方法，当用户敲击一次键盘，触发 KeyEvent 事件，自动调用该方法响应用户操作。

## 12.4.4 案例代码实现

RelayRaceWindow 类是 javax.swing 包中 JFrame 类的子类,实现了接口 KeyListener 和接口 Runnable,负责 GUI 界面的实现和模拟跑步接力。

首先需要进行 GUI 界面设计。RelayRaceWindow 窗口对象的布局是 BorderLayout 型,在"Center"位置添加一个 JPanel 类型的面板,设置该 JPanel 类对象的布局为 null,往该对象中按顺序添加 4 个不同颜色的按钮;在"South"位置添加一个标签对象 message,用它来停止用户如何开始演示跑步接力和如何退出应用程序。

接下来的工作是定义线程体,让它来控制 4 个按钮,模拟 $4 \times 100$ 米跑步接力。在具体实现的过程中,设计了一个同步方法 move(JButton)来控制 JButton 按钮的移动,并使用了 wait() 和 notifyAll() 方法来协助工作的完成。

GUI 界面设计的工作较简单,线程的创建及其状态的变迁前面案例也叙述过了,下面重点来了解下面线程的同步知识。

### 1. 线程的同步与资源共享

在多线程的程序中,当多个线程并发执行时,虽然各个线程中语句的执行顺序是确定的,但线程的相对执行顺序是不确定的。有些情况下,这种因多线程并发执行而引起的执行顺序的不确定性是无害的,不影响程序运行的结果。但在有些情况下如多线程对共享数据操作时,这种线程运行顺序的不确定性将会产生执行结果的不确定性,使共享数据的一致性被破坏,因此在某些应用程序中必须对线程的并发操作进行控制。

1) 线程的同步与 synchronized 关键字

为了确保可以在线程之间以受控方式共享数据,Java 语言提供了两个关键字:synchronized 和 volatile。

synchronized 有两个重要的含义:它确保了一次只有一个线程可以执行代码的受保护部分,而且它确保了一个线程更改的数据对于其他线程是可见的。

一个程序的各个并发线程中对同一个对象进行访问的代码段,称为临界区(Critical Sections)。在 Java 语言中,临界区可以是一个语句块或是一个方法,并且用 synchronized 关键字标识。

synchronized 关键字可以实现多个线程对临界区的互斥访问;为所有对临界区进行访问的方法加上 synchronized 关键字,这样就可以使得多个线程对临界区进行互斥访问。

另外,synchronized 关键字还可以用于线程的同步,因为它不仅可以确保同时只有一个线程访问对象的 synchronized 方法,而且还确保了一个线程更改的数据对其他所有线程是可见的。

synchronized 关键字保证了数据操作的原子性,否则,数据很容易处于不一致状态。例如,一个线程在对一个结构体的各个域进行修改,但当修改了一部分的时候,另一个线程对该结构进行读访问,那么该结构体中的数据就处于错误状态。而 synchronized 关键字保证了数据操作的原始性;当一个线程对数据的操作没有完成时,其他线程均不能访问数据。

volatile 比 synchronized 更简单,但只适合于控制对基本变量(整数、布尔变量等)的单个实例的访问。当一个变量被声明成 volatile,任何对该变量的写操作都会绕过高速缓存,直接写入主内存,而任何对该变量的读取也都绕过高速缓存,直接取自主内存。这表示所有线程在任何时候看到的 volatile 变量值都相同。如果没有正确的同步,线程可能会看到旧的变量值,或者引起其他形式的数据损坏。

volatile 对于确保每个线程看到最新的变量值非常有用,但如果需要保护比较大的代码片段,如涉及更新多个变量的片段,这时需要应用监听器或锁的概念,以协调对特定代码块的访问。

每个 Java 对象都有一个相关的锁,但是同一时间只能有一个线程持有 Java 锁。当线程进入 synchronized 代码块时,线程会阻塞并等待,直到锁可用,当它可用时,就会获得这个锁,然后执行代码块。当线程退出受保护的代码块时,它就会释放该锁。这样,每次只有一个线程可用执行受保护的代码块。从其他线程的角度看,该代码块可以看作是原子的,它要么全部执行,要么根本不执行。

可以使用 synchronized 块来将一组相关更新作为一个集合来执行,而不必担心其他线程中断或看到计算的中间结果。

2) 线程的等待与唤醒

当线程在继续执行前需要等待一个条件,仅有 synchronized 关键字是不够的。虽然 synchronized 关键字阻止并发访问一个对象,但它没有实现线程之间的通信。为此,Object 提供了 wait()、notify() 和 notifyAll() 方法。

获得对象锁的线程可以通过调用该对象的 wait() 方法主动释放锁,等待在该对象的线程等待队列上,此时其他线程可以得到锁从而访问该对象,之后可以通过调用 notify() 或 notifyAll() 方法来唤醒先前因调用 wait() 方法而等待的线程。一般情况下,对于 wait()、notify() 和 notifyAll() 方法的调用都是根据一定的条件来进行的。

当线程即将结束 synchronized 方法时,则调用 notify 方法或 notifyAll 方法使处于等待状态的线程进入就绪状态,以便使这些线程参与临界区的竞争。

notifyAll() 方法和 notify() 方法的区别是:notifyAll() 唤醒所有由于使用这个同步方法而处于等待的线程结束等待,曾中断的线程就会从刚才中断处继续执行这个同步方法,并遵循"先中断先继续"的原则;notify() 方法,只是通知处于等待中的某一个线程结束等待。

本模拟跑步接力的案例中,每个线程负责模拟一个按钮跑一段路程,并且必须是有序的。这就可以在同步方法 move(JButton) 中设定一定的条件,决定它们什么时候处于等待状态,什么时候轮到它们跑了。

**2. 代码(RelayRaceWindow.java)**

```
package caida.xinxi.TestThread4;

import java.awt.BorderLayout;
import java.awt.Color;
import java.awt.Font;
import java.awt.event.KeyEvent;
import java.awt.event.KeyListener;

import javax.swing.JButton;
import javax.swing.JFrame;
```

```java
import javax.swing.JLabel;
import javax.swing.JPanel;

@SuppressWarnings("serial")
public class RelayRaceWindow extends JFrame implements Runnable,KeyListener{
    Thread first,second,third,fourth;
    JButton redButton,greenButton,blueButton,yellowButton;
    int distance;
    JPanel centerPanel;

    //main方法,程序的入口方法
    public static void main(String[] args) {
        new RelayRaceWindow();
    }

    //构造方法,负责初始化窗口,并创建线程
    public RelayRaceWindow(){
        setTitle("线程同步——4×100米接力");
        //分别创建4个线程,处于新建状态
        first = new Thread(this);
        second = new Thread(this);
        third = new Thread(this);
        fourth = new Thread(this);
        //创建面板对象和标签对象,分别添加到窗口中
        centerPanel = new JPanel();
        JLabel message = new JLabel("按'S'键开始接力,按'Q'键退出程序",JLabel.CENTER);
        message.setFont(new Font("宋体",Font.BOLD,15));
        add(centerPanel,BorderLayout.CENTER );
        add(message,BorderLayout.SOUTH );
        //分别创建4个按钮,设置其背景色、大小、位置,并添加到面板中
        redButton = new JButton();
        redButton.setBackground(Color.red);
        greenButton = new JButton();
        greenButton.setBackground(Color.green);
        blueButton = new JButton();
        blueButton.setBackground(Color.blue);
        yellowButton = new JButton();
        yellowButton.setBackground(Color.yellow );
        centerPanel.setLayout(null);
        centerPanel.setBackground(Color.white );
        centerPanel.add(redButton);
        redButton.setBounds(0,60,15,15);
        centerPanel.add(greenButton);
        greenButton.setBounds(100,60,15,15);
        centerPanel.add(blueButton);
        blueButton.setBounds(200,60,15,15);
        centerPanel.add(yellowButton);
        yellowButton.setBounds(300,60,15,15);

        //让窗口获得焦点,要不然窗口无法响应按键事件
        this.setFocusable(true);
```

```java
        //给窗口注册 KeyEvent 事件监听器
        addKeyListener(this);
        //设置窗口的一些常见属性
        setBounds(100,100,430,200);
        setVisible(true);
        this.setDefaultCloseOperation(JFrame.EXIT_ON_CLOSE );
        validate();
    }

    //定义线程体,实现 4 个线程模拟 4 个按钮跑步接力的过程
    public void run() {
        while(true){
            //如果当前占用 CPU 资源的线程是 first 线程,则移动 redButton 按钮
            if(Thread.currentThread() == first){
                move(redButton);
                try {
                    //红色按钮每移动一次暂停 0.03 秒
                    Thread.sleep(30);
                } catch (InterruptedException e) {
                    e.printStackTrace();
                }
            }
            //如果当前占用 CPU 资源的线程是 second 线程,则移动 greenButton 按钮
            if(Thread.currentThread() == second){
                move(greenButton);
                try {
                    //绿色按钮每移动一次暂停 0.06 秒
                    Thread.sleep(60);
                } catch (InterruptedException e) {
                    e.printStackTrace();
                }
            }
            //如果当前占用 CPU 资源的线程是 third,则负责移动 blueButton 按钮
            if(Thread.currentThread() == third){
                move(blueButton);
                try {
                    //蓝色按钮每移动一次暂停 0.05 秒
                    Thread.sleep(50);
                } catch (InterruptedException e){
                    e.printStackTrace();
                }
            }
            //如果当前占用 CPU 资源的线程是 fourth 线程,则移动黄色按钮
            if(Thread.currentThread() == fourth){
                move(yellowButton);
                try {
                    //黄色按钮每移动一次,暂停 0.03 秒
                    Thread.sleep(30);
                } catch (InterruptedException e) {
                    e.printStackTrace();
                }
```

```java
            }
        }
    }

    //同步方法,实现接力
    public synchronized void move(JButton button) {
        if(Thread.currentThread() == first){
            //如果不是first线程该跑的路程,线程必须等待,直到轮到它跑
            while(!(distance>=0&&distance<=100)){
                try {
                    wait();
                } catch (InterruptedException e){
                    e.printStackTrace();
                }
            }
            //如果轮到first线程跑,则设置其跑步步长
            distance = distance + 1;
            button.setLocation(distance, 60);
            // 如果first线程跑完它负责的100米,则回到它的初始处
            //并通知其他处于等待状态的线程
            if(distance>=100){
                button.setLocation(0,60);
                //通知等待的线程
                notifyAll();
            }
        }
        if(Thread.currentThread() == second){
            while(!(distance>=100&&distance<=200)){
                try {
                    wait();
                } catch (InterruptedException e) {
                    e.printStackTrace();
                }
            }
            distance = distance + 1;
            button.setLocation(distance, 60);
            if(distance>=200){
                button.setLocation(100,60);
                notifyAll();
            }
        }
        if(Thread.currentThread() == third){
            while(!(distance>=200&&distance<=300)){
                try {
                    wait();
                } catch (InterruptedException e) {
                    e.printStackTrace();
                }
            }
            distance = distance + 1;
            button.setLocation(distance, 60);
```

```java
            if(distance >= 300){
                button.setLocation(200,60);
                notifyAll();
            }
        }
        if(Thread.currentThread() == fourth){
            while(!(distance >= 300&&distance <= 400)){
                try {
                    wait();
                } catch (InterruptedException e) {
                    e.printStackTrace();
                }
            }
            distance = distance + 1;
            button.setLocation(distance, 60);
            if(distance >= 400){
                distance = 0;
                button.setLocation(300,60);
                notifyAll();
            }
        }
    }

    //重写接口 KeyListener 中的方法,响应用户的按键行为
    public void keyPressed(KeyEvent e) {
        //当用户按下 S 键,则启动线程,开始演示跑步接力
        if(e.getKeyCode() == KeyEvent.VK_S){
            first.start();
            second.start();
            third.start();
            fourth.start();
        }
        //当用户按下 Q 键,则退出应用程序
        else if(e.getKeyCode() == KeyEvent.VK_Q){
            System.exit(0);
        }
    }
    public void keyReleased(KeyEvent e) {
    }
    public void keyTyped(KeyEvent e) {
    }
}
```

## 12.4.5 案例练习题目

(1) 请读者删掉 RelayRaceWindow.java 代码中的 public synchronized void move (JButton button)方法的修饰词 synchronized,多次调试程序观察运行结果,解释其中的原因,理解关键字 synchronized 的作用。

(2) 如果要把 public synchronized void move(JButton button)方法中 4 个 if 分支中的

代码 while()语句改成 if 语句,可以吗?会出现什么现象,为什么?

(3) 由于是使用了同步方法,程序中的各线程保持了同步,现在请不要使用同步方法,而使用同步语句块来保持线程间的同步。

(4) 请设计程序,模拟两个队伍参加 4×100 米的接力竞赛,确定优胜者。

(5) 请设计程序模拟一个银行账户存款的过程。

(6) 请设计程序模拟火车多窗口售票的过程。

(7) 请设计程序模拟生产者-消费者的例子。

# 第13章 综合案例：拼图游戏

## 13.1 设计要求

拼图游戏是一款非常经典的智力小游戏。它简单有趣，无论男女老少都喜欢玩。拼图的设计对每一位 Java 程序设计者进行语言提高和进阶都是一个非常好的锻炼机会。通过本章的综合案例，读者可以用到前面学习过的 Swing 图形界面设计、Java I/O 流、Java 异常处理及多媒体程序设计、多线程设计、集合等知识。

本游戏的具体要求如下：

(1) 在开始游戏之前，用户可以按照默认的游戏等级、默认的图像开始游戏，也可以通过菜单选择游戏的等级和拼图图像。

(2) 拼图游戏分为两个等级：普通级别、高级级别。对于普通级别游戏，将一幅图像分成 3×3 幅小图像，将最后一幅小图像(图像的右下角)移到游戏界面的副窗口，将其余各幅小图像打乱顺序后放在拼图面板的按钮上，最终目标是通过移动方块恢复原始图像。同理，对于高级级别游戏，是将一幅图像分成 4×4 幅小图像，后面的要求同上。

(3) 用鼠标单击任何与空格子水平或垂直相邻的方块就可以把该方块移入到空格子，而当前方块移动之前所在的格子成为空格子。通过不断地移动方块可以将方块一行一行地按图像本来的顺序拼好。

(4) 当用户按要求拼好方块后，弹出对话框，提示用户成功的消息，并自动保存用户的成绩。用户单击菜单上相应级别的游戏排行榜，可以看到自己的成绩。

(5) 用户还可以单击菜单上的播放背景音乐命令，弹出音乐控制器，播放背景音乐。

(6) 拼图游戏提供三幅图像，用户可以使用这些图像来玩拼图游戏。用户也可以使用菜单选择本地的一幅新图像，然后使用这个新的图像来玩拼图游戏。

(7) 游戏的简单规则在游戏界面的"帮助"菜单内的"游戏说明"信息框内已经简单介绍了。游戏前可以先预览看看。

拼图游戏的运行效果如图 13-1 所示。

图 13-1 拼图游戏运行效果

## 13.2 总体设计

在设计拼图游戏时，用到了 8 个 Java 源文件：Cell.java、ControlGamePanel.java、GameWindow.java、MusicDialog.java、Player.java、PuzzlePanel.java、ResultRecordDialog.java、SplitImage.java。

拼图游戏所用到的一些重要的类及其之间的组合关系如图 13-2 所示。

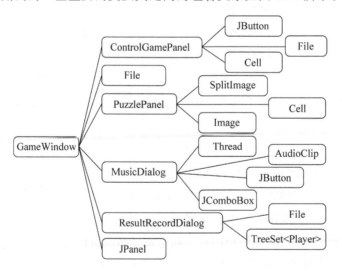

图 13-2 类之间的组合关系

以下是 8 个源文件的总体设计。

### 13.2.1　GameWindow.java

GameWindow 类负责创建拼图游戏的主窗口,该类含有 main 方法,拼图游戏从该类开始执行。GameWindow 类主要负责创建游戏的菜单,并响应用户对菜单的操作。整个窗口由 4 个面板组成:"North"方向是显示玩家信息的面板,是 JPanel 类对象;"Center"方向是拼图界面,是 PuzzlePanel 类对象;"East"方向是图像面板,显示被移去的右下角的小图像,是 JPanel 类对象;"South"方向是控制游戏的功能面板,是 ControlGamePanel 类对象。除此以外 GameWindow 类还有两个重要的对象,分别是 MusicDialog 类对象,用来控制背景音乐播放;ResultRecordDialog 类对象,用来显示玩家游戏排行榜的信息。

GameWindow 类的主要成员的作用将在后面的详细设计中阐述。

### 13.2.2　PuzzlePanel.java

PuzzlePanel 类是主类 GameWindow 窗口中的一个 JPanel 容器,所创建的对象叫作"拼图面板",这个"拼图面板"被添加到 GameWindow 窗口的中心。PuzzlePanel 类主要负责根据游戏的等级和玩家所选图像初始化拼图面板,并将图像分割成 row×column 个小图像,再在拼图面板去添加 row×column 个带小图像的单元格;另外可以判断拼图游戏什么时候成功结束。PuzzlePanel 类的主要成员的作用将在后面的详细设计中阐述。

### 13.2.3　Cell.java

Cell 类是 JButton 的一个子类,创建的对象是 PuzzlePanel 类的重要成员之一,用来表示拼图面板区中的单元格。通过该对象可以设置和获取单元格的位置和所显示的图像。

### 13.2.4　ControlGamePanel.java

ControlGamePanel 类是 JPanel 的一个子类,创建的对象是 GameWindow 类的重要成员之一,用来控制游戏的运行。通过该控制面板,可以开始新游戏、在拼图的过程中预览全图、保存未完成的游戏状态、提取游戏继续以前未完成的游戏以及结束当前游戏等功能。该类的主要成员的作用将在后面的详细设计中阐述。

### 13.2.5　SplitImage.java

SplitImage 类所创建的对象负责将一幅图像分割成若干个更小的小图像,并将小图像保存到一个 Image 类型的二维数组中。

### 13.2.6　MusicDialog.java

MusicDialog 类的对象是一个对话框,负责控制拼图游戏背景音乐的播放。它是利用线程技术实现的。MusicDialog 类具体的成员和实现请看后面的详细设计。

### 13.2.7 Player.java

Player 对象用来封装玩家的姓名和成绩,以便 ResultRecordDialog 对象可以按成绩的高低排序玩家。

### 13.2.8 ResultRecordDialog.java

ResultRecordDialog 类是 javax.swing 包中 JDialog 对话框的子类,当用户单击有关查看排行榜的菜单项时,会弹出对话框,显示目前所有成功完成游戏的玩家的成绩排行。

## 13.3 详细设计

### 13.3.1 GameWindow 类

**1. 效果图**

GameWindow 创建的窗口效果如图 13-3 所示,菜单条上有"游戏"和"帮助"两个菜单。从图 13-3 可以看到默认级别是"普通级别 3×3",默认图像是"QQ 图片"。

图 13-3　GameWindow 创建的窗口

在"选择图像"级联菜单下,可在游戏提供的三个备选图像之间进行选择,如果选择"花图片",效果可见图 13-4。从图 13-4 中可以看到整个游戏主窗口由 4 个部分组成(messagePanel、imagePanel、puzzlePanel、controlPanel)。如果选择"从本地选择图片",可打开"打开图像文件"对话框(见图 13-5),用户可以从本地选一幅自己喜欢的图片进行拼图游戏。

如果用户对游戏的规则不熟悉,可以单击"帮助"菜单下的"游戏说明"命令,程序会弹出消息对话框(见图 13-6),简单介绍游戏的规则。

图 13-4 选择"花图片"后的游戏初始效果

图 13-5 打开图像文件对话框

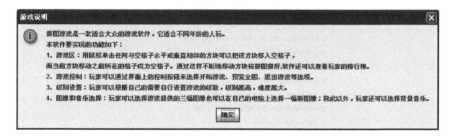

图 13-6 "游戏说明"对话框

2．UML 图

GameWindow 类是 javax.swing 包中的 JFrame 的一个子类，并实现了 ActionListener 接口。该类的主要成员变量和方法如图 13-7 所示。

以下是 UML 图中有关数据和方法的详细说明。

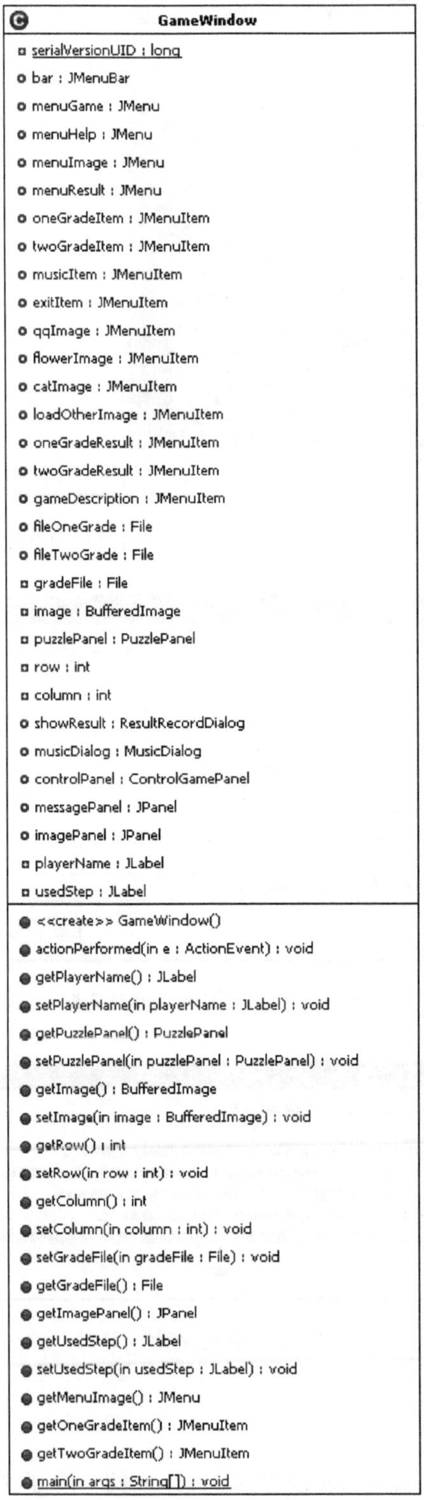

```
┌─────────────────────────────────────────────────┐
│ Ⓒ              GameWindow                       │
├─────────────────────────────────────────────────┤
│ □ serialVersionUID : long                       │
│ ○ bar : JMenuBar                                │
│ ○ menuGame : JMenu                              │
│ ○ menuHelp : JMenu                              │
│ ○ menuImage : JMenu                             │
│ ○ menuResult : JMenu                            │
│ ○ oneGradeItem : JMenuItem                      │
│ ○ twoGradeItem : JMenuItem                      │
│ ○ musicItem : JMenuItem                         │
│ ○ exitItem : JMenuItem                          │
│ ○ qqImage : JMenuItem                           │
│ ○ flowerImage : JMenuItem                       │
│ ○ catImage : JMenuItem                          │
│ ○ loadOtherImage : JMenuItem                    │
│ ○ oneGradeResult : JMenuItem                    │
│ ○ twoGradeResult : JMenuItem                    │
│ ○ gameDescription : JMenuItem                   │
│ ○ fileOneGrade : File                           │
│ ○ fileTwoGrade : File                           │
│ □ gradeFile : File                              │
│ □ image : BufferedImage                         │
│ □ puzzlePanel : PuzzlePanel                     │
│ □ row : int                                     │
│ □ column : int                                  │
│ ○ showResult : ResultRecordDialog               │
│ ○ musicDialog : MusicDialog                     │
│ ○ controlPanel : ControlGamePanel               │
│ ○ messagePanel : JPanel                         │
│ ○ imagePanel : JPanel                           │
│ □ playerName : JLabel                           │
│ □ usedStep : JLabel                             │
├─────────────────────────────────────────────────┤
│ ● <<create>> GameWindow()                       │
│ ● actionPerformed(in e : ActionEvent) : void    │
│ ● getPlayerName() : JLabel                      │
│ ● setPlayerName(in playerName : JLabel) : void  │
│ ● getPuzzlePanel() : PuzzlePanel                │
│ ● setPuzzlePanel(in puzzlePanel : PuzzlePanel) : void │
│ ● getImage() : BufferedImage                    │
│ ● setImage(in image : BufferedImage) : void     │
│ ● getRow() : int                                │
│ ● setRow(in row : int) : void                   │
│ ● getColumn() : int                             │
│ ● setColumn(in column : int) : void             │
│ ● setGradeFile(in gradeFile : File) : void      │
│ ● getGradeFile() : File                         │
│ ● getImagePanel() : JPanel                      │
│ ● getUsedStep() : JLabel                        │
│ ● setUsedStep(in usedStep : JLabel) : void      │
│ ● getMenuImage() : JMenu                        │
│ ● getOneGradeItem() : JMenuItem                 │
│ ● getTwoGradeItem() : JMenuItem                 │
│ ● main(in args : String[]) : void               │
└─────────────────────────────────────────────────┘
```

图 13-7　GameWindow 类的 UML 图

1) 成员变量
- bar 是 JMenuBar 类创建的菜单条，可以向 bar 中添加菜单。
- menuGame、menuHelp、menuImage、menuResult 是 JMenu 类创建的 4 个菜单。这 4 个菜单的名称分别是"游戏"、"帮助"、"选择图像"和"查看排行榜"。
- oneGradeItem、twoGradeItem、musicItem、exitItem 是 JMenuItem 类创建的 4 个菜单项，被添加到 menuGame 菜单中。这 4 个菜单项的名称分别是"普通级别 3×3"、"高级级别 4×4"、"背景音乐播放控制"和"退出"。
- qqImage、catImage、flowerImage、loadOtherImage 是 JMenuItem 类创建的 4 个菜单项，被添加到 menuImage 中。这 4 个菜单项的名称分别是"QQ 图片"、"猫图片"、"花图片"和"从本地选择图片"。
- oneGradeResult、twoGradeResult、gameDescription 是 JMenuItem 类创建的 3 个菜单项。gameDescription 被添加到 menuHelp 中，其余两个被添加到 menuResult 中。这 3 个菜单项的名称分别是："普通级别排行"、"高级级别排行"和"游戏说明"。
- fileOneGrade、fileTwoGrade、gradeFile 是 File 类创建的三个文件对象。fileOneGrade、fileTwoGrade 两个文件对象所引用的文件分别是"普通级别 游戏排行榜.txt"和"高级级别 游戏排行榜.txt"，分别用来写入两个级别的成绩。gradeFile 可以是 fileOneGrade 和 fileTwoGrade 中的某一个。
- image 是 BufferedImage 对象用来表示加载的图像。
- puzzlePanel 是 PuzzlePanel 类对象，表示拼图界面，放置在主窗口的中心位置。
- row 和 column 是 int 型数据，其值用来确定拼图区的方块数。
- showResult 是 ResultRecordDialog 对象，用来显示成绩的对话框。
- musicDialog 是 MusicDialog 对象，负责显示播放背景音乐的对话框。
- controlPanel 是 ControlGamePanel 对象，负责控制游戏。
- messagePanel 和 imagePanel 分别是 JPanel 类对象。messagePanel 面板放置在主窗口的"North"位置，显示玩家姓名和玩家拼图所走过的步数；imagePanel 面板放置在主窗口的"East"位置，用来显示被移去的小图像。
- playerName 和 usedStep 是 JLabel 对象。playerName 的文本内容用来显示当前玩家姓名；usedStep 的文本内容用来显示玩家所走过的步数。

2) 成员方法
- GameWindow() 是构造方法，负责完成窗口的初始化。
- actionPerformed(ActionEvent) 方法是 GameWindow 类实现的 ActionListener 接口中的方法，负责执行菜单项发出的有关命令。用户选择菜单中的菜单项可以触发 ActionEvent 事件，导致 actionPerformed(ActionEvent) 方法执行相应的操作。
- main(String[]) 方法是软件运行的入口方法。
- getPlayerName() 和 setPlayerName() 方法分别用来获得和设置 playerName。
- 同理其余方法也都是 GameWindow 类的重要属性的 setter 方法和 getter 方法。

### 3. 代码（GameWindow.java）

```java
package caida.xinxi.jigsaw;

import java.awt.BorderLayout;
import java.awt.Color;
import java.awt.Dimension;
import java.awt.FlowLayout;
import java.awt.event.ActionEvent;
import java.awt.event.ActionListener;
import java.awt.image.BufferedImage;
import java.io.File;
import java.io.IOException;

import javax.imageio.ImageIO;
import javax.swing.ButtonGroup;
import javax.swing.JFileChooser;
import javax.swing.JFrame;
import javax.swing.JLabel;
import javax.swing.JMenu;
import javax.swing.JMenuBar;
import javax.swing.JMenuItem;
import javax.swing.JOptionPane;
import javax.swing.JPanel;
import javax.swing.JRadioButtonMenuItem;
import javax.swing.border.EtchedBorder;
import javax.swing.filechooser.FileNameExtensionFilter;
public class GameWindow extends JFrame implements ActionListener{
    private static final long serialVersionUID = 1L;
    JMenuBar bar;
    JMenu menuGame,menuHelp;
    JMenu menuImage;
    JMenu menuResult;
    JMenuItem oneGradeItem,twoGradeItem;
    JMenuItem musicItem,exitItem;
    JMenuItem qqImage,flowerImage,catImage,loadOtherImage;
    JMenuItem oneGradeResult,twoGradeResult;
    JMenuItem gameDescription;
    File fileOneGrade,fileTwoGrade;
    private File gradeFile;
    private BufferedImage image;
    private PuzzlePanel puzzlePanel;
    private int row = 3,column = 3;           //默认级别是初级,3×3的拼图游戏
    ResultRecordDialog showResult;            //显示游戏结果(排行榜)对话框
    MusicDialog musicDialog;                  //播放背景音乐对话框
    ControlGamePanel controlPanel;
    JPanel messagePanel,imagePanel;
    private JLabel playerName,usedStep;
```

```java
//GameWindow类构造方法,初始化窗口
public GameWindow(){
    setTitle("拼图小游戏");

    /** 设计游戏主窗口界面菜单条,有"游戏"和"帮助"两个菜单。"游戏"菜单可帮助选择图
    像、设置游戏级别,选择背景音乐,查看排行榜及退出游戏;"帮助"菜单可弹出消息窗口,显示游戏运
    行规则 */
    bar = new JMenuBar();
    setJMenuBar(bar);
    menuGame = new JMenu("游戏");
    menuHelp = new JMenu("帮助");
    bar.add(menuGame);
    bar.add(menuHelp);
    /** 设计"选择图像"的级联菜单项,游戏自带三张可选图像,默认选项为"QQ"图片,玩家也
    可单击"从本地选择图片"从本机选择图片开始游戏 */
    menuImage = new JMenu("选择图像");
    this.qqImage = new JRadioButtonMenuItem("QQ 图片",true);
    this.qqImage.addActionListener(this);
    this.flowerImage = new JRadioButtonMenuItem("花图片");
    this.flowerImage.addActionListener(this);
    this.catImage = new JRadioButtonMenuItem("猫图片");
    this.catImage.addActionListener(this);
    this.loadOtherImage = new JMenuItem("从本地选择图片");
    this.loadOtherImage.addActionListener(this);
    //单选按钮菜单并不会自动实现互斥特性,下面具体实现互斥特性
    ButtonGroup group1 = new ButtonGroup();
    group1.add(qqImage);
    group1.add(flowerImage);
    group1.add(catImage);
    menuImage.add(qqImage);
    menuImage.add(flowerImage);
    menuImage.add(catImage);
    menuImage.addSeparator();
    menuImage.add(loadOtherImage);
    //设置游戏难度菜单项,默认为 3×3 的普通级别
    oneGradeItem = new JRadioButtonMenuItem("普通级别 3×3",true);
    this.oneGradeItem.addActionListener(this);
    twoGradeItem = new JRadioButtonMenuItem("高级级别 4×4",false);
    this.twoGradeItem.addActionListener(this);
    ButtonGroup group2 = new ButtonGroup();
    group2.add(oneGradeItem);
    group2.add(twoGradeItem);
    //控制背景音乐播放菜单设置
    musicItem = new JMenuItem("背景音乐播放控制");
    this.musicItem.addActionListener(this);
    //查看排行榜级联菜单设计
    menuResult = new JMenu("查看排行榜");
    this.oneGradeResult = new JMenuItem("普通级别排行");
    this.oneGradeResult.addActionListener(this);
    this.twoGradeResult = new JMenuItem("高级级别排行");
    this.twoGradeResult.addActionListener(this);
```

```java
            this.menuResult.add(this.oneGradeResult);
            this.menuResult.add(this.twoGradeResult);
            //退出应用程序菜单设计
            exitItem = new JMenuItem("退出");
            this.exitItem.addActionListener(this);
            //给"游戏"菜单添加各菜单项
            menuGame.add(this.menuImage);
            menuGame.addSeparator();
            menuGame.add(this.oneGradeItem);
            menuGame.add(this.twoGradeItem);
            menuGame.addSeparator();
            menuGame.add(this.musicItem);
            menuGame.addSeparator();
            menuGame.add(this.menuResult);
            menuGame.addSeparator();
            menuGame.add(this.exitItem);
            this.gameDescription = new JMenuItem("游戏说明");
            this.gameDescription.addActionListener(this);
            this.menuHelp.add(this.gameDescription);

            //创建用于保存玩家成绩的文件及文件的引用
            fileOneGrade = new File("普通级别 游戏排行榜.txt");
            fileTwoGrade = new File("高级级别 游戏排行榜.txt");
            //设置普通级别游戏排行榜为默认选项
            setGradeFile(fileOneGrade);
            //如果保存排行榜的文件不存在,则创建它
            if(!fileOneGrade.exists()){
                try {
                    fileOneGrade.createNewFile();
                } catch (IOException e) {
                    e.printStackTrace();
                }
            }
            if(!fileTwoGrade.exists()){
                try {
                    fileTwoGrade.createNewFile();
                } catch (IOException e) {
                    e.printStackTrace();
                }
            }

            showResult = new ResultRecordDialog();
            musicDialog = new MusicDialog();

            //设置初始默认图像为QQ图像
            try {
                image = ImageIO.read(new File("image/qq.jpg"));
            } catch (IOException e1) {
            }

            //游戏区,上部——显示游戏信息的面板
```

```java
        messagePanel = new JPanel();
        messagePanel.setLayout(new FlowLayout(FlowLayout.CENTER));
        messagePanel.add(new JLabel("当前玩家："));
        playerName = new JLabel(" ");
        //JLabel 对象默认是透明的，必须先取消其透明度，才可以设置颜色
        playerName.setOpaque(true);
        playerName.setBackground(Color.yellow );
        messagePanel.add(playerName);
        messagePanel.add(new JLabel("您完成游戏所走过的步数："));
        usedStep = new JLabel("0");
        usedStep.setOpaque(true);
        usedStep.setBackground(Color.yellow );
        messagePanel.add(usedStep);
        //创建拼图游戏主界面，并按默认设置初始化界面
        puzzlePanel = new PuzzlePanel(this);
        puzzlePanel.initPanel();
        //创建图像面板，显示（放置）分割后的图片最后一个（右下角的）小图像
        imagePanel = new JPanel();
        imagePanel.setBackground(Color.white );
        imagePanel.setBorder(new EtchedBorder());
        imagePanel.setPreferredSize(new Dimension(190, Cell.HEIGHT * row + 120));
        //创建控制面板，用于控制游戏
        controlPanel = new ControlGamePanel(this);
        // 游戏主窗口中添加面板
        add(messagePanel, BorderLayout.NORTH );
        add(puzzlePanel, BorderLayout.CENTER );
        add(imagePanel, BorderLayout.EAST );
        add(controlPanel, BorderLayout.SOUTH );
        //设置游戏主窗口的常规属性
        setLocation(50,50);
        this.setSize(new Dimension(Cell.WIDTH * column + 200,Cell.HEIGHT * row + 128));
        this.setDefaultCloseOperation(JFrame.EXIT_ON_CLOSE );
        this.setVisible(true);
        this.setResizable(false);
        validate();
    }

    //实现 ActionListener 接口中的方法，负责响应用户单击菜单的行为
    public void actionPerformed(ActionEvent e) {
        if(e.getSource() == this.qqImage ){
            //设置拼图图像为 QQ，并初始化拼图界面
            try {
                image = ImageIO.read(new File("image/qq.jpg"));
                puzzlePanel.initPanel();
            } catch (IOException e1) {
                e1.printStackTrace();
            }
        }
        else if(e.getSource() == this.flowerImage ){
            //设置拼图图像为 flower，并初始化拼图界面
            try {
```

```java
                    image = ImageIO.read(new File("image/flower.jpg"));
                    puzzlePanel.initPanel();
                } catch (IOException e1) {
                    e1.printStackTrace();
                }
            }
            else if(e.getSource() == this.catImage ){
                //设置拼图图像为cat,并初始化拼图界面
                try {
                    image = ImageIO.read(new File("image/cat.jpg"));
                    puzzlePanel.initPanel();
                } catch (IOException e1) {
                    e1.printStackTrace();
                }
            }
            // 响应用户单击"从本地选择图片"菜单项的行为
            else if(e.getSource() == this.loadOtherImage ){
                //创建图像文件选择器
                JFileChooser chooser = new JFileChooser();
                //图片过滤器
                FileNameExtensionFilter filter = new FileNameExtensionFilter("jpg&jpeg&gif&png images","jpg","gif","png","jpeg");
                //设置文件过滤器
                chooser.setFileFilter(filter);
                //设置该文件选择器只能选择文件
                chooser.setFileSelectionMode(JFileChooser.FILES_ONLY);
                chooser.setDialogTitle("打开图像文件");
                int result = chooser.showOpenDialog(this);
                //得到目录下的指定类型的文件
                File file = chooser.getSelectedFile();
                if(file!= null&&result == JFileChooser.APPROVE_OPTION){
                    //加载图像,将新图像设置为拼图对象,并初始化拼图界面
                    try {
                        image = ImageIO.read(file);
                        puzzlePanel.initPanel();
                    } catch (IOException e1) {
                        e1.printStackTrace();
                    }
                }
            }
            //响应用户单击"普通级别"的操作
            else if(e.getSource() == this.oneGradeItem ){
                row = 3;
                column = 3;
                setGradeFile(fileOneGrade);
                this.setSize(new Dimension(Cell.WIDTH * column + 200,Cell.HEIGHT * row + 128));
                puzzlePanel.initPanel();
            }
            //响应用户单击"高级级别"的操作
            else if(e.getSource() == this.twoGradeItem ){
                row = 4;
```

```java
            column = 4;
            setGradeFile(fileTwoGrade);
            this.setSize(new Dimension(Cell.WIDTH * column + 200,Cell.HEIGHT * row + 128));
            puzzlePanel.initPanel();
        }
        //响应用户设置背景音乐的操作
        else if(e.getSource() == this.musicItem ){
            musicDialog.setVisible(true);


        }
        //响应用户单击"普通级别排行榜"菜单的行为
        else if(e.getSource() == this.oneGradeResult ){
            showResult.setGradeFile(fileOneGrade);
            showResult.showRecord();
            showResult.setVisible(true);
        }
        //响应用户单击"高级级别排行榜"菜单的行为
        else if(e.getSource() == this.twoGradeResult ){
            showResult.setGradeFile(fileTwoGrade);
            showResult.showRecord();
            showResult.setVisible(true);
        }
        //退出程序
        else if(e.getSource() == this.exitItem ){
            System.exit(0);
        }
        //响应用户单击"游戏说明"菜单的操作,简单显示游戏规则
        else if(e.getSource() == this.gameDescription ){
            JOptionPane.showMessageDialog(null, "拼图游戏是一款适合大众的游戏软件,它适合不同年龄的人玩。" + "\n本软件要实现的功能如下: " + "\n1、游戏区：用鼠标单击任何与空格子水平或垂直相邻的方块可以把该方块移入空格子,\n而当前方块移动之前所在的格子成为空格子。通过这样不断地移动方块将拼图拼好,软件还可以查看玩家的排行榜。" + "\n2、游戏控制：玩家可以通过界面上的控制按钮来选择开始游戏、预览全图、退出游戏等选项。" + "\n3、级别设置：玩家可以根据自己的需要自行设置游戏的级数,级别越高,难度越大。" + "\n4、图像和音乐选择：玩家可以选择游戏提供的三幅图像也可以在自己的电脑中选择一幅新图像；除此以外,玩家还可以选择背景音乐。","游戏说明",JOptionPane.INFORMATION_MESSAGE);
        }
    }

    public JLabel getPlayerName() {
        return playerName;
    }
    public void setPlayerName(JLabel playerName) {
        this.playerName = playerName;
    }

    public PuzzlePanel getPuzzlePanel() {
        return puzzlePanel;
    }
    public void setPuzzlePanel(PuzzlePanel puzzlePanel) {
        this.puzzlePanel = puzzlePanel;
```

```java
    }

    public BufferedImage getImage() {
        return image;
    }
    public void setImage(BufferedImage image) {
        this.image = image;
    }

    public int getRow() {
        return row;
    }
    public void setRow(int row) {
        this.row = row;
    }

    public int getColumn() {
        return column;
    }
    public void setColumn(int column) {
        this.column = column;
    }

    public void setGradeFile(File gradeFile) {
        this.gradeFile = gradeFile;
    }
    public File getGradeFile() {
        return gradeFile;
    }

    public JPanel getImagePanel() {
        return imagePanel;
    }

    public JLabel getUsedStep() {
        return usedStep;
    }
    public void setUsedStep(JLabel usedStep) {
        this.usedStep = usedStep;
    }

    public JMenu getMenuImage() {
        return menuImage;
    }

    public JMenuItem getOneGradeItem() {
        return oneGradeItem;
    }

    public JMenuItem getTwoGradeItem() {
        return twoGradeItem;
```

        }
        public static void main(String[] args) {
            new GameWindow();
        }
    }

## 13.3.2 PuzzlePanel 类

### 1. 效果图

PuzzlePanel 类创建的游戏面板区的效果如图 13-8 所示，这是一个高级级别 4×4 的一个测试效果。

图 13-8　PuzzlePanel 创建的拼图面板

### 2. UML 图

PuzzlePanel 类是 swing 包中 JPanel 容器的子类，所创建的对象 puzzlePanel 是 GameWindow 类中的重要成员之一，作为一个容器添加到 GameWindow 窗口的中心位置。表明 PuzzlePanel 类的主要成员变量和方法如图 13-9 所示。

以下是 UML 图中有关数据和方法的详细说明。

1）成员变量

- cells 是 Cell 类型的二维数组，数组元素的个数是 row 和 column 的乘积，它的每一个单元存放着一个 Cell 对象。PuzzlePanel 对象在调用 initPanel() 方法时完成对 cells 单元的初始化，即创建有图像和位置坐标的单元格对象。
- gameWin 是 GameWindow 类对象，表示游戏主窗口。
- image 是 BufferedImage 对象，表示当前拼图面板所选图像。
- row 和 column 是 int 型数据，表示拼图面板区单元格的行数和列数。

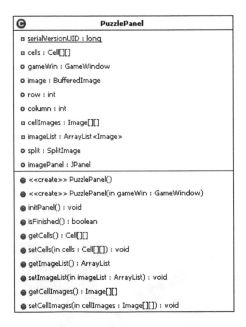

图 13-9　PuzzlePanel 类的 UML 图

- cellImage 是 Image 类型的二维数组，保存图像分割后的所有小图像。元素个数是 row × column。
- split 是 SplitImage 对象，是图像分割器。
- imagePanel 是 JPanel 容器对象，面板区显示被移走的小图像。

2）成员方法

- PuzzlePanel()是构造方法。
- PuzzlePanel(GameWindow)是构造方法。
- initPanel()方法负责初始化游戏面板，确定游戏选定的图像和级别，负责分割图像，并将分割后的小图像保存到二维图像数组和数组链表中；还负责创建带图像的单元格，并添加到拼图面板中。
- isFinished()方法负责判断拼图游戏是否成功完成。
- getCells()方法和 setCells(Cell[][])方法分别用来获取和设置保存拼图面板区中所有单元格的数组对象。
- getImageList()方法和 setImageList(ArrayList(Image))方法分别用来获取和设置存放拼图面板区中所有图像的链表对象。
- getCellImage()方法和 setCellImages(Image[][])方法分别用来获取和设置保存所有分割后的小图像的图像数组。

3．代码（puzzlePanel.java）

```
package caida.xinxi.jigsaw;

import java.awt.Image;
```

```java
import java.awt.Point;
import java.awt.image.BufferedImage;
import java.util.ArrayList;
import javax.swing.JPanel;
public class PuzzlePanel extends JPanel {
    private static final long serialVersionUID = 1L;
    private Cell[][] cells;
    GameWindow gameWin;
    BufferedImage image;
    int row,column;
    private Image[][] cellImages;         //所有分割成的小图片都保存在这个二维图像数组里
    private ArrayList<Image> imageList;
                                          //把除了最后一个小图像的所有图像保存到该集合中
    SplitImage split;
    JPanel imagePanel;

    //PuzzlePanel 类构造方法
    public PuzzlePanel(){}
    public PuzzlePanel(GameWindow gameWin){
        this.gameWin = gameWin;
        imagePanel = gameWin.getImagePanel();
        setLayout(null);
    }

    //此方法可以初始化游戏主面板,以全图显示图片
    public void initPanel() {
        //每次初始化前,要把以前的内容给清除掉
        removeAll();
        image = gameWin.getImage();
        row = gameWin.getRow();
        column = gameWin.getColumn();
        //设置游戏面板的大小
        setSize(Cell.WIDTH * column,Cell.HEIGHT * row);
        cells = new Cell[row][column];
        //获得图片分割器 split
        SplitImage split = new SplitImage();
        //分割图片,并保存到图像数组中
        cellImages = split.getImages(image,row,column);
        //创建图像链表
        imageList = new ArrayList<Image>();
        //将分割后的小图像放入数组链表中.(除了最后一幅小图像,它不参与拼图)
        for(int i = 0;i < row;i++){
            for(int j = 0;j < column;j++){
                imageList.add(cellImages[i][j]);
            }
        }
        imageList.remove(row * column - 1);

        //在游戏面板中添加 row×column 个有图像的单元格
        for(int i = 0;i < row;i++){
            for(int j = 0;j < column;j++){
```

/** 创建单元格对象,单元格上有图片,有 Point 对象标记——用它来标记单元格的位置坐标,以后可用此 Point 对象来判断当前玩家单击单元格与空图像单元格是否水平相邻或垂直相邻 */

```java
                cells[i][j] = new Cell(cellImages[i][j],new Point(i,j));
                cells[i][j].setBorder(null);
                add(cells[i][j]);
                cells[i][j].setLocation(j * Cell.WIDTH , i * Cell.HEIGHT );
            }
        }
    }

    //判断拼图游戏是否正确完成的方法
    public boolean isFinished(){
        boolean boo = true;
        mark:for(int i = 0;i < row;i++){
            if(i < row - 1){
                for(int j = 0;j < column;j++){
                    if(!(cells[i][j].getButtonImage() == cellImages[i][j])){
                        boo = false;
                        break mark;
                    }
                }
            }
            else{
                for(int j = 0;j < column - 1;j++){
                    if(!(cells[i][j].getButtonImage() == cellImages[i][j])){
                        boo = false;
                        break mark;
                    }
                }
            }
        }
        return boo;
    }

    public Cell[][] getCells() {
        return cells;
    }
    public void setCells(Cell[][] cells) {
        this.cells = cells;
    }
    public ArrayList<Image> getImageList() {
        return imageList;
    }
    public void setImageList(ArrayList<Image> imageList) {
        this.imageList = imageList;
    }
    public Image[][] getCellImages() {
        return cellImages;
    }
    public void setCellImages(Image[][] cellImages) {
```

```
            this.cellImages = cellImages;
    }
}
```

### 13.3.3 Cell 类

**1. 效果图**

Cell 创建的对象效果如图 13-10 所示。

图 13-10　Cell 创建的对象效果

**2. UML 图**

Cell 类是 javax.swing 包中 JButton 类的一个子类，创建的对象是二维数组 cells 中的元素。Cell 对象是 PuzzlePanel 类的重要成员之一。标明 Cell 类的主要成员及方法的 UML 图如图 13-11 所示。

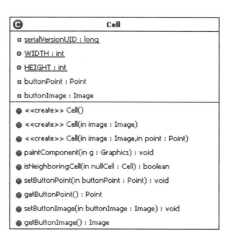

图 13-11　Cell 类创建的 UML 图

以下是 UML 图中有关数据和方法的详细说明。

1）成员变量
- WIDTH 是 int 型常量，表示 Cell 类对象（即单元格）的宽。
- HEIGHT 是 int 型常量，表示 Cell 类对象的高。
- buttonPoint 是 Point 类对象，表示 Cell 对象在 PuzzlePanel 游戏面板中的位置。
- buttonImage 是 Image 类对象，表示 Cell 对象上所显示的图像。

2）成员方法
- Cell()、Cell(Image) 和 Cell(Image,Point) 是 Cell 类的构造方法。

- paintComponent(Graphics)方法重写 JComponent 中的方法,重绘图像。
- isNeighboringCell(Cell)方法判断当前单元格与无图像(空)单元格是否水平相邻或垂直相邻。
- setButtonPoint(Point)方法和 getButtonPoint()方法分别用来设置和获取单元格在游戏面板中的位置。
- setButtonImage(Image)方法和 getButtonImage()方法分别用来设置和获取单元格上所显示的图像。

### 3. 代码(Cell.java)

```java
package caida.xinxi.jigsaw;

import java.awt.Color;
import java.awt.Graphics;
import java.awt.Image;
import java.awt.Point;
import javax.swing.JButton;

//单元格类
public class Cell extends JButton{
    private static final long serialVersionUID = 1L;
    public static final int WIDTH = 120;
    public static final int HEIGHT = 120;
    private Point buttonPoint;
    private Image buttonImage;

    //Cell 类构造方法
    public Cell(){
        super();
    }
    public Cell(Image image){
        buttonImage = image;
        setSize(WIDTH,HEIGHT);
        setBackground(Color.blue);
        repaint();
    }
    public Cell(Image image,Point point){
        this(image);
        buttonPoint = point;
    }

    public void paintComponent(Graphics g){
        super.paintComponent(g);
        g.drawImage(buttonImage,0,0,WIDTH,HEIGHT,this);
    }

    //该方法判断当前单元格与无图像单元格是否水平相邻或垂直相邻
    public boolean isNeighboringCell(Cell nullCell){
```

```
            Point currentCellPoint = this.getButtonPoint();
            Point nullCellPoint = nullCell.getButtonPoint();
            boolean condition1 = Math.abs(currentCellPoint.getX() - nullCellPoint.getX()) ==
1&&currentCellPoint.getY() == nullCellPoint.getY();
            boolean condition2 = Math.abs(currentCellPoint.getY() - nullCellPoint.getY()) ==
1&&currentCellPoint.getX() == nullCellPoint.getX();
            if(condition1||condition2){
                return true;
            }
            else
                return false;
        }

        public void setButtonPoint(Point buttonPoint) {
            this.buttonPoint = buttonPoint;
        }
        public Point getButtonPoint() {
            return buttonPoint;
        }
        public void setButtonImage(Image buttonImage) {
            this.buttonImage = buttonImage;
        }
        public Image getButtonImage() {
            return buttonImage;
        }
}
```

### 13.3.4 ControlGamePanel 类

**1. 效果图**

ControlGamePanel 类创建的对象效果如图 13-12 所示。

图 13-12 ControlGamePanel 创建的对象

一旦玩家单击"开始新游戏"后,会弹出登录对话框,要求输入玩家的姓名;输入玩家姓名后,拼图面板区打乱图形显示,要求玩家完成拼图;在游戏主窗口的右侧面板显示右下角被移去的图像,如图 13-13 所示。

在如图 13-13 所示效果的基础上,玩家可按照游戏规则移动单元格,如果在移动过程中需要预览原图,可单击"预览全图"按钮,即可预览原图;若想返回拼图界面,则单击"返回"按钮,即可实现,具体效果可见图 13-14。

如果玩家成功拼好图像,则会弹出消息对话框,并将玩家成绩自动保存(见图 13-15);玩家单击"结束当前游戏"按钮则可结束此次游戏,玩家可开始新游戏。

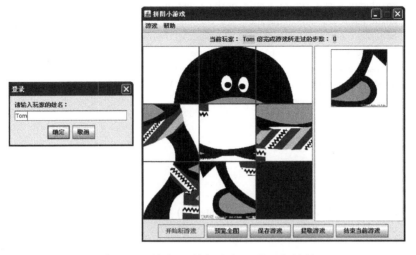

图 13-13 单击"开始新游戏"后的运行效果

图 13-14 "预览全图"按钮效果

图 13-15 成功完成游戏效果

2. UML 图

ControlGamePanel 类是 javax.swing 包中 JPanel 容器的一个子类，并实现了 ActionListener 接口，创建的对象 controlPanel 是 GameWindow 类的重要成员之一，负责控制游戏。ControlGamePanel 类的主要成员及方法如图 13-16 所示。

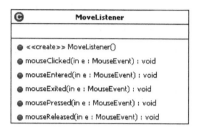

图 13-16　ControlGamePanel 类的 UML 图

以下是 UML 图中有关数据和方法的详细说明。

1）成员变量

- gameWin 是控制面板所在的 GameWindow 类窗口对象。
- buttonStart、buttonPreview、buttonSave、buttonDraw 和 buttonExit 是 JButton 类对象，用来控制游戏，分别给它们注册了动作监听器，响应玩家的操作。按钮上的文本分别是"开始新游戏"、"预览全图"、"保存游戏"、"提取游戏"、"结束当前游戏"。
- puzzlePanel 是 PuzzelPanel 类对象。
- row、column 是 int 型数据，表示拼图区的单元格的行数和列数。
- gradeFile 是 File 类对象，将玩家成绩写入 gradeFile 指向的文件。
- imageList 是 ArrayList<Image> 链表对象，其中存储拼图区中的所有图像。
- cells 是 Cell 类型的二维数组，用来存放拼图区的所有单元格。
- lastCell 是 Cell 类型的按钮，被添加到图像面板 imagePanel 中。
- noImageCell 是 Cell 类型的单元格，表示拼图区中的空单元格。

- imagePanel 是 JPanel 类对象，被添加到 gameWin 窗口中，显示被移去的单元格。
- cellImages 是 Image 类型的二维数组，用来保存最原始的图像分割后的所有小图像。
- imagesState 是 Image 类型的二维数组，用来保存拼图游戏某一过程中的图像对象。
- playerName 和 usedStep 是 JLabel 对象，它们的文本内容分别表示玩家姓名和玩家到目前为止所走过的步数。
- step 是 int 型数据，表示玩家玩拼图游戏所走过的步数。
- menuImage 是 JMenu 对象，其名称为"选择图像"；oneGradeItem 和 twoGradeItem 是 JMenuItem 对象，其名称分别为"普通级别 3×3"和"高级级别 4×4"。

2) 成员方法
- ControlGamePanel(GameWindow)是构造方法。
- actionPerformed(ActionEvent)方法是 ControlGamePanel 类实现的 ActionListener 接口中的方法，负责响应用户单击按钮的行为，实现程序控制游戏的功能。

3) 内部类 MoveListener
- MoveListener()是构造方法。
- 其余的 5 个方法是实现 MouseListener 接口中的方法。其中，mousePressed(MouseEvent)方法负责实现单元格与相邻空单元格的交换；mouseReleased(MouseEvent)方法负责实现当玩家拼图成功时，弹出消息框，自动保存玩家成绩的功能。

3. 代码（ControlGamePanel.java）

```
package caida.xinxi.jigsaw;

import java.awt.Dimension;
import java.awt.Image;
import java.awt.event.ActionEvent;
import java.awt.event.ActionListener;
import java.awt.event.MouseEvent;
import java.awt.event.MouseListener;
import java.io.File;
import java.io.RandomAccessFile;
import java.util.ArrayList;
import java.util.Collections;
import javax.swing.JButton;
import javax.swing.JLabel;
import javax.swing.JMenu;
import javax.swing.JMenuItem;
import javax.swing.JOptionPane;
import javax.swing.JPanel;
public class ControlGamePanel extends JPanel implements ActionListener {
    private static final long serialVersionUID = 1L;
    GameWindow gameWin;
    JButton buttonStart,buttonPreview,buttonSave,buttonDraw,buttonExit;
    PuzzlePanel puzzlePanel;
```

```java
int row,column;
File gradeFile;
ArrayList<Image> imageList;
Cell[][] cells;
Cell lastCell,noImageCell;
JPanel imagePanel;
Image[][] cellImages;
Image[][] imagesState;
JLabel playerName,usedStep;
int step;
JMenu menuImage;
JMenuItem oneGradeItem,twoGradeItem;

//ControlGamePanel 构造方法
public ControlGamePanel(GameWindow gameWin){
    this.gameWin = gameWin;
    gradeFile = gameWin.getGradeFile();
    //创建控制游戏的一系列控制按钮,并添加到面板中,给按钮添加动作监听器
    buttonStart = new JButton("开始新游戏");
    buttonPreview = new JButton("预览全图");
    buttonSave = new JButton("保存游戏");
    buttonDraw = new JButton("提取游戏");
    buttonExit = new JButton("结束当前游戏");
    this.buttonStart.addActionListener(this);
    this.buttonPreview.addActionListener(this);
    this.buttonSave.addActionListener(this);
    this.buttonDraw.addActionListener(this);
    this.buttonExit.addActionListener(this);
    add(this.buttonStart);
    add(this.buttonPreview );
    add(this.buttonSave );
    add(this.buttonDraw );
    add(this.buttonExit );

    imagePanel = gameWin.getImagePanel();
    usedStep = gameWin.getUsedStep();
    step = Integer.parseInt(usedStep.getText());
    menuImage = gameWin.getMenuImage();
    oneGradeItem = gameWin.getOneGradeItem();
    twoGradeItem = gameWin.getTwoGradeItem();
}

//实现 ActionListener 接口中的方法,负责响应用户单击按钮的操作
public void actionPerformed(ActionEvent e) {
    //响应用户单击"开始新游戏"按钮的操作
    if(e.getSource() == buttonStart){
        String name = JOptionPane.showInputDialog(null,"请输入玩家的姓名：","登录",JOptionPane.PLAIN_MESSAGE);
        playerName = gameWin.getPlayerName();
        playerName.setText(name);
        //设置"开始新游戏"按钮、"选择图像"及选择游戏级别的菜单项为不可用
```

```java
            buttonStart.setEnabled(false);
            menuImage.setEnabled(false);
            oneGradeItem.setEnabled(false);
            twoGradeItem.setEnabled(false);
            //获取当前拼图面板的行数和列数
            row = gameWin.getRow();
            column = gameWin.getColumn();
            puzzlePanel = gameWin.getPuzzlePanel();
            cells = puzzlePanel.getCells();
            cellImages = puzzlePanel.getCellImages();
            imageList = puzzlePanel.getImageList();
            /** 用 imagesState 保存游戏当前的图像状态;在"预览全图"中需要利用它来恢复拼
图面板界面 */
            imagesState = new Image[row][column];
            noImageCell = cells[row - 1][column - 1];
            //得到右下角单元格上显示的图像
            Image cellImagePanel = noImageCell.getButtonImage();

            //在右边的 imagePanel 面板上显示右下角的单元格上显示的图像
            lastCell = new Cell(cellImagePanel);
            lastCell.setPreferredSize(new Dimension(Cell.WIDTH ,Cell.HEIGHT ));
            imagePanel.add(lastCell);
            imagePanel.validate();

            noImageCell.setButtonImage(null);
            noImageCell.updateUI();

            Collections.shuffle(imageList);   //打乱图片顺序
            MoveListener l = new MoveListener();
            //给除了右下角的所有按钮重新显示打乱的图片,并给它们注册监听器 l
            int k = 0;
            for(int i = 0;i < row;i++){
                for(int j = 0;j < column;j++){
                    if(i == row - 1&&j == column - 1){
                        break;
                    }
                    cells[i][j].setButtonImage(imageList.get(k));
                    cells[i][j].repaint();
                    cells[i][j].updateUI();
                    cells[i][j].addMouseListener(l);
                    k++;
                }
            }
            //给右下角无图像的单元格注册动作监听器
            noImageCell.addMouseListener(l);
            puzzlePanel.validate();
        }

        //响应用户单击"预览全图"按钮的操作
        else if(e.getSource() == buttonPreview){
            //在拼图面板预览原图(完整的)
```

```java
        if(e.getActionCommand().equals("预览全图")){
            //在预览以前,保存当前拼图界面的状况到 imagesState 中
            for(int i = 0;i < row;i++){
                for(int j = 0;j < column;j++){
                    imagesState[i][j] = cells[i][j].getButtonImage();
                }
            }
            //显示(预览)原图
            for(int i = 0;i < row;i++){
                for(int j = 0;j < column;j++){
                    cells[i][j].setButtonImage(cellImages[i][j]);
                    cells[i][j].repaint();
                    cells[i][j].setBorder(null);
                }
            }
            buttonPreview.setText("返回");
        }
        //预览结束后,返回拼图界面,恢复拼图状态
        else if(e.getActionCommand().equals("返回")){
            for(int i = 0;i < row;i++){
                for(int j = 0;j < column;j++){
                    cells[i][j].setButtonImage(imagesState[i][j]);
                    cells[i][j].repaint();
                    cells[i][j].updateUI();
                }
            }
            buttonPreview.setText("预览全图");
        }
    }

    //响应用户"保存游戏"按钮的操作,保存未完成的游戏,以便下次有时间接着玩
    else if(e.getSource() == buttonSave){
        /** 在此处编写代码实现功能:保存当前未完成游戏状态,供下次接着玩 */
    }

    //响应用户单击"提取游戏"按钮操作,接着玩上次未完成的游戏
    else if(e.getSource() == buttonDraw){
        /** 在此处编程代码实现功能:提取上次未完成的游戏状态,接着玩 */
    }

    // 响应用户单击"结束当前游戏"按钮的操作
    //有两种情况:一种是玩家未完成游戏,不愿意继续玩了;一种是成功完成了
    else if(e.getSource() == buttonExit){
        //重新初始化游戏面板,即回到游戏的最初界面
        puzzlePanel.initPanel();
        imagePanel.remove(lastCell);
        usedStep.setText("0");
        playerName.setText(" ");
        step = 0;
        //设置"开始新游戏"按钮及等级菜单项和图像菜单项为可用
```

```java
                this.buttonStart.setEnabled(true);
                menuImage.setEnabled(true);
                oneGradeItem.setEnabled(true);
                twoGradeItem.setEnabled(true);
            }
        }

        //内部类,负责处理鼠标事件
        public class MoveListener implements MouseListener{
            public MoveListener(){}
            public void mouseClicked(MouseEvent e) {
            }
            public void mouseEntered(MouseEvent)
            }
            public void mouseExited(MouseEvent e) {
            }

            //交换相邻单元格的图像
            public void mousePressed(MouseEvent e) {
                Cell currentCell = (Cell)e.getSource();
                //如果单击的单元格与空单元相邻——水平相邻或垂直相邻,则交换二者图像
                if(currentCell.isNeighboringCell(noImageCell)){
                    Image img = currentCell.getButtonImage();
                    noImageCell.setButtonImage(img);
                    noImageCell.requestFocus();
                    noImageCell = currentCell;
                    noImageCell.setButtonImage(null);
                    usedStep.setText(String.valueOf(++step));
                }
            }

            //当成功完成游戏时,弹出消息窗口,恭喜玩家,并将结果写入排行榜
            public void mouseReleased(MouseEvent e) {
                if(puzzlePanel.isFinished()){
                    //游戏自动将最后一张图补全
                    cells[row-1][column-1].setButtonImage(lastCell.getButtonImage());
                    puzzlePanel.requestFocus();
                    //显示消息对话框
                    JOptionPane.showMessageDialog(null,"恭喜你完成了拼图!玩家成绩将自动保存到排行榜中");
                    //将成绩写入排行榜,即保存玩家成绩
                    try {
                        RandomAccessFile out = new RandomAccessFile(gradeFile,"rw");
                        long length = gradeFile.length();
                        out.seek(length);
                        out.writeUTF(playerName.getText());
                        out.writeInt(step);
                        out.close();
```

```
                } catch (Exception e1) {
                    e1.printStackTrace();
                }
            }
        }
    }
}
```

### 13.3.5　SplitImage 类

**1. 效果图**

SplitImage 创建的对象负责将一幅图像分解成若干个小的图像,没有可显示的效果图。

**2. UML 图**

SplitImage 类是 javax.swing 包中 JComponent 类的一个子类,创建对象 split 负责将一幅图像分解成若干个小的图像,是 PuzzlePanel 类的重要成员之一。SplitImage 类的主要成员及 UML 图如图 13-17 所示。

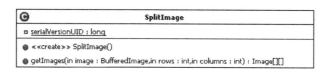

图 13-17　SplitImage 类的 UML 图

以下是 UML 图中有关方法的详细说明。
成员方法：
- SplitImage()方法是构造方法。
- getImages(BufferedImage,int,int)方法将参数指定的 BufferedImage 对象分解成若干个更小的图像,并将这些图像存放到一个 Image 类型的二维数组中,然后返回该数组。

**3. 代码**(SplitImage.java)

```java
package caida.xinxi.jigsaw;

import java.awt.Image;
import java.awt.image.BufferedImage;
import javax.swing.JComponent;
public class SplitImage extends JComponent{
    private static final long serialVersionUID = 1L;
    //构造方法
    public SplitImage(){
        super();
    }
```

```java
    public Image[][] getImages(BufferedImage image,int rows, int columns){
        //图像分割后的小图像放在此二维数组 cellImages 中
        Image[][] cellImages = new Image[rows][columns];
        int imageWidth = image.getWidth();              //获得全图的宽
        int imageHeight = image.getHeight();            //获得全图的高
        int imageCellWidth = imageWidth/columns;        // 获得分割后的图像的宽
        int imageCellHeight = imageHeight/rows;         //获得分割后的图像的高
        for(int i = 0;i<rows;i++){
            for(int j = 0;j<rows;j++){
    cellImages[i][j] = image.getSubimage(j*imageCellWidth,i*imageCellHeight,imageCellWidth,
imageCellHeight);
            }
        }
        return cellImages;
    }
}
```

### 13.3.6　MusicDialog 类

**1. 效果图**

MusicDialog 类创建的对话框的效果如图 13-18 所示。

图 13-18　MusicDialog 类创建的对话框效果

**2. UML 图**

MusicDialog 类是 javax.swing 包中 JDialog 类的子类，实现了 Runnable 接口、ActionListener 接口和 ItemListener 接口。当用户单击"背景音乐播放控制"菜单项时，会弹出该控制窗口。MusicDialog 类的主要成员及 UML 图如图 13-19 所示。

以下是 UML 图中有关数据和方法的详细说明。

1) 成员变量

- musicThread 是 Thread 类对象，该线程负责控制音乐的播放。
- musicSelector 是 JComboBox 类对象，是一个下拉列表对象，负责显示所有可供选择的歌曲清单。
- buttonPlay、buttonLoop、buttonStop 和 buttonExit 是 JButton 类对象，按钮分别用于控制音乐的播放，每个按钮都注册了 ActionEvent 监听器，负责响应用户单击按钮的行为。按钮上的文本依次为"播放"、"循环"、"停止"、"退出"。
- curMusicName 是 String 类对象，表示当前正播放的背景音乐的名字。
- music 是 AudioClip 对象。
- url 是 URL 类对象。
- isLoop 和 isPlay 是 boolean 型变量，分别用来标识当前音乐是否循环播放、当前是

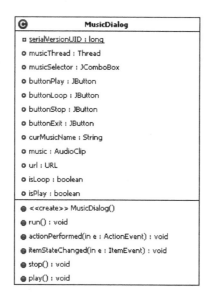

图 13-19　MusicDialog 类的 UML 图

否有音乐在播放。

2) 成员方法

- MusicDialog()是构造方法,初始化音乐对话框。
- run()方法是接口 Runnable 中的方法,是线程体,负责控制音乐的播放。
- actionPerformed(ActionEvent)方法是 MusicDialog 类实现的接口 ActionListener 中的方法,负责响应用户单击按钮的行为。当用户单击按钮,该方法自动调用执行。
- itemStateChanged(ItemEvent)方法是 MusicDialog 类实现的接口 ItemListener 中的方法,当用户在下拉列表中选择某个歌曲选项时,触发 ItemEvent 事件,itemStateChanged(ItemEvent)被自动调用执行。
- stop()方法用于停止播放音乐。
- play()方法用于播放音乐。

3. 代码(MusicDialog.java)

```
package caida.xinxi.jigsaw;

import java.applet.Applet;
import java.applet.AudioClip;
import java.awt.Color;
import java.awt.FlowLayout;
import java.awt.event.ActionEvent;
import java.awt.event.ActionListener;
import java.awt.event.ItemEvent;
import java.awt.event.ItemListener;
import java.io.File;
import java.net.MalformedURLException;
import java.net.URL;
```

```java
import java.util.Vector;
import javax.swing.JButton;
import javax.swing.JComboBox;
import javax.swing.JDialog;
public class MusicDialog extends JDialog implements Runnable, ActionListener, ItemListener {
    private static final long serialVersionUID = 1L;
    Thread musicThread;
    JComboBox musicSelector;
    JButton buttonPlay,buttonLoop,buttonStop,buttonExit;
    String curMusicName;
    AudioClip music;
    URL url;
    boolean isLoop;                                          //标识当前音乐是否循环播放
    boolean isPlay;                                          //标识当前是否有歌曲播放

    //构造方法
    public MusicDialog(){
        setTitle("背景音乐播放窗口");
        musicThread = null;
        isPlay = false;
        isLoop = false;

        //设置音乐列表
        Vector<String> musicCollection = new Vector<String>();
        musicCollection.add("buzaiyouyu.wav");
        musicCollection.add("girlfriend.wav");
        musicCollection.add("inno.wav");
        musicCollection.add("kiss me.wav");
        musicCollection.add("zhandou.wav");
        musicSelector = new JComboBox(musicCollection);
        //也可用下面的方式来设置音乐列表,见注释/**  */中的语句
        /** musicSelector = new JComboBox();
        musicSelector.addItem("girlfriend.wav");
        musicSelector.addItem("inno.wav");
        musicSelector.addItem("kiss me.wav");
        musicSelector.addItem("zhandou.wav");
        musicSelector.addItem("buzaiyouyu.wav"); */
        musicSelector.addItemListener(this);
        musicSelector.setBackground(Color.white);
        musicSelector.setForeground(Color.black );

        //创建控制播放的按钮,并给它们注册动作监听器
        buttonPlay = new JButton("播放");
        buttonLoop = new JButton("循环");
        buttonStop = new JButton("停止");
        buttonExit = new JButton("退出");
        this.buttonPlay.addActionListener(this);
        this.buttonLoop.addActionListener(this);
        this.buttonStop.addActionListener(this);
        this.buttonExit.addActionListener(this);
        //给对话框中依次添加组件
```

```java
        setLayout(new FlowLayout());
        add(musicSelector);
        add(buttonPlay);
        add(buttonLoop);
        add(buttonStop);
        add(buttonExit);
        setBounds(200,5,500,80);
        this.getContentPane().setBackground(Color.pink);
    }

    //用线程来控制音乐的播放——播放一次和循环播放
    public void run() {
        if(!curMusicName.equals(null)){
            File file = new File("music/" + curMusicName);
            try {
                url = file.toURI().toURL();
                music = Applet.newAudioClip(url);
            } catch (MalformedURLException e) {
                e.printStackTrace();
            }
            if(isLoop == false&&isPlay == true){
                music.play();
            }
            else if(isLoop == true&&isPlay == true){
                music.loop();
            }
        }
    }

    //实现接口 ActionListener 中的方法,响应用户单击按钮的行为
    public void actionPerformed(ActionEvent e) {
        if(e.getSource() == buttonPlay){
            isLoop = false;                              //当前播放状态为:播放一次
            play();
        }
        else if(e.getSource() == buttonLoop){
            isLoop = true;                               //播放状态为:循环播放
            play();
        }
        else if(e.getSource() == buttonStop){
            stop();                                      //停止播放背景音乐
        }
        else if(e.getSource() == buttonExit){
            setVisible(false);                           //退出对话框,返回游戏主窗口
        }
    }

    //实现接口 ItemListener 中的方法,响应用户选择不同音乐文件的操作
    public void itemStateChanged(ItemEvent e) {
        if(e.getStateChange() == ItemEvent.SELECTED){
            if(isPlay == true){
```

```
                    stop();              //如果程序当前正播放背景音乐,则停止当前正在播放的音乐
                }
                //获取用户选择的当前要播放的歌曲名称
                curMusicName = (String)musicSelector.getSelectedItem();
            }
        }

        //停止音乐
        public void stop() {
            if(curMusicName!= null){
                buttonPlay.setEnabled(true);
                buttonLoop.setEnabled(true);
                buttonStop.setEnabled(false);
                isPlay = false;                              //设置播放状态为假
                music.stop();                                //停止播放音乐
                //结束播放背景音乐的线程
                if(musicThread!= null){
                    musicThread = null;
                }
            }
        }

        //播放音乐
        public void play() {
            if(curMusicName!= null){
                isPlay = true;                               //设置播放状态为真
                buttonPlay.setEnabled(false);
                buttonLoop.setEnabled(false);
                buttonStop.setEnabled(true);
                //创建控制播放音乐的线程,并启动它
                if(musicThread == null){
                    musicThread = new Thread(this);
                    musicThread.start();
                }
            }
        }
    }
```

## 13.3.7 Player 类

### 1. 效果图

Player 类创建的对象中封装了玩家的姓名和成绩,它被显示在 ResultRecordDialog 对话框中。其效果见 ResultRecordDialog 类的效果图。

### 2. UML 图

Player 对象中的数据由 ResultRecordDialog 对话框从 gradeFile 文件中读取的玩家姓名和成绩所构成。ResultRecordDialog 对话框将 Player 对象作为其 treeSet 树集上的节点,

以便按照成绩由高到低排列 Player 对象。Player 类涉及的主要成员及 UML 图如图 13-20 所示。

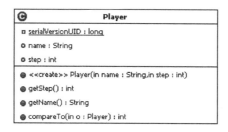

图 13-20　Player 类的 UML 图

以下是 UML 图中有关数据和方法的详细说明。

1) 成员变量
- name 是 String 类型变量,表示玩家姓名。
- step 是 int 型变量,表示玩家完成某次游戏所走过的步数。

2) 成员方法
- Player(String,int)是构造方法。
- getStep()方法返回用户成功完成某次游戏所走过的步数。
- getName()方法返回玩家的姓名。

compareTo(Player)是 Comparable 接口中的方法,其操作是确定 Player 对象的大小关系。

### 3. 代码（Player. java）

```java
package caida.xinxi.jigsaw;

import java.io.Serializable;
public class Player implements Comparable<Player>,Serializable{
    private static final long serialVersionUID = 1L;
    String name;
    int step = 0;
    //构造方法
    public Player(String name, int step) {
        super();
        this.name = name;
        this.step = step;
    }

    public int getStep(){
        return step;
    }
    public String getName(){
        return name;
    }
```

```
        //重写接口中的方法
        public int compareTo(Player o) {
            //如果两个玩家完成游戏所走过的步数相同
            if((this.step - o.step ) == 0){
                return 1;
            }
            else{
                return (this.step - o.step );
            }
        }
    }
```

### 13.3.8　ResultRecordDialog 类

**1. 效果图**

ResultRecordDialog 创建的对话框如图 13-21 所示。

图 13-21　ResultRecordDialog 对话框

**2. UML 图**

ResultRecordDialog 类是 javax.swing 包中 JDialog 的子类，该类创建的对象是 GameWindow 窗口的成员之一。当用户选择窗口上的菜单"查看排行榜"中的某个菜单项时，该对话框可见，并显示相应级别文件中存储的成绩。ResultRecordDialog 类的主要成员和 UML 图如图 13-22 所示。

图 13-22　ResultRecordDialog 类的 UML 图

以下是 UML 图中有关数据和方法的详细说明。

1) 成员变量

- gradeFile 是 File 类对象，是对话框要读取的文件，该文件存储成绩。
- showArea 是 JTextArea 类对象。

- treeSet 是 TreeSet<Player>对象,负责将成绩按由高到低排列。

2) 成员方法
- ResultRecordDialog()是构造方法,初始化对话框。
- setGradeFile(File)方法。ResultRecordDialog 类创建的排行榜对话框是主类 GameWindow 窗口中的一个成员。当用户选择窗口上的"查看排行榜"菜单中的某个菜单项时,ResultRecordDialog 对话框调用 setGradeFile(File)方法将相应的级别传递给 gradeFile。
- showRecord()方法。ResultRecordDialog 对话框调用该方法读取 gradeFile 文件中的成绩,为了将成绩由高到低顺序地显示在 showArea 文本区中,showRecord()方法根据读取的玩家姓名和该玩家的成绩创建一个 Player 对象,并将该对象存放在了 treeSet 树集中。

3. 代码(ResultRecordDialog.java)

```java
package caida.xinxi.jigsaw;

import java.awt.Font;
import java.awt.event.WindowAdapter;
import java.awt.event.WindowEvent;
import java.io.File;
import java.io.RandomAccessFile;
import java.util.Iterator;
import java.util.TreeSet;
import javax.swing.JDialog;
import javax.swing.JScrollPane;
import javax.swing.JTextArea;
public class ResultRecordDialog extends JDialog{
    private static final long serialVersionUID = 1L;
    File gradeFile;
    JTextArea showArea = null;
    TreeSet<Player> treeSet;            //TreeSet 对象中元素按照升序排序

    //构造方法,初始化对话框
    public ResultRecordDialog(){
        treeSet = new TreeSet<Player>();
        showArea = new JTextArea(10,8);
        showArea.setFont(new Font("宋体",Font.BOLD ,18));
        add(new JScrollPane(showArea),"Center");
        setBounds(100,100,300,200);
        setModal(true);
        addWindowListener(new WindowAdapter(){
            public void windowClosing(WindowEvent e){
                setVisible(false);
            }
        }
        );
```

```java
            }

            public void setGradeFile(File file) {
                gradeFile = file;
                setTitle(file.getName());
            }

            public void showRecord() {
                showArea.setText(null);
                treeSet.clear();
                //从指定等级的文件中读取信息,创建 Player 对象,并添加到集合中
                try {
                    RandomAccessFile in = new RandomAccessFile(gradeFile,"rw");
                    long length = in.length();
                    long filePoint = 0;
                    while(filePoint < length){
                        String name = in.readUTF();
                        int step = in.readInt();
                        filePoint = in.getFilePointer();
                        Player player = new Player(name,step);
                        treeSet.add(player);
                    }
                    in.close();
                    //按玩家成绩由高到低在文本区中显示排行榜
                    Iterator<Player> iterator = treeSet.iterator();
                    while(iterator.hasNext()){
                        Player p = iterator.next();
                        showArea.append("玩家:" + p.getName() + ",成绩:" + p.getStep() + "步");
                        showArea.append("\n");
                    }
                } catch (Exception e) {
                    e.printStackTrace();
                }
            }
        }
```

## 13.4 案例练习题目

(1) 改进程序。对本案例提供的拼图游戏测试后,可以发现在"开始新游戏"时,玩家登录窗口无法实现"玩家姓名不能为空"这样的功能,且如果单击登录窗口的"取消"按钮时,"取消"按钮失效,程序仍将开始一个新游戏。本题希望读者能修改程序,避免出现这样的状况。

(2) 在练习(1)的基础上改进程序。要求实现如果玩家姓名重复,则在保存玩家成绩时,选择玩家历史记录中的最高成绩存入文件,在排行榜中显示玩家的最好成绩。

(3) 补全程序。实现本案例程序中 ControlGamePanel 中的"保存游戏"和"提取游戏"的功能。(备注:可以使用 Java I/O 流的知识来保存游戏状态和恢复游戏状态。需要注意

的是，Image类不可以序列化，不能直接用ObjectOutputStream和ObjectInputStream来写入和读出图像)

(4) 修改排行榜。本案例衡量玩家成绩好坏是采用比较玩家所走过的步数进行的。请读者换一种标准，不采用所用的步数，而是玩家成功完成游戏所花费的时间来排行。

(5) 改进排行榜。要求：能同时统计玩家完成游戏所花费的时间和所走过的步数，请读者设计一个综合这两种因素的衡量标准，改进排行榜。

(6) 请读者换一种思路编写程序，实现打乱拼图面板中的单元格的功能。并注意排除游戏出现无解的情况。

(7) 请用你自己的思路重新编写拼图游戏程序。

(8) 编写游戏：俄罗斯方块。

(9) 编写游戏：贪吃蛇。

(10) 编写游戏：龟兔赛跑。

(11) 尝试编写程序解决生活中或学习中遇到的问题或经常处理的事务。

# 第3部分

# 基于Java的综合课程设计

第 14 章　Java 与数据库：资料室图书管理系统
第 15 章　Java 与网络：P2P 聊天系统
第 16 章　Java 与网络：Web 服务器与浏览器
第 17 章　Java 与网络、数据库：基于 B/S 的用户登录管理系统

# 第14章 Java与数据库：资料室图书管理系统

## 14.1 资料室图书管理系统需求分析

资料室图书管理信息系统需实现的功能如下。
（1）图书管理：包括录入、查询、修改和删除图书信息。
（2）借书：包括借阅图书和查看借书记录。
（3）还书：包括还书和查看还书记录。
为了保证系统安全，进入系统之前，应先通过登录验证。

## 14.2 资料室图书管理系统设计

### 14.2.1 数据库设计

根据以上需求分析，设计数据库myBooks，包含4个表：图书信息表（books）、借书记录表（lendRecord）、还书记录表（returnRecord）和用户信息表（users）。表数据结构分别如表14-1～表14-4所示。

表14-1 图书信息表（books）结构

| 字段名 | 数据类型 | 备注 |
| --- | --- | --- |
| id | int | 图书顺序号，主键，自增 |
| no | 字符串 | ISBN号 |
| name | 字符串 | 书名 |
| author | 字符串 | 作者 |
| publisher | 字符串 | 出版社 |
| price | 字符串 | 价格 |
| pubDate | 字符串 | 出版日期 |
| deposit | 字符串 | 存放位置 |
| quantity | int | 数量 |
| lend | int | 借出数量 |
| imgFile | 字符串 | 封面图像文件名 |

表 14-2　借书记录表(lendRecord)结构

| 字段名 | 数据类型 | 备注 |
| --- | --- | --- |
| id | int | 借书记录顺序号,主键,自增 |
| bookId | int | 图书顺序号 |
| borrower | 字符串 | 借书人 |
| borrowerUnit | 字符串 | 借书人所在单位 |
| userId | int | 系统操作用户编号 |
| borrowDate | 字符串 | 借书日期 |
| state | bit | 还书状态,已还:true |

表 14-3　还书记录表(returnRecord)结构

| 字段名 | 数据类型 | 备注 |
| --- | --- | --- |
| id | int | 还书记录顺序号,主键,自增 |
| bookId | int | 图书顺序号 |
| returner | 字符串 | 还书人 |
| returnerUnit | 字符串 | 还书人所在单位 |
| userId | int | 系统操作用户编号 |
| returnDate | 字符串 | 还书日期 |

表 14-4　用户信息表(users)结构

| 字段名 | 数据类型 | 备注 |
| --- | --- | --- |
| id | int | 系统操作用户编号,主键,自增 |
| userName | 字符串 | 用户名 |
| password | 字符串 | 密码 |

## 14.2.2　系统功能设计

根据需求分析,设计系统功能如下。

(1) 登录:显示登录界面,用户输入用户名和密码,单击"登录"按钮后,系统验证用户名和密码是否正确,如正确则进入主界面。

(2) 主界面:显示菜单,菜单结构如图 14-1 所示。

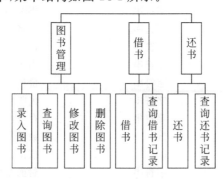

图 14-1　资料室图书管理系统主菜单结构

(3) 图书管理：共 4 项子功能，分别说明如下。

① 录入图书：用户在主界面菜单中选择此功能时，进入录入界面。用户在录入界面中输入各项图书信息后，单击"提交"按钮，系统将采集界面中的图书信息，将其存入数据库的图书信息表中。

② 查询图书：用户在主界面菜单中选择此功能时，进入查询界面。系统为用户提供完全查询和模糊查询功能。用户在查询界面中输入书名、作者和出版社信息（或任意的一部分信息，甚至没有输入任何信息）后，选择查询条件为"完全一致"或"模糊查询"，最后单击"查询"按钮，系统按要求查询图书信息表，查询出图书信息显示于列表中。如用户想查看某一本书的图片，可以在列表中选择图书后单击"查看图片"按钮，系统为用户弹出该书图片。

③ 修改图书：用户在主界面菜单中选择此功能时，进入修改界面一。首先根据用户输入信息完成图书信息的查询（详细请参见(3)图书管理②查询图书），用户在列表中选择需修改的图书后，进入修改界面二。用户在修改界面二中完成图书信息的修改后，单击"修改"按钮，系统采集界面二中的信息后，修改图书信息表中的相应记录。

④ 删除图书：用户在主界面菜单中选择此功能时，进入删除界面。首先根据用户输入信息完成图书信息的查询（详细请参见(3)图书管理②查询图书），用户在列表中选择需删除的图书后，单击"删除"按钮，系统删除图书信息表中的相应记录。

(4) 借书：共两项子功能，分别说明如下。

① 借书：用户在主界面菜单中选择此功能时，进入借阅图书界面。首先根据用户输入信息完成图书信息的查询（详细请参见(3)图书管理②查询图书），用户在列表中选择需借阅的图书后，单击"借书"按钮，系统修改图书信息表中的相应记录（将借出数量加一，如借出数量已经大于该图书的数量，则不加修改，且提示用户该图书已经全部借出），并在借书记录表中登记一条记录（如借书不成功则不予登记）。

② 查询借书记录：用户在主界面菜单中选择此功能时，进入借书记录界面。系统根据用户输入的信息完成借书记录的查询（查询涉及三个表：借书记录表、图书信息表、用户表），如用户在查出的借书记录列表中选择一条借书记录，单击"查看详细信息"按钮，系统为用户查出所借图书的详细信息。

(5) 还书：共两项子功能，分别说明如下。

① 还书：用户在主界面菜单中选择此功能时，进入还书界面。首先根据用户输入信息完成借书记录的查询（详细请参见(4)借书②查询借书记录），用户在列表中选择需还的图书后，单击"还书"按钮，系统修改图书信息表中的相应记录（将借出数量减一），且修改借书记录表中相应记录的"还书状态"为 true，并在还书记录表中登记一条记录。

② 查询还书记录：用户在主界面菜单中选择此功能时，进入还书记录界面。系统根据用户输入信息完成还书记录的查询（涉及三个表：还书记录表、图书信息表、用户表），如用户在查出的还书记录列表中选择一条还书记录，单击"查看详细信息"按钮，系统为用户查出所还图书的详细信息。

## 14.3 资料室图书管理系统实现思路

为避免图形用户界面中对按钮事件处理的方法中代码堆砌太多，造成程序的结构性太差，系统实现时采用清晰的分层结构模型，如图 14-2 所示。

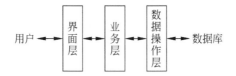

图 14-2　本系统采用的三层结构模型

　　用户通过界面输入信息或单击按钮，界面层通过事件处理控制转向业务层的不同处理方法进行处理，当需要对数据库进行操作时，业务层调用数据操作层的相应数据操作方法完成；之后数据操作层将结果返回业务层继续处理，业务层处理完毕，则再将结果返回到界面层，界面层显示结果，用户由此可进一步在界面上进行操作。

## 14.4　资料室图书管理系统实现

　　在资料室图书管理信息系统中，根据以上设计和实现思路，共写了 8 个类。其中界面层 5 个类：LoginManager 类、BooksManager 类、UpdatePanel 类、QueryPanel 类和 CardPanel 类；业务层 1 个类：Service 类；数据库操作层 1 个类：DataOperator 类；另外还有一个附加类：MD5 类，该类用于添加用户时以及登录时对密码进行加密处理。

　　在界面层中，LoginManager 类用于登录界面；BooksManager 类是主界面；UpdatePanel 类用于"录入"和"修改"图书功能所需界面；QueryPanel 类用于检索图书界面，还用于"修改"或"删除"图书前、"借书"前查询图书信息所需界面，以及查看"借书记录"所需界面、"还书"前查询借书记录所需界面、查看"还书记录"所需界面；CardPanel 类用于"修改"图书时所需卡片布局界面，该界面将 QueryPanel 面板和 UpdatePanel 面板摞起来，首先显示 QueryPanel 面板查询出的图书信息，挑选一本书后，将信息显示在 UpdatePanel 面板上进行修改。

　　在业务层中，Service 类封装了界面层中除主界面之外，各个界面中按钮事件处理时所需调用的业务处理方法，这些方法都是静态的。

　　在数据库操作层中，DataOperator 对象封装了所有对数据库进行操作的方法，这些方法供业务层中的方法调用。

　　以下根据系统功能分类对其实现细节进行分析。

### 14.4.1　建立数据库表

　　在 SQL Server 数据库管理系统下，建立数据库 myBooks，在该数据库中创建 4 个表，表名分别为"boos"、"lendRecord"、"returnRecord"和"users"，表结构参见 14.2.1 节。参照 14.4.2 节中，3. DataOperator 对象的 connect() 方法的源代码，在 SQL Server 数据库管理系统下，建立登录用户名"sa1"，密码"123"，默认操作的数据库为 myBooks。

## 14.4.2 登录功能的实现

**1. LoginManager 界面类**

1) 效果图(图 14-3)

图 14-3 "登录"功能运行效果

2) UML 图(图 14-4)

图 14-4 标识出了 LoginManager 类的主要成员。

LoginManager 类是 JFrame 类的子类,并实现了 ActionListener 接口。

封装在 LoginManager 类中的 main(String)方法是系统运行的入口方法。在该方法中,创建 LoginManager 窗口对象,并为 LoginManager 窗口注册窗口监听器(使用匿名类对象),最后显示 LoginManager 窗口。

LoginManager()是构造方法,负责完成登录窗口的初始化,并为 login 按钮注册动作事件监听器。

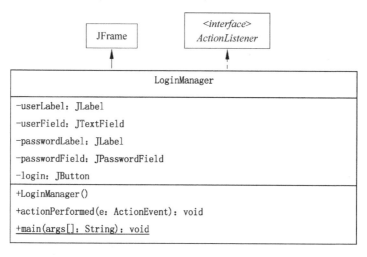

图 14-4 LoginManager 类的 UML 图

actionPerformed(ActionEvent)是 ActionListener 接口定义的方法。在 LoginManager 窗口中,当用户输入了用户名和密码,单击"登录"按钮时,执行该方法。在该方法中,取得用户在 userField 和 passwordField 输入的用户名与密码,对密码进行加密后,调用 Service 类的 login(String,string)方法,完成登录操作。如接收 login(String,string) 方法返回值大于 0(返回值为该用户在 users 表中对应存储的顺序号或标识号),登录成功,则创建主界面 BooksManager 窗口对象(同时将用户标识号传递给主界面对象),并为主界面窗口注册窗口监听器(使用匿名类对象),最后显示主界面。

不得不说明的是，当用户关闭 LoginManager 窗口时，会执行 main(String) 方法中为 LoginManager 窗口注册的匿名窗口监听器对象中的 WindowClosing(WindowEvent) 方法，该方法首先调用 Service 类的 quit() 方法关闭数据库连接，然后退出应用程序；而当用户关闭 BooksManager 主界面窗口时，同样会执行上述 actionPerformed(ActionEven) 方法中为创建的主界面窗口注册的匿名窗口监听器对象中的 WindowClosing(WindowEvent) 方法，该方法执行同样的操作。

3）源代码（LoginManager.java）

```java
import java.awt.Container;
import java.awt.GridLayout;
import java.awt.event.*;
import java.sql.*;
import javax.swing.*;
public class LoginManager extends JFrame implements ActionListener {
    private JLabel userLabel;
    private JTextField userField;
    private JLabel passwordLabel;
    private JPasswordField passwordField;
    private JButton login;
    public LoginManager() {
        super("登录管理");
        userLabel = new JLabel("登录名");
        userField = new JTextField(23);
        passwordLabel = new JLabel("密 码");
        passwordField = new JPasswordField(23);
        passwordField.setEchoChar('*');
        login = new JButton("登录");
        login.addActionListener(this);
        JPanel p1 = new JPanel(new GridLayout(2, 1));
        JPanel p2 = new JPanel(new GridLayout(2, 1));
        p1.add(userLabel);
        p1.add(passwordLabel);
        p2.add(userField);
        p2.add(passwordField);
        Box box1 = new Box(BoxLayout.X_AXIS);
        Box box2 = new Box(BoxLayout.X_AXIS);
        box1.add(p1);
        box1.add(p2);
        box2.add(login);
        Box box = new Box(BoxLayout.Y_AXIS);
        box.add(box1);
        box.add(box2);
        Container c = getContentPane();
        c.add(box);
    }
    public void actionPerformed(ActionEvent e) {
        String userName = userField.getText().trim();
        String password = MD5.GetMD5Code( passwordField.getText().trim());
        int id;
```

```java
            if ((id = Service.login(userName,password)) > 0){
                //登录成功,id 为返回的用户标识号
                JFrame app = new BooksManager(id);
                app.addWindowListener(new WindowAdapter(){
                    public void windowClosing(WindowEvent e){
                        Service.quit();      //结束业务操作,关闭数据库连接
                        System.exit(0);      //退出应用程序
                    }
                });
                app.setVisible(true);
                setVisible(false);
            } else if (id == 0){
                JOptionPane.showMessageDialog(this, "用户名或密码不正确!");
            } else {
                JOptionPane.showMessageDialog(this, "查询用户表出错!");
            }
        }
        public static void main(String[] args) {
            JFrame loginManager = new LoginManager();
            loginManager.addWindowListener(new WindowAdapter(){
                public void windowClosing(WindowEvent e){
                    Service.quit();      //结束业务操作,关闭数据库连接
                    System.exit(0);      //退出应用程序
                }
            });
            loginManager.pack();
            loginManager.setVisible(true);
        }
    }
```

## 2. Service 类的 login（String,String）方法和 quit()方法

在 login（String,String）方法中,依次调用了 DataOperator 对象的 loadDatabaseDriver() 方法、connect()方法、addSuperUser()方法和 userQuery(userName,password)方法,分别完成"加载 SQL Server 数据库的 JDBC 驱动程序"、"连接 myBooks 数据库"、"在 users 表中添加 admin 用户（如果表 users 为空）"、"查询用户表,核对用户名和密码是否正确"的操作。如 login（String,String）方法返回值大于 0,则登录成功,且返回值即为 users 表中存储的用户顺序号。

在 quit()方法中,调用了 DataOperator 对象的 disconnect()方法,关闭数据库连接。

具体源代码如下：

```java
import java.util.Vector;
import javax.swing.JOptionPane;
import java.awt.Component;
public class Service {
    private static DataOperator dataOperate = new DataOperator();
    public static int login(String userName, String password) {
        dataOperate.loadDatabaseDriver();
```

```java
            dataOperate.connect();
            dataOperate.addSuperUser();        //如果表 users 为空,添加 admin 用户
            return dataOperate.userQuery(userName, password);
        }
        public static void quit(){
            dataOperate.disconnect();
        }
    }
```

3. **DataOperator 对象的 loadDatabaseDriver( )方法、connect( )方法、addSuperUser( ) 方法、userQuery(userName,password)方法和 disconnect( )方法**

```java
import java.sql.*;
import java.util.Vector;
public class DataOperator {
    Connection con;
    private PreparedStatement pstmt;
    private String sql;
    public void loadDatabaseDriver() {
        try {
            Class.forName("com.microsoft.sqlserver.jdbc.SQLServerDriver");
        } catch (ClassNotFoundException e) {
            System.err.println("加载数据库驱动失败!");
            System.err.println(e);
        }
    }
    public void connect(){
        try {
            String connectString = "jdbc:sqlserver://localhost:1433; DatabaseName=myBooks";
            con = DriverManager.getConnection(connectString, "sa1", "123");
        } catch (SQLException e) {
            System.err.println("数据库连接出错!");
            System.err.println(e);
        }
    }
    public void addSuperUser(){           //如果表 users 为空,添加 Admin 用户
        try {
            sql = "SELECT * from users";
            pstmt = con.prepareStatement(sql);
            ResultSet rs = pstmt.executeQuery();
            if (!rs.next()) {
                String userName = "Admin";
                String password = MD5.GetMD5Code("123456");
                sql = "INSERT into users VALUES (?,?)";
                pstmt = con.prepareStatement(sql);
                pstmt.setString(1, userName);
                pstmt.setString(2, password);
                pstmt.executeUpdate();
            }
        } catch (SQLException e) {
```

```
            System.err.println("添加超级用户出错!");
            System.err.println(e);
        }
    }
    public int userQuery(String userName, String password){
        //查询用户表,核对用户名和密码是否正确
        try {
            sql = "SELECT id from users WHERE userName = ? AND password = ?";
            pstmt = con.prepareStatement(sql);
            pstmt.setString(1, userName);
            pstmt.setString(2, password);
            ResultSet rs = pstmt.executeQuery();
            if (rs.next()) return rs.getInt(1);    //核对正确,返回用户标识号
            return 0;                              //用户名或密码不正确,返回0
        } catch (SQLException se) {
            System.err.println("查询用户表出错!");
            System.err.println(se);
            return -1;                             //查询出错,返回-1
        }
    }
    public void disconnect(){
        try {
            if (con != null)
                con.close();
        } catch (SQLException e) {
            System.err.println("关闭数据库连接出错!");
            System.err.println(e);
        }
    }
}
```

### 14.4.3　主界面类 BooksManager 的实现

1. 效果图（图 14-5）

图 14-5　主界面类 BooksManager 运行效果

2. UML 图（图 14-6）

图 14-6 所示的 UML 图标识出了 BooksManager 类的主要成员。

主界面 BooksManager 类是 JFrame 类的子类,并实现了 ActionListener 接口。

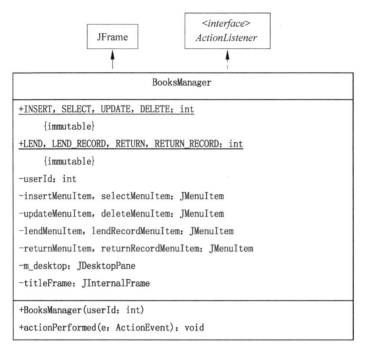

图 14-6　BooksManager 类的 UML 图

  BooksManager 类中封装了 8 个静态常量：INSERT、SELECT、UPDATE、DELETE、LEND、LEND_RECORD、RETURN、RETURN_RECORD。这些常量对应于 8 个菜单项，当某个菜单项被用户选中时，所对应的常量作为操作标记，被传递给 QueryPanel 面板或 UpdatePanel 面板，用来标识不同的操作。

  userId 接收从 LoginManager 窗口对象传递进来的参数，表示当前用户的标识号。

  insertMenuItem、selectMenuItem、updateMenuItem、deleteMenuItem、lendMenuItem、lendRecordMenuItem、returnMenuItem、returnRecordMenuItem 为定义的 8 个菜单项。

  m_desktop 为定义的桌面容器。当用户选择一个菜单项时，会创建一个内部窗口框架（JInternalFrame）放在 m_desktop 桌面容器上。桌面容器上放的第一个内部窗口框架是 titleFrame——显示标题的内部窗口框架，如图 14-5 所示。

  BooksManager(int) 是构造方法，负责完成主界面窗口的初始化，并为各菜单项注册动作监听器。其中 int 类型的参数是从登录窗口对象传递进来的用户标识号（userId）。

  actionPerformed(ActionEvent) 方法是 ActionListener 接口定义的方法。当用户在主界面中选中不同的菜单项时，执行该方法。在该方法中，创建 QueryPanel 面板或 UpdatePanel 面板对象，并且根据不同的事件源（用户选中的菜单项），向 QueryPanel 面板或 UpdatePanel 面板对象传递不同的操作标记（用 BooksManager 类定义的常量表示），同时也向 QueryPanel 面板或 UpdatePanel 面板对象传递用户标识号（userId）。

3．源代码（BooksManager.java）

```
import java.awt.*;
```

```java
import java.awt.event.*;
import javax.swing.*;
import javax.swing.event.*;
public class BooksManager extends JFrame implements ActionListener {
    public final static int INSERT = 0;
    public final static int SELECT = 1;
    public final static int UPDATE = 2;
    public final static int DELETE = 3;
    public final static int LEND = 4;
    public final static int LEND_RECORD = 5;
    public final static int RETURN = 6;
    public final static int RETURN_RECORD = 7;
    private JDesktopPane m_desktop = new JDesktopPane();
    private JMenuItem insertMenuItem;
    private JMenuItem selectMenuItem;
    private JMenuItem updateMenuItem;
    private JMenuItem deleteMenuItem;
    private JMenuItem lendMenuItem;
    private JMenuItem lendRecordMenuItem;
    private JMenuItem returnMenuItem;
    private JMenuItem returnRecordMenuItem;
    private JInternalFrame titleFrame;
    private int numberOfInternalFrame = 0;
    private int userId;
    public BooksManager(int userId) {
        super("资料室图书管理信息系统");
        this.userId = userId;
        JMenuBar theMenuBar = new JMenuBar();
        JMenu manageMenu = new JMenu("图书管理");
        JMenu lendMenu = new JMenu("借书");
        JMenu returnMenu = new JMenu("还书");
        insertMenuItem = new JMenuItem("录入");
        selectMenuItem = new JMenuItem("检索");
        updateMenuItem = new JMenuItem("修改");
        deleteMenuItem = new JMenuItem("删除");
        lendMenuItem = new JMenuItem("借书");
        lendRecordMenuItem = new JMenuItem("借书记录");
        returnMenuItem = new JMenuItem("还书");
        returnRecordMenuItem = new JMenuItem("还书记录");
        setJMenuBar(theMenuBar);
        theMenuBar.add(manageMenu);
        theMenuBar.add(lendMenu);
        theMenuBar.add(returnMenu);
        manageMenu.add(insertMenuItem);
        manageMenu.add(selectMenuItem);
        manageMenu.add(updateMenuItem);
        manageMenu.add(deleteMenuItem);
        lendMenu.add(lendMenuItem);
        lendMenu.add(lendRecordMenuItem);
        returnMenu.add(returnMenuItem);
        returnMenu.add(returnRecordMenuItem);
```

```java
            Container theContainer = getContentPane();
            theContainer.add(m_desktop);
            JLabel label = new JLabel("资料室图书管理信息系统", JLabel.CENTER);
            label.setFont(new Font("隶书", Font.BOLD, 30));
            label.setForeground(Color.blue);
            titleFrame = new JInternalFrame(null, true);
            Container c = titleFrame.getContentPane();
            c.add(label, BorderLayout.CENTER);
            titleFrame.setSize(500, 300);
            m_desktop.add(titleFrame);
            titleFrame.setVisible(true);
            setSize(505, 358);
            insertMenuItem.addActionListener(this);
            selectMenuItem.addActionListener(this);
            updateMenuItem.addActionListener(this);
            deleteMenuItem.addActionListener(this);
            lendMenuItem.addActionListener(this);
            lendRecordMenuItem.addActionListener(this);
            returnMenuItem.addActionListener(this);
            returnRecordMenuItem.addActionListener(this);
        }
        public void actionPerformed(ActionEvent e) {
            JPanel panel = null;
            String s = null;
            Object source = e.getSource();
            if (source == insertMenuItem) {
                s = "录入图书";
                panel = new UpdatePanel(INSERT);
            }
            if (source == selectMenuItem) {
                s = "查询图书";
                panel = new QueryPanel(SELECT, userId);
            }
            if (source == updateMenuItem) {
                s = "修改图书信息";
                panel = new CardPanel(userId);
            }
            if (source == deleteMenuItem) {
                s = "删除图书";
                panel = new QueryPanel(DELETE, userId);
            }
            if (source == lendMenuItem) {
                s = "借阅图书";
                panel = new QueryPanel(LEND, userId);
            }
            if (source == returnMenuItem) {
                s = "还书";
                panel = new QueryPanel(RETURN, userId);
            }
            if (source == lendRecordMenuItem) {
                s = "借书记录";
```

```java
            panel = new QueryPanel(LEND_RECORD, userId);
        }
        if (source == returnRecordMenuItem) {
            s = "还书记录";
            panel = new QueryPanel(RETURN_RECORD, userId);
        }
        JInternalFrame internalFrame = new JInternalFrame(s, true, true, true, true);
        numberOfInternalFrame ++;
        Container c = internalFrame.getContentPane();
        c.setLayout(new BorderLayout());
        c.add(panel, BorderLayout.CENTER);
        internalFrame.pack();
        m_desktop.add(internalFrame);

        Dimension d = internalFrame.getSize();
        setSize(new Dimension((int) d.getWidth() + 5, (int) d.getHeight() + 58));
        internalFrame.setVisible(true);
        MenuItemsEnabled(false);
        internalFrame.addInternalFrameListener(new InternalFrameHandler());
    }
    private void MenuItemsEnabled(boolean b) {
        insertMenuItem.setEnabled(b);
        selectMenuItem.setEnabled(b);
        updateMenuItem.setEnabled(b);
        deleteMenuItem.setEnabled(b);
        lendMenuItem.setEnabled(b);
        returnMenuItem.setEnabled(b);
    }
    /* 以下是内部窗口框架监听器类, 被定义为 BooksManager 类中的内部类, 用来处理内部窗口被
关闭的事件 */
    private class InternalFrameHandler extends InternalFrameAdapter {
        public void internalFrameClosing(InternalFrameEvent e) {
            numberOfInternalFrame --;
            if (numberOfInternalFrame == 0)
                MenuItemsEnabled(true);
            Dimension d1 = titleFrame.getSize();
            setSize(new Dimension((int) d1.getWidth(),
                (int) d1.getHeight() + 55));
        }
    }
}
```

## 14.4.4 录入图书功能的实现

### 1. UpdatePanel 面板类

在主界面，当用户选择"图书管理"菜单下的"录入"菜单项时，创建该面板类对象，将其放在主界面新建的内部窗口框架中。该面板从主界面 BooksManager 对象接收到操作标记（operateFlag）等参数。

1) 效果图(图 14-7)

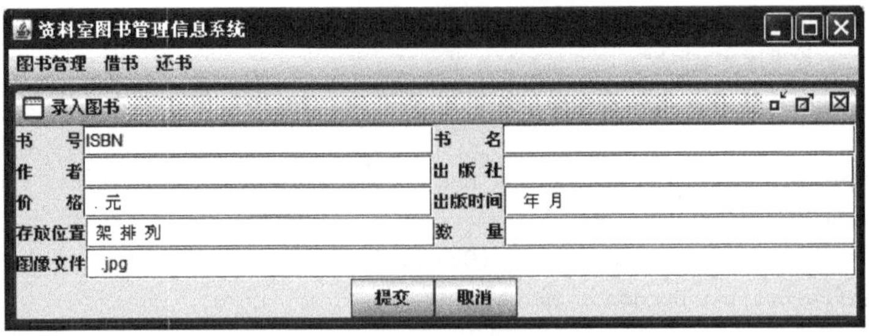

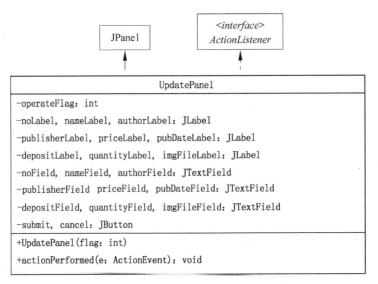

图 14-7 "录入图书"功能运行效果

2) UML 图

图 14-8 所示的 UML 图标识出了 UpdatePanel 类中完成"录入图书"操作的主要成员。UpdatePanel 类是 JPanel 类的子类,并实现了 ActionListener 接口。

operateFlag 变量接收从 BooksManager 对象传递过来的操作标记——BooksManager.INSERT,表示"录入图书"。

图 14-8 UpdatePanel 类的 UML 图

UpdatePanel(int)是构造方法,负责完成录入图书面板的初始化,并为按钮 submit 和按钮 cancel 注册动作监听器。其中 int 类型的参数是从主界面窗口对象传递进来的操作标记。

actionPerformed(ActionEvent)方法是 ActionListener 接口定义的方法。当用户在面板上的文本框中输入了图书信息,单击"提交"(submit)按钮时,执行该方法。在该方法中,取得用户在文本框中输入的图书信息,调用 Service 类的 addBook(Vector<String>)方法,完成在数据库中添加图书的操作。如接收 addBook(Vector<String>)返回值等于 0,则操作成功,否则操作失败。

3) 源代码(UpdatePanel.java)

```java
import java.awt.*;
import java.awt.event.*;
import java.sql.*;
import java.util.Vector;
import javax.swing.*;
public class UpdatePanel extends JPanel implements ActionListener {
    private int operateFlag;
    private JLabel noLabel,nameLabel,authorLabel;
    private JLabel publisherLabel,priceLabel,pubDateLabel;
    private JLabel depositLabel,quantityLabel,imgFileLabel;
    private JTextField noField,nameField,authorField;
    private JTextField publisherField,priceField,pubDateField;
    private JTextField depositField,quantityField,imgFileField;
    private JButton submit, cancel;
    private Container container;
    private CardLayout card;                        //修改图书信息时从 CardPanel 面板容器接收该参数
    private int updatedBookID;                      //修改图书信息时,存储"需修改图书"的标识号
    public UpdatePanel(int flag) {
        operateFlag = flag;                         // 设置操作标记
        // 初始化组件
        String buttonString = null;
        if (operateFlag == BooksManager.INSERT)
            buttonString = "提交";                  //录入图书信息时用此做按钮标签
        if (operateFlag == BooksManager.UPDATE)
            buttonString = "修改";                  //修改图书信息时用此做按钮标签
        noLabel = new JLabel("书        号");
        noField = new JTextField("ISBN ", 23);
        nameLabel = new JLabel("书        名");
        nameField = new JTextField(23);
        authorLabel = new JLabel("作        者");
        authorField = new JTextField(23);
        publisherLabel = new JLabel("出    版    社");
        publisherField = new JTextField(23);
        priceLabel = new JLabel("价        格");
        priceField = new JTextField("   .   元", 23);
        pubDateLabel = new JLabel("出版时间");
        pubDateField = new JTextField("    年    月", 23);
        depositLabel = new JLabel("存放位置");
        depositField = new JTextField("  架    排    列", 23);
        quantityLabel = new JLabel("数        量");
        quantityField = new JTextField(23);
        imgFileLabel = new JLabel("图像文件");
        imgFileField = new JTextField("    .jpg", 23);
        submit = new JButton(buttonString);
        cancel = new JButton("取消");
        submit.addActionListener(this);
        cancel.addActionListener(this);
        // 设置组件
        Box box1 = new Box(BoxLayout.X_AXIS);
        Box box2 = new Box(BoxLayout.X_AXIS);
        Box box3 = new Box(BoxLayout.X_AXIS);
        Box box4 = new Box(BoxLayout.X_AXIS);
```

```java
                Box box5 = new Box(BoxLayout.X_AXIS);
                Box box6 = new Box(BoxLayout.X_AXIS);
                box1.add(noLabel);
                box1.add(noField);
                box1.add(nameLabel);
                box1.add(nameField);
                box2.add(authorLabel);
                box2.add(authorField);
                box2.add(publisherLabel);
                box2.add(publisherField);
                box3.add(priceLabel);
                box3.add(priceField);
                box3.add(pubDateLabel);
                box3.add(pubDateField);
                box4.add(depositLabel);
                box4.add(depositField);
                box4.add(quantityLabel);
                box4.add(quantityField);
                box5.add(imgFileLabel);
                box5.add(imgFileField);
                box6.add(submit);
                box6.add(cancel);
                Box box = new Box(BoxLayout.Y_AXIS);
                box.add(box1);
                box.add(box2);
                box.add(box3);
                box.add(box4);
                box.add(box5);
                box.add(box6);
                setLayout(new BorderLayout());
                add(box, BorderLayout.CENTER);
            }
        public UpdatePanel(int flag, Container c, CardLayout card) {
        //该构造方法在修改图书信息时被调用
                this(flag);
                container = c;                              // 获取父组件
                this.card = card;                           // 获取父组件的卡片布局引用
            }
    //下面的方法将"被修改书的信息"显示在修改面板上,该方法在"修改图书"时被调用
            public void setUpdatedBookInfo(String updatedBookInfo) {
                if (updatedBookInfo == null)
                    return;
                String str[] = new String[11];
                int index = -1;
                for (int i = 0; i < 10; i++) {
                    index = updatedBookInfo.indexOf(',');
                    str[i] = updatedBookInfo.substring(0, index);
                    updatedBookInfo = updatedBookInfo.substring(index + 1);
                }
                str[10] = updatedBookInfo;
                updatedBookID = Integer.parseInt(str[0]);   //设置需修改书的标识
                //将查询出的需修改书的信息显示于界面,用户修改后提交保存
                noField.setText(str[1]);
                nameField.setText(str[2]);
```

```java
            authorField.setText(str[3]);
            publisherField.setText(str[4]);
            priceField.setText(str[5]);
            pubDateField.setText(str[6]);
            depositField.setText(str[7]);
            quantityField.setText(str[8]);
            imgFileField.setText(str[10]);
        }
        public void clearField() {                          // 清除文本框内容
            noField.setText("ISBN ");
            nameField.setText("");
            authorField.setText("");
            publisherField.setText("");
            priceField.setText("   .  元");
            pubDateField.setText("   年   月");
            depositField.setText("  架   排   列");
            quantityField.setText("");
            imgFileField.setText("   .jpg");
        }
        public void actionPerformed(ActionEvent e) {
            if (e.getSource() == submit) {//添加或修改图书信息
                //获取界面上用户输入或修改后的信息
                String no = noField.getText().trim();
                String name = nameField.getText().trim();
                String author = authorField.getText().trim();
                String publisher = publisherField.getText().trim();
                String price = priceField.getText().trim();
                String pubDate = pubDateField.getText().trim();
                String deposit = depositField.getText().trim();
                String quantityString = quantityField.getText().trim();
                String imgFile = imgFileField.getText().trim();
                Vector<String> bookInfo = new Vector<String>();
                //将获取到的信息字符串存入集合
                bookInfo.add(no);      bookInfo.add(name);
                bookInfo.add(author);  bookInfo.add(publisher);
                bookInfo.add(price);   bookInfo.add(pubDate);
                bookInfo.add(deposit); bookInfo.add(quantityString);
                bookInfo.add(imgFile);
                if (operateFlag == BooksManager.INSERT &&
                        Service.addBook(bookInfo) == 0)//添加图书
                    JOptionPane.showMessageDialog(null, "添加成功!");
                if (operateFlag == BooksManager.UPDATE &&
                        Service.modifyBook(updatedBookID, bookInfo) == 0) {
                    //修改图书,updatedBookID 为修改图书的标识
                    card.next(container);
                    JOptionPane.showMessageDialog(null, "修改成功!请单击\"查询\"按钮查看修改结果.");
                }
            }
            clearField();
        }
    }
}
```

## 2. Service 类的 addBook(Vector<String>)方法

在 addBook(Vector<String>)方法中,Vector<String>类型的参数接收从 UpdatePanel 面板输入后传递的图书信息。该方法调用 DataOperator 对象的 insert(Vector<String>)方法,完成"在 books 表中添加图书信息"的操作。如接收到 insert(Vector<String>)方法返回值等于 0,则 addBook(Vector<String>)方法返回 0,表示添加成功,否则添加失败。

具体源代码如下:

```java
public class Service {
    private static DataOperator dataOperate = new DataOperator();
    public static int addBook(Vector<String> bookInfo){ //添加图书
        if (dataOperate.insert(bookInfo) == -1)
            return -1;
        return 0;
    }
}
```

## 3. DataOperator 对象的 insert(Vector<String>)方法

在 insert(Vector<String>)方法中,Vector<String>类型的参数接收从 Service 类的 addBook(Vector<String>)方法传递的图书信息。该方法完成"在 books 表中添加图书信息"的操作。如添加成功,该方法返回 0;否则添加失败,返回-1。

具体源代码如下:

```java
public class DataOperator {
    Connection con;
    private PreparedStatement pstmt;
    private String sql;
    public int insert(Vector<String> bookInfo){
        try {
            sql = "INSERT into books VALUES (?,?,?,?,?,?,?,?,0,?)";
            pstmt = con.prepareStatement(sql);
            for(int i = 1; i <= bookInfo.size(); i++){
                if(i == 8)
                    pstmt.setInt(i, Integer.parseInt(bookInfo.elementAt(i-1)));
                else
                    pstmt.setString(i, bookInfo.elementAt(i-1));
            }
            pstmt.executeUpdate();
        } catch (SQLException se) {
            System.err.println("数据库增加记录出错!");
            System.err.println(se);
            return -1;
        }
        return 0;
    }
}
```

## 14.4.5 检索图书功能的实现

**1. QueryPanel 面板类**

在主界面,当用户选择"图书管理"菜单下的"检索"菜单项时,创建该面板类对象,将其放在主界面新建的内部窗口框架中。该面板从主界面 BooksManager 对象接收到操作标记(operateFlag)、用户标识号(userId)等参数。

1)效果图(图 14-9)

图 14-9 "检索图书"功能运行效果

2) UML 图

图 14-10 所示的 UML 图标识出了 QueryPanel 类中完成"检索图书"操作的主要成员。

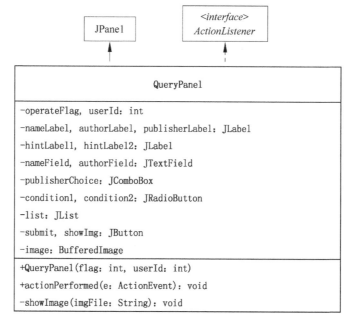

图 14-10 QueryPanel 类(完成"检索图书"操作)的 UML 图

QueryPanel 类是 JPanel 类的子类,并实现了 ActionListener 接口。

operateFlag 变量接收从 BooksManager 对象传递过来的操作标记——BooksManager.SELECT,表示"检索图书"。

userId 接收从 BooksManager 对象传递过来的用户标识号。

submit 是"查询"按钮,showImg 是"显示图片"按钮。

QueryPanel(int,int)是构造方法,负责完成查询面板的初始化,并为按钮 submit 和按钮 showImg 注册动作监听器。两个 int 类型的参数分别是从主界面窗口对象传递进来的操作标记和用户标识号。初始化查询面板时,先调用 Service 类的 publishers()方法,从 books 表中读取所有的出版社信息,放在组合框中供用户选择。

actionPerformed(ActionEvent)是 ActionListener 接口定义的方法。当用户在面板上单击"查询"按钮或"查看图片"按钮时,执行该方法。

如用户单击"查询"按钮,在 actionPerformed 方法中,取得用户在文本框、组合框以及单选按钮输入的查询条件,调用 Service 类的 seek(int, String, String, String, String)方法,完成在数据库中检索图书的操作,最后将 seek 方法返回的查询结果集合显示于 list 变量表示的列表中。

如用户单击"查看图片"按钮,在 actionPerformed 方法中,取得用户在 list 列表中选择的图书信息,调用 Service 类的 getImgFile (int, String) 方法,获取这本书的图像文件名字符串,再调用 showImage(String)方法,在弹出的窗口中显示图片。

showImage(String)是私有方法,该方法弹出一个窗口,在窗口的内容面板上显示用户选择图书的图像文件。在该方法中,将内部类 ImgPanel 的对象设置为弹出窗口的内容面板。

3) 源代码(QueryPanel.java)

```java
import java.awt. * ;
import java.awt.event. * ;
import java.awt.image.BufferedImage;
import java.io. * ;
import java.sql. * ;
import java.util.Vector;
import javax.imageio.ImageIO;
import javax.swing. * ;
public class QueryPanel extends JPanel implements ActionListener {
    private JLabel nameLabel, authorLabel, publisherLabel;
    private JLabel hintLabel1;              //查询条件选择提示标签
    private JLabel hintLabel2;              //列表显示内容提示标签
    private JTextField nameField, authorField;
    private JComboBox publisherChoice;
    private JRadioButton condition1, condition2;
    private ButtonGroup group;
    private JList list;
    private JButton submit, showImg, showDetails, update, delete, lend, returnB;
    private BufferedImage image;
    private int operateFlag;
    private Container container;
                //如为"修改图书"前的查询,该变量获取父组件引用,即 CardPanel 对象
```

```java
    private CardLayout card;         //如为"修改图书"前的查询,该变量获取父组件的卡片布局引用
    private UpdatePanel updatePanel; //如为"修改图书"前的查询,该变量获取修改面板引用
    private int userId;
    public QueryPanel(int flag, int userId) {
        this.userId = userId;
        operateFlag = flag;              // 设置操作标志
        // 调用 Service 类的 publishers 方法,读取出版社信息
        Vector<String> publisherInfo = Service.publishers();
        // 初始化组件
        nameLabel = new JLabel("书 名");
        nameField = new JTextField(23);
        authorLabel = new JLabel("作 者");
        authorField = new JTextField(10);
        publisherLabel = new JLabel("出版社");
        publisherChoice = new JComboBox(publisherInfo);
        hintLabel1 = new JLabel("查询条件");
        condition1 = new JRadioButton("完全一致", true);
        condition2 = new JRadioButton("模糊查询");
        group = new ButtonGroup();
        group.add(condition1);
        group.add(condition2);
        String s = null;
        if (operateFlag == BooksManager.LEND_RECORD
                || operateFlag == BooksManager.RETURN)              //查询借书记录
            s = "查询结果(借书记录):";
        else if (operateFlag == BooksManager.RETURN_RECORD)         //查询还书记录
            s = "查询结果(还书记录):";
        else
            s = "查询结果(图书信息):";
        hintLabel2 = new JLabel(s, JLabel.LEFT);
        list = new JList();
        list.setSelectionMode(ListSelectionModel.SINGLE_SELECTION);
        JScrollPane scroll = new JScrollPane();
        scroll.getViewport().setView(list);
        submit = new JButton("查询");
        submit.addActionListener(this);
        // 设置组件
        Box box1 = new Box(BoxLayout.X_AXIS);
        box1.add(nameLabel);
        box1.add(nameField);
        Box box2 = new Box(BoxLayout.X_AXIS);
        box2.add(authorLabel);
        box2.add(authorField);
        Box box3 = new Box(BoxLayout.X_AXIS);
        box3.add(publisherLabel);
        box3.add(publisherChoice);
        Box box4 = new Box(BoxLayout.X_AXIS);
        box4.add(hintLabel1);
        box4.add(condition1);
        box4.add(condition2);
        box4.add(submit);
```

```java
            Box box5 = new Box(BoxLayout.X_AXIS);
            box5.add(hintLabel2);
            Box box6 = new Box(BoxLayout.X_AXIS);
            if (operateFlag == BooksManager.LEND_RECORD
                    || operateFlag == BooksManager.RETURN_RECORD) {
                showDetails = new JButton("查看详细信息");
                                        //该按钮只在查看"借书记录"或"还书记录"时用到
                showDetails.addActionListener(this);
                box6.add(showDetails);
            } else {
                showImg = new JButton("查看图片");
                showImg.addActionListener(this);
                box6.add(showImg);
            }
            if (operateFlag == BooksManager.UPDATE) {
                update = new JButton("修改");          //该按钮只在"修改图书"时用到
                update.addActionListener(this);
                box6.add(update);
            }
            if (operateFlag == BooksManager.DELETE) {
                delete = new JButton("删除");          //该按钮只在"删除图书"时用到
                delete.addActionListener(this);
                box6.add(delete);
            }
            if (operateFlag == BooksManager.LEND) {
                lend = new JButton("借书");            //该按钮只在"借书"时用到
                lend.addActionListener(this);
                box6.add(lend);
            } if (operateFlag == BooksManager.RETURN) {
                returnB = new JButton("还书");         //该按钮只在"还书"时用到
                returnB.addActionListener(this);
                box6.add(returnB);
            }
            Box box = new Box(BoxLayout.Y_AXIS);
            box.add(box1);
            box.add(box2);
            box.add(box3);
            box.add(box4);
            box.add(box5);
            box.add(scroll);
            box.add(box6);
            setLayout(new BorderLayout());
            add(box, BorderLayout.CENTER);
        }
        //下面的构造方法只在"修改图书"时用到
        public QueryPanel(int flag, Container c, CardLayout card,
                UpdatePanel updatePanel, int userId) {
            this(flag, userId);
            container = c;                          // 获取父组件
            this.card = card;                       // 获取父组件的卡片布局引用
            this.updatePanel = updatePanel;         // 获取修改面板引用
```

```java
        }
        public void actionPerformed(ActionEvent e) {
            Object source = e.getSource();
            if (source == submit) {
            /*查询按钮：当完成"图书管理"中的"检索"、"修改"、"删除"，
             * 以及"借书"功能时，均需先查询全部图书信息；
             * 而当查看"借书记录"或完成"还书"操作时，需查询全部借书记录信息 */
                Vector<String> infoStringCollection = new Vector<String>();
                //上面的语句定义存放查询结果的集合
                list.setListData(infoStringCollection);    //清除显示查询结果的列表
                // 下面从界面获取用户输入或选择的查询条件
                String name = nameField.getText().trim();
                String author = authorField.getText().trim();
                String publisher = publisherChoice.getSelectedItem().toString();
                String condition = null;
                if (condition1.isSelected())
                    condition = condition1.getText().trim();
                if (condition2.isSelected())
                    condition = condition2.getText().trim();
                /* 调用 Service 类的 seek 方法返回查询结果集合 */
                infoStringCollection = Service.seek(operateFlag, name, author, publisher,
condition);
                list.setListData(infoStringCollection);// 将查询结果显示于列表
                //清除界面
                nameField.setText("");
                authorField.setText("");
                publisherChoice.setSelectedIndex(0);
                return;
            }
            String book = (String) list.getSelectedValue();
            //从列表中获取用户选择的一本书的字符串信息进行处理
            if (book == null) {                        //用户没有在列表显示的查询结果中选择
                String str = null;
                if (operateFlag == BooksManager.LEND_RECORD
                        || operateFlag == BooksManager.RETURN)
                    str = "请先在查询结果中选择一条借书记录!";         //只用于查询借书记录
                else if (operateFlag == BooksManager.RETURN_RECORD)
                    str = "请先在查询结果中选择一条还书记录!";         //只用于查询还书记录
                else
                    str = "请先在查询结果中选择一本书!";
                JOptionPane.showMessageDialog(this, str);                //显示提示信息
                return;
            }
            if (source == showImg) {              // 单击了"显示图片"按钮，显示所选图书的图片
                String imgFile = Service.getImgFile(operateFlag, book);
                showImage(imgFile);
            }
            if (source == showDetails) {              //只用于查询借书记录或还书记录
                // 单击了"查看详细信息"按钮，显示所选借书或还书记录的详细信息
                String details = Service.detailsOfBook(operateFlag, book);
                JOptionPane.showMessageDialog(this, details);
```

```java
        }
        if (source == update) {                    // 单击了对图书信息的"修改"按钮,修改所选图书
            //对修改面板设置修改图书标识,将图书原来的信息显示在修改面板上,以供修改
            updatePanel.setUpdatedBookInfo(book);
            card.next(container);                   //通过卡片布局转向修改面板
        } //该 if 块中代码只用于"修改图书"功能
        if (source == delete) {                    // 单击了对图书信息的"删除"按钮,删除所选图书
            if (Service.deleteBook(this, book) == 0)
                JOptionPane.showMessageDialog(this, "删除成功!单击\"查询\"按钮可查看结果。");
        } //该 if 块中代码只用于"删除图书"功能
        if (source == lend) {                      // 单击了"借书"按钮
            StringBuffer hintMessage = new StringBuffer("");
            int lendFlag = Service.lendBook(this, userId, book, hintMessage);
            if (lendFlag == 0)
                JOptionPane.showMessageDialog(this,"借书成功!单击\"查询\"按钮可查看结果.");
            if (lendFlag == 2)
                JOptionPane.showMessageDialog(this, hintMessage + "该书已全部被借出!对不起.");
        } //该 if 块中代码只用于"借书"功能
        if (source == returnB) {                   // 单击了"还书"按钮
            if (Service.returnBook(this, userId, book) == 0)
                JOptionPane
                    .showMessageDialog(this, "还书成功!单击\"查询\"按钮可查看结果.");
        }   //该 if 块中代码只用于"还书"功能
    }
    private void showImage(String imgFile) {       //显示图片
        try {
            image = ImageIO.read(new File(imgFile));
        } catch (Exception ee) {
            JOptionPane.showMessageDialog(this, "该书没有照片文件!");
            System.out.println("读取图像文件出错!" + "\n" + ee);
            return;
        }
        int width = image.getWidth(this);
        int height = image.getHeight(this);
        ImgPanel imgPanel = new ImgPanel();         //创建显示图片面板
        JFrame popupWindow = new JFrame();          //弹出窗口
        popupWindow.setContentPane(imgPanel);       //将显示图片面板设置为弹出窗口的内容面板
        popupWindow.setSize(width, height);
        popupWindow.setVisible(true);
    }
    private class ImgPanel extends JPanel {        //内部类,定义显示图片面板
        public void paintComponent(Graphics g) {
            super.paintComponent(g);
            g.drawImage(image, 0, 0, null);
        }
    }
}
```

## 2. Service 类的 publishers()方法、seek(int，String，String，String，String)方法和 getImgFile (int，String) 方法

在 publishers()方法中，调用 DataOperator 对象的 publishersQuery()方法，从 books 表查询所有的出版社信息，存储于结果集合中，返回。

在 seek(int，String，String，String，String)方法中，int 参数表示从 QueryPanel 面板对象传递过来的操作标记(operateFlag 为 BooksManager.SELECT)，4 个字符串参数表示从 QueryPanel 面板对象传递过来的书名、作者、出版社、是否为模糊查询的查询条件信息。根据查询条件，首先组织查询字符串，然后调用 DataOperator 对象的 generalQuery(int，String，int，String，String，String)方法，从 books 表查询所有符合条件的图书信息，存储于结果集合中，返回。

在 getImgFile (int，String) 方法中，int 参数表示从 QueryPanel 面板对象传递过来的操作标记(operateFlag 为 BooksManager.SELECT)，字符串参数表示从 QueryPanel 面板对象传递过来的(用户选择的)一本书的信息。对参数字符串进行处理，从中找出这本书的图像文件名，返回。

```
public class Service {
    private static DataOperator dataOperate = new DataOperator();
    public static Vector<String> publishers(){       // 获取出版社信息
        return dataOperate.publishersQuery();
    }
    public static Vector<String> seek(int operateFlag, String name, String author, String publisher, String condition) {
//该方法根据查询条件,查询图书信息或借书记录
        // 开始组织查询语句,sql 为查询语句字符串
        String sql = "SELECT * from books";
        int selectFlag = 0;                          // 查询语句为"SELECT * from books"
        if (name != null && !name.equals("")) {
            sql += " WHERE name LIKE ?";
            selectFlag = 1;
            // 查询语句为"SELECT * from books WHERE name LIKE ?"
            if (author != null && !author.equals("")) {
                sql += " AND author LIKE ?";
                selectFlag = 2;
// 查询语句为"SELECT * from books WHERE name LIKE ? AND author LIKE ?"
                if (!publisher.equals("")) {
                    sql += " AND publisher LIKE ?";
                    selectFlag = 3;
// 查询语句为"SELECT * from books WHERE name LIKE ? AND author LIKE ? AND
//publisher LIKE ?"
                }
            } else {
                if (!publisher.equals("")) {
                    sql += " AND publisher LIKE ?";
                    selectFlag = 4;
// 查询语句为"SELECT * from books WHERE name LIKE ? AND publisher LIKE ?"
```

```java
                        }
                    }
                } else {
                    if (author != null && !author.equals("")) {
                        sql += " WHERE author LIKE ?";
                        selectFlag = 5;
                         // 查询语句为"SELECT * from books WHERE author LIKE ?";
                        if (!publisher.equals("")) {
                            sql += " AND publisher LIKE ?";
                            selectFlag = 6;
// 查询语句为"SELECT * from books WHERE author LIKE ? AND publisher LIKE ?"
                        }
                    } else {
                        if (!publisher.equals("")) {
                            sql += " WHERE publisher LIKE ?";
                            selectFlag = 7;
                            // 查询语句为"SELECT * from books WHERE publisher LIKE ?"
                        }
                    }
                }
                /* 以上组织了查询资料室书库(表名books)的字符串,下面修改为查询借书记录的字符串,查询条
                件不变,查询内容涉及3个表:lendRecord,books,users,该操作只用于"查看借书记录"或"还书" */
                if (operateFlag == BooksManager.LEND_RECORD
                        || operateFlag == BooksManager.RETURN) {
                    StringBuffer sb = new StringBuffer(sql);
                    sb.replace(7,20,"lendRecord.id, books.no, books.name, "
                        + "books.author, books.publisher, books.pubDate, "
                        + "lendRecord.borrower, lendRecord.borrowerUnit, "
                        + "users.userName, lendRecord.borrowDate, lendRecord.state "
                        + "FROM lendRecord "
                        + "INNER JOIN books ON lendRecord.bookId = books.id "
                        + "INNER JOIN users ON lendRecord.userId = users.id ");
                    sql = sb.toString();
                }
                /* 下面将查询资料室书库(表名books)的字符串修改为查询还书记录的字符串,查询条件不变,查
                询内容涉及3个表:returnRecord,books,users,该操作只用于"查看还书记录" */
                if (operateFlag == BooksManager.RETURN_RECORD) {
                    StringBuffer sb = new StringBuffer(sql);
                    sb.replace(7,20,"returnRecord.id, books.no, books.name, "
                        + "books.author, books.publisher, books.pubDate, "
                        + "returnRecord.returner, returnRecord.returnerUnit, "
                        + "users.userName, returnRecord.returnDate "
                        + "FROM returnRecord "
                        + "INNER JOIN books ON returnRecord.bookId = books.id "
                        + "INNER JOIN users ON returnRecord.userId = users.id ");
                    sql = sb.toString();
                }
                if (condition.equals("模糊查询")) {
                    name = "%" + name + "%";
                    author = "%" + author + "%";
                    publisher = "%" + publisher + "%";
```

```
            }
            Vector < String > infoStringCollection = dataOperate.generalQuery(operateFlag, sql, 
    selectFlag, name, author, publisher);
            return infoStringCollection;
        }
        public static String getImgFile(int operateFlag, String book){
            String imgFile = null;
            if (operateFlag == BooksManager.RETURN) {
                /* 以从借书记录中选择的一本书为索引,查找其图片并显示.只用于"还书" */
                int index = book.indexOf(',');
                book = book.substring(0, index);
                index = Integer.parseInt(book);
                imgFile = dataOperate.imgFileQuery(index);
            } else {
                /*以从图书信息中选择的一本书为索引,查找其图片并显示. */
                if (book.endsWith(".jpg") || book.endsWith(".jpeg")) {
                    int index = book.lastIndexOf(",");
                    imgFile = book.substring(index + 1);
                }
            }
            return imgFile;
        }// getImgFile 方法定义结束
}//Service 类定义结束
```

## 3. DataOperator 对象的 publishersQuery()方法和 generalQuery(int,string,int, string,string,string)方法

publishersQuery()方法查询 books 表中所有的出版社信息,存储于集合中,返回。

在 generalQuery(int,String,int,String,String,String)方法中,第一个 int 参数接收从 Service 类的 seek 方法传递的操作标记(为 BooksManager.SELECT),第二个 int 参数接收从 Service 类的 seek 方法传递的查询语句种类标记(在 0～7 范围内),4 个字符串参数分别接收从 Service 类的 seek 方法传递的查询语句字符串、书名、作者、出版社。该方法查询 books 表中所有符合条件的图书信息,存储于集合中,返回。

具体代码如下:

```
public Vector < String > publishersQuery(){              // 连接数据库,读取出版社信息
    Vector < String > publisherInfo = new Vector < String >();
    try{
    sql = "SELECT publisher from books UNION SELECT publisher from books";
        pstmt = con.prepareStatement(sql);
        ResultSet rs = pstmt.executeQuery();
        publisherInfo.add("");
        while (rs.next()) {
            publisherInfo.add(rs.getString(1));
        }
    } catch (SQLException e) {
        System.err.println("数据库查询出错!");
        System.err.println(e);
```

```java
        }
        return publisherInfo;
    }
    public Vector<String> generalQuery(int operateFlag, String sql,
            int selectFlag, String name, String author, String publisher) {
        Vector<String> infoStringCollection = new Vector<String>();
        try {
            pstmt = con.prepareStatement(sql);
            switch (selectFlag) {
                case 0:
                    break;
                case 1:
                    pstmt.setString(1, name);
                    break;
                case 2:
                    pstmt.setString(1, name);
                    pstmt.setString(2, author);
                    break;
                case 3:
                    pstmt.setString(1, name);
                    pstmt.setString(2, author);
                    pstmt.setString(3, publisher);
                    break;
                case 4:
                    pstmt.setString(1, name);
                    pstmt.setString(2, publisher);
                    break;
                case 5:
                    pstmt.setString(1, author);
                    break;
                case 6:
                    pstmt.setString(1, author);
                    pstmt.setString(2, publisher);
                    break;
                case 7:
                    pstmt.setString(1, publisher);
            }
            ResultSet rs = pstmt.executeQuery();
            String infoString = null;
            while (rs.next()) { // 将查询结果记录组织为字符串,并存入集合
                infoString = new String();
                infoString += rs.getInt(1) + ",";
                int count = 12;
                if (operateFlag == BooksManager.LEND_RECORD
                        || operateFlag == BooksManager.RETURN)
                    count = 12;
                if (operateFlag == BooksManager.RETURN_RECORD)
                    count = 11;
```

```
                for (int i = 2; i < count; i++)
                    infoString += rs.getString(i).trim() + ",";
                infoString = infoString.substring(0, infoString.length() - 1);
                    infoStringCollection.add(infoString);
            }
        } catch (SQLException se) {
            System.err.println("查询数据库出错!");
            System.err.println(se);
            se.printStackTrace(System.err);
        }
        return infoStringCollection;
    }
```

## 14.4.6 修改图书功能的实现

**1. 界面类（CardPanel 面板类、QueryPanel 面板类、UpdatePanel 面板类）**

在主界面，当用户选择"图书管理"菜单下的"修改"菜单项时，创建 CardPanel 面板类对象，将其放在主界面新建的内部窗口框架中。

CardPanel 面板对象从主界面 BooksManager 对象接收到用户标识号（userId）等参数后，创建 QueryPanel 面板类对象和 UpdatePanel 面板类对象，并以卡片布局形式放置在 CardPanel 面板上；首先显示 QueryPanel 面板，由用户给出查询条件并单击"查询"按钮后，actionPerformed(ActionEvent)方法查询出符合条件的图书信息，显示于 QueryPanel 面板的列表中；用户在 QueryPanel 面板的列表中选择一本书，单击"修改"按钮，执行 QueryPanel 面板对象的 actionPerformed(ActionEvent)方法，该方法获取用户在列表中选择的一本图书的信息，将其传递给 UpdatePanel 面板对象，控制卡片布局显示出 UpdatePanel 面板；UpdatePanel 面板对象将得到的图书信息显示于界面，供用户修改；用户在 UpdatePanel 面板上对图书信息进行修改后，单击"修改"按钮，执行 UpdatePanel 面板对象的 actionPerformed(ActionEvent)方法，该方法获取用户在 UpdatePanel 面板上修改的信息，完成对数据库中图书信息表 books 中相应记录的修改。

1）效果图（图 14-11 和图 14-12）

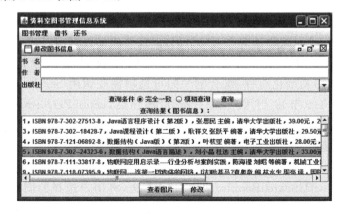

图 14-11 "修改图书"功能运行效果——QueryPanel 面板

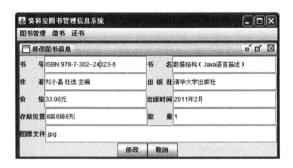

图 14-12 "修改图书"功能运行效果——UpdatePanel 面板

2) UML 图(图 14-13～图 14-15)

图 14-13 的 UML 图标识出 CardPanel 类的主要成员,该类是 JPanel 的子类。

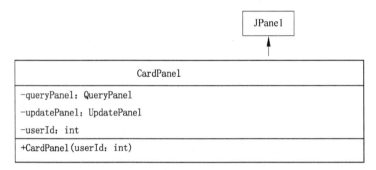

图 14-13 CardPanel 类的 UML 图

CardPanel(int)是构造方法,int 参数为从主界面 BooksManager 对象传递过来的用户标识号。在该方法中,用参数初始化成员变量 userId,设置 CardPanel 面板使用卡片布局,创建 QueryPanel 面板对象和 UpdatePanel 面板对象,分别赋值给成员变量 queryPanel 和 updatePanel,将这两块面板添加到 CardPanel 面板上,并设置首先显示查询面板。

图 14-14 的 UML 图标识出 QueryPanel 类中完成"修改图书"操作的主要成员。与图 14-10 相比,多了 update("修改"按钮)、container(表示父组件,即 CardPanel)、card(父组件的卡片布局引用)、updatePanel(修改面板引用)这些成员变量;相应地,构造方法多了 Container、CardLayout、UpdatePanel 这些类型的参数,用于接收从 CardPanel 面板对象传递的父组件 CardPanel 对象和其卡片布局引用,以及修改面板引用。

QueryPanel 面板对象从 CardPanel 面板对象接收到操作标记(BooksManager. UPDATE)等参数后,初始化组件(组合框中的可选项需要调用 Service 类的 publishers 方法获取,具体见 14.4.5 节);当用户单击"查询"按钮后,actionPerformed(ActionEvent)方法获取用户在界面给出的条件,调用 Service 类的 seek 方法(具体见 14.4.5 节)查询出符合条件的图书信息,显示于列表中;当用户在列表中选择了一本书,单击"修改"按钮,执行 actionPerformed(ActionEvent)方法的相关操作。这时,在该方法中首先调用 updatePanel (修改面板引用)的 setUpdatedBookInfo(String)方法,将用户选择的图书信息放在修改面板的文本框中以供修改,也将用户选择的图书信息中包含的图书标识号设置给修改面板;

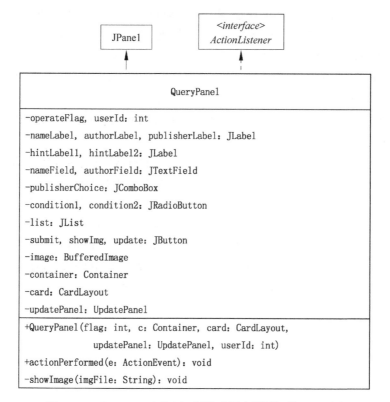

图 14-14　QueryPanel 类(完成"修改图书"操作)的 UML 图

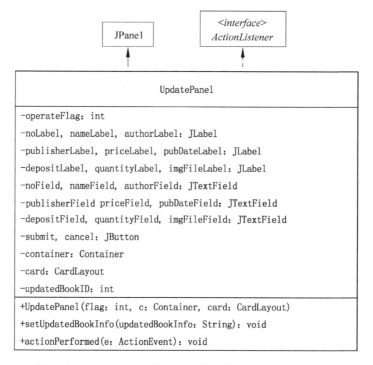

图 14-15　UpdatePanel 类(完成"修改图书"操作)的 UML 图

然后调用 card（父组件的卡片布局引用）的 next(Container) 方法，将焦点转向父组件 CardPanel 面板上的另一组件——修改面板。具体见 14.4.5 节中源代码 QueryPanel.java。

图 14-15 的 UML 图标识出 UpdatePanel 类中完成"修改图书"操作的主要成员。与图 14-8 相比，多了 container（表示父组件，即 CardPanel）、card（父组件的卡片布局引用）、updatedBookID（被修改图书标识号）这些成员变量；相应地，构造方法多了 Container、CardLayout 两种类型的参数，用于接收从 CardPanel 面板对象传递的父组件 CardPanel 对象和其卡片布局引用。

setUpdatedBookInfo(String) 方法由之前显示的 QueryPanel 面板对象调用，方法的字符串参数接收从 QueryPanel 面板对象传递的需修改的图书信息。该方法将被修改书的信息做处理后，显示在修改面板上；并从图书信息中提取出书的标识号，赋给成员变量 updatedBookID，以供执行数据库修改操作时定位要修改的图书。

当用户在 UpdatePanel 面板单击"修改"按钮时执行 actionPerformed(ActionEvent) 方法的相关操作。这时，在该方法中首先收集用户在文本框中修改好的信息；然后调用 Service 类的 modifyBook(int，Vector<String>) 方法修改图书；最后通过父组件的卡片布局引用，控制父组件显示查询面板，以便用户可以通过查询证实图书信息是否被修改。具体见 14.4.4 节中源代码 UpdatePand.java。

3) 源代码（CardPanel.java）

说明：以下是 CardPanel.java 源代码，QueryPanel.java 源代码请参考 14.4.5 节，UpdatePanel.java 源代码请参考 14.4.4 节。

```java
import java.awt.CardLayout;
import javax.swing.JPanel;
public class CardPanel extends JPanel {
    private QueryPanel queryPanel;
    private UpdatePanel updatePanel;
    private int userId;
    public CardPanel(int userId) {
        this.userId = userId;
        CardLayout card = new CardLayout();
        setLayout(card);
        // 初始化、设置组件
        updatePanel = new UpdatePanel(BooksManager.UPDATE, this, card);
        queryPanel = new QueryPanel(BooksManager.UPDATE, this, card,
                updatePanel, userId);
        add("query", queryPanel);
        add("update", updatePanel);
        card.show(this, "query");
    }
}
```

**2. Service 类的 modifyBook(int，Vector<String>) 方法**

该方法是用户在 UpdatePanel 面板单击"修改"按钮后，程序执行 actionPerformed

(ActionEvent)方法时调用的,int 参数接收被修改图书标识号,Vector＜String＞参数接收被修改图书信息。modifyBook 会调用 DateOperator 对象的 update(int,Vector＜String＞)方法完成修改的数据库操作。具体源代码如下:

```java
public static int modifyBook(int updatedBookID, Vector<String> bookInfo)
{
    if (dataOperate.update(updatedBookID, bookInfo) == -1)
        return -1;
    return 0;
}
```

**3. DateOperator 对象的 update(int,Vector＜String＞)方法**

该方法完成修改图书信息的数据库操作,修改成功返回 0,修改失败返回-1。

```java
public int update(int updatedBookID, Vector<String> bookInfo){
    try {
        sql = "UPDATE books SET no = ?,name = ?,author = ?,"
            + "publisher = ?,price = ?,pubDate = ?,deposit = ?, "
            + "quantity = ?,imgFile = ? WHERE id = ?";
        pstmt = con.prepareStatement(sql);
        int number = bookInfo.size();
        for(int i = 1; i <= number; i++){
            if(i == 8)
                pstmt.setInt(i, Integer.parseInt(bookInfo.elementAt(i-1)));
            else
                pstmt.setString(i, bookInfo.elementAt(i-1));
        }
        pstmt.setInt(number + 1, updatedBookID);
        pstmt.executeUpdate();
    } catch (SQLException se) {
        System.err.println("数据库修改记录出错!");
        System.err.println(se);
        return -1;
    }
    return 0;
}
```

### 14.4.7 删除图书功能的实现

**1. 界面类(QueryPanel 面板类)**

在主界面,当用户选择"图书管理"菜单下的"删除"菜单项时,创建 QueryPanel 面板类对象,将其放在主界面新建的内部窗口框架中。在该面板上,用户首先查询符合条件的图书信息显示于列表,然后从列表选择一本书,可以通过查看图片的方式确认需要删除的图书,最后单击"删除"按钮执行删除操作。

1) 效果图(图 14-16)

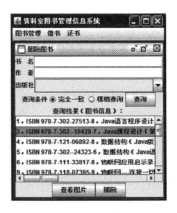

图 14-16 "删除图书"功能运行效果——QueryPanel 面板

2) UML 图(图 14-17)

图 14-17 的 UML 图标识出 QueryPanel 类中完成"删除图书"操作的主要成员。与图 14-10 相比,多了 delete("删除"按钮)这一成员变量。

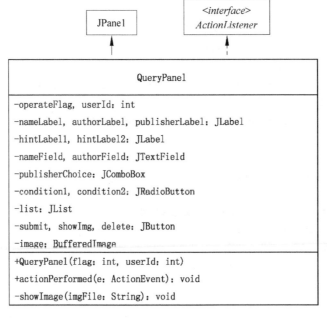

图 14-17 QueryPanel 类(完成"删除图书"操作)的 UML 图

QueryPanel 面板对象从主界面 BooksManager 对象接收到操作标记(BooksManager.DELETE)、用户标识号两个参数后,初始化组件(组合框中的可选项需要调用 Service 类的 publishers 方法获取,具体见 14.4.5 节);当用户单击"查询"按钮后,actionPerformed(ActionEvent)方法获取用户在界面给出的条件,调用 Service 类的 seek 方法(具体见 14.4.5 节),查询出符合条件的图书信息,显示于列表中;当用户在列表中选择一本书,单击"删除"按钮时,actionPerformed(ActionEvent)方法获取用户在列表中选择的图书信息字符串,调用 Service 类的 deleteBook(Container，String)方法，完成删除操作。删除成功后，显示提示信息，用户在该面板单击"查询"按钮可以在列表中看到删除后的显示结果。

3）源代码（QueryPanel.java，请参考 14.4.5 节）

### 2．Service 类的 deleteBook(Component，String)方法

该方法是用户在 QueryPanel 面板单击"删除"按钮后，程序执行 actionPerformed (ActionEvent) 方法时调用的，Component 参数接收 QueryPanel 面板引用，用于在 deleteBook 方法中用标准对话框显示信息，String 参数接收被删除的图书信息。方法返回 0 表示删除成功，返回 −1 表示删除失败。

deleteBook 方法首先对接收的图书信息字符串进行处理，然后使用标准对话框显示出图书信息，询问用户是否确认要删除，如果用户确认，则调用 DateOperator 对象的 delete (int)方法完成删除的数据库操作。具体代码如下：

```java
public static int deleteBook(Component c, String book) {
    String str[] = new String[8];
    int index = -1;
    for (int i = 0; i < 8; i++) {
        index = book.indexOf(',');
        str[i] = book.substring(0, index);
        book = book.substring(index + 1);
    }
    int confirm = JOptionPane.showConfirmDialog(c,
"您决定要删除的一本书的信息如下：\n"
        + "书号：" + str[1] + "\n" + "书名：" + str[2] + "\n"
        + "作者：" + str[3] + "\n" + "出版社：" + str[4] + "\n"
        + "价格：" + str[5] + "\n" + "出版时间：" + str[6] + "\n"
        + "存放位置：" + str[7] + "\n" + "确实需要删除吗?");
    if (confirm > 0)
        return 1;
    int deletedBookID = Integer.parseInt(str[0]);
    if (dataOperate.delete(deletedBookID) == -1)
        return -1;
    return 0;
}
```

### 3．DateOperator 对象的 delete（int)方法

该方法完成删除图书信息的数据库操作，删除成功返回 0，删除失败返回 −1。int 参数接收从 Service 类的 deleteBook 方法传递的图书标识号。具体代码如下：

```java
public int delete(int deletedBookID){
    try {
        sql = "DELETE from books WHERE id = ?";
        pstmt = con.prepareStatement(sql);
        pstmt.setInt(1, deletedBookID);
```

```
            pstmt.executeUpdate();
        } catch (SQLException se) {
            System.err.println("数据库删除记录出错!");
            System.err.println(se);
            return -1;
        }
        return 0;
    }
```

### 14.4.8 借书功能的实现

**1. 界面类（QueryPanel 面板类）**

在主界面,当用户选择"借书"菜单下的"借书"菜单项时,创建 QueryPanel 面板类对象,将其放在主界面新建的内部窗口框架中。在该面板上,用户首先查询符合条件的图书信息显示于列表,然后从列表选择一本书,可以通过查看图片的方式确认要借阅的图书,最后单击"借书"按钮执行借书操作。

1) 效果图(图 14-18)

图 14-18 "借书"功能运行效果——QueryPanel 面板

2) UML 图(图 14-19)

图 14-19 的 UML 图标识出 QueryPanel 类中完成"借书"操作的主要成员。与图 14-10 相比,多了 lend("借书"按钮)这一成员变量。

QueryPanel 面板对象从主界面 BooksManager 对象接收到操作标记(BooksManager. LEND)、用户标识号两个参数后,初始化组件(组合框中的可选项需要调用 Service 类的 publishers 方法获取,具体见 14.4.5 节);当用户单击"查询"按钮后,actionPerformed (ActionEvent)方法获取用户在界面给出的条件,调用 Service 类的 seek 方法(具体见 14.4.5 节),查询出符合条件的图书信息,显示于列表中;当用户在列表中选择一本书,单击"借书"按钮后,actionPerformed(ActionEvent)方法获取用户在列表中选择的一本图书的信息,调用 Service 的 lendBook 方法,完成借书操作。

3) 源代码(QueryPanel.java,请参考 14.4.5 节)

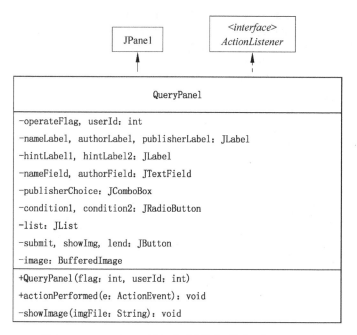

图 14-19　QueryPanel 类(完成"借书"操作)的 UML 图

### 2. Service 类的 lendBook(Component，int，String，StringBuffer)方法

该方法是用户在 QueryPanel 面板单击"借书"按钮后，程序执行 actionPerformed (ActionEvent)方法时调用的，Component 参数接收 QueryPanel 面板引用，用于在 lendBook 方法中用标准对话框显示信息，int 参数接收用户标识号，String 参数接收用户选择的要借阅的图书信息，StringBuffer 参数接收 QueryPanel 面板给出的字符串缓冲区引用，编辑字符串后再返回给 QueryPanel 面板，以便于 QueryPanel 面板显示借书结果信息。方法返回 0 表示借书成功，返回 1 表示未借书(用户通过对话框取消了借书)，返回 −1 表示数据库操作失败，返回 2 表示库存为 0。

lendBook 方法首先对接收的图书信息字符串进行处理，计算库存，并组织提示信息字符串；然后测试库存，如大于 0，使用标准对话框显示出提示信息，询问用户是否确认要借阅，如果用户确认，则两次弹出标准对话框，收集用户输入的借书人姓名、借书人所在单位；最后调用 DateOperator 对象的 lend 方法完成借书的数据库操作。具体代码如下：

```java
public static int lendBook(Component c, int userId, String book, StringBuffer hintMessage) {
    String str[] = new String[10];
    int index = −1;
    for (int i = 0; i < 10; i++) {
        index = book.indexOf(',');
        str[i] = book.substring(0, index);
        book = book.substring(index + 1);
    }
    int remainder = Integer.parseInt(str[8]) − Integer.parseInt(str[9]);
    hintMessage.append("您决定要借阅的一本书的信息如下：\n"
        + "书号：" + str[1] + "\n" + "书名：" + str[2] + "\n"
```

```java
            + "作者:" + str[3] + "\n" + "出版社:" + str[4] + "\n"
            + "价格:" + str[5] + "\n" + "出版时间:" + str[6] + "\n"
            + "存放位置:" + str[7] + "\n" + "库存数量:" + remainder + "\n");
        if (remainder > 0) {    // 有库存
            int confirm = JOptionPane.showConfirmDialog(c,
                         hintMessage + "确定借阅吗?");
            if (confirm > 0)
                return 1;
            int id = Integer.parseInt(str[0]);
            int lentQuantity = Integer.parseInt(str[9]) + 1;
            String s1 = JOptionPane.showInputDialog(c, "请输入借书人姓名");
            String s2 = JOptionPane.showInputDialog(c, "请输入借书人所在单位");
            if (dataOperate.lend(id, lentQuantity, s1, s2, userId) == -1)
                return -1;
            return 0;
        } else {    // 库存为 0
            return 2;
        }
    }
}
```

3. DateOperator 对象的 lend(int，int，String，String，int)方法、updateStock(int，int)方法、insertLendRecord(int，String，String，int)方法和 rollback()方法

lend 方法完成借书的数据库操作,借书成功返回 0,借书失败返回-1。3 个 int 参数分别接收从 Service 类的 lendBook 方法传递的图书标识号、修改库存时应存储的借出数量、用户标识号,两个 String 参数分别接收从 Service 类的 lendBook 方法传递的借书人姓名、借书人所在单位。

在 lend 方法中,首先关闭数据库事务自动更新模式,然后调用 updateStock 方法修改库存,调用 insertLendRecord 方法保存借书记录,最后提交事务,恢复数据库事务自动更新模式。如在数据库操作中发生异常,则调用 rollback 方法撤销事务。

updateStock(int，int)方法完成 books 表中借阅图书的借出数量(加 1 后)字段的存储。两个 int 参数分别接收借阅图书的标识号、借出数量(加 1 后)。

insertLendRecord(int，String，String，int)方法将借书信息存储于借书记录表(lendRecord)中。两个 int 参数分别接收借阅图书的标识号、用户标识号,两个字符串参数分别接收借阅人姓名、借阅人所在单位。

具体代码如下:

```java
public int lend(int id, int lentQuantity, String s1, String s2,
                int userId){
    try {
        con.setAutoCommit(false);                    // 关闭数据库事务自动更新模式
        if (updateStock(id, lentQuantity) == -1)     // 修改库存
            return -1;
        if (insertLendRecord(id, s1, s2, userId) == -1)              // 存储借书记录
            return -1;
        con.commit();
```

```java
            con.setAutoCommit(true);
            return 0;
        } catch (SQLException e){
            System.err.println("事务提交或设置事务自动提交出错!");
            System.err.println(e);
            rollback();
            return -1;
        }
    }
    public int updateStock(int id, int lentQuantity){      // 修改库存
        try {
            sql = "UPDATE books SET lend = ? WHERE id = ?";                    //修改库存
            pstmt = con.prepareStatement(sql);
            pstmt.setInt(1, lentQuantity);
            pstmt.setInt(2, id);
            pstmt.executeUpdate();
        } catch (SQLException se) {
            System.err.println("修改库存记录出错!");
            System.err.println(se);
            rollback();
            return -1;
        }
        return 0;
    }
    public int insertLendRecord(int id, String s1, String s2, int userId) {
        // 借书时,存储借书记录
        try {
            sql = "INSERT into lendRecord Values (?,?,?,?,?,?)";
            pstmt = con.prepareStatement(sql);
            pstmt.setInt(1, id);
            pstmt.setString(2, s1);
            pstmt.setString(3, s2);
            pstmt.setInt(4, userId);
            java.util.Date d = new java.util.Date();
            pstmt.setString(5, d.toString());
            pstmt.setBoolean(6, false);
            pstmt.executeUpdate();
        } catch (SQLException se) {
            System.err.println("存储借书记录出错!");
            System.err.println(se);
            rollback();
            return -1;
        }
        return 0;
    }
```

```
public void rollback(){                              //撤销事务
    if (con == null) return;
    try {
        System.err.println("发生异常,正在撤销事务------");
        con.rollback();
    } catch(SQLException e) {
        System.err.println(e.getMessage());
    }
}
```

### 14.4.9 查看借书记录功能的实现

**1. 界面类(QueryPanel 面板类)**

在主界面,当用户选择"借书"菜单下的"借书记录"菜单项时,创建 QueryPanel 面板类对象,将其放在主界面新建的内部窗口框架中。在该面板上,用户查询符合条件的借书记录显示于列表,然后从列表选择一条记录,可以查看该条借书记录的详细信息。

1) 效果图(图 14-20)

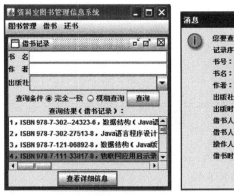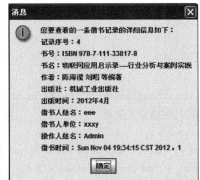

图 14-20 "查看借书记录"功能运行效果——QueryPanel 面板及弹出的信息框

2) UML 图(图 14-21)

图 14-21 所示的 UML 图标识出 QueryPanel 类中完成"查看借书记录"操作的主要成员。与图 14-10 相比,成员变量 showImg("查看图片"按钮)换成了 showDetails("查看详细信息"按钮)。

QueryPanel 面板对象从主界面 BooksManager 对象接收到操作标记(BooksManager.LEND_RECORD)、用户标识号两个参数后,初始化组件(组合框中的可选项需要调用 Service 类的 publishers 方法获取,具体见 14.4.5 节),并为按钮 submit、showDetails 注册动作监听器;当用户单击"查询"按钮后,actionPerformed(ActionEvent)方法获取用户在界面给出的条件,调用 Service 类的 seek 方法查询出符合条件的借书记录,显示于列表中;当用户在列表中选择一条记录,单击"查看详细信息"按钮后,actionPerformed(ActionEvent)方法获取用户在列表中选择的一条借书记录字符串,调用 Service 类的 detailsOfBook 方法

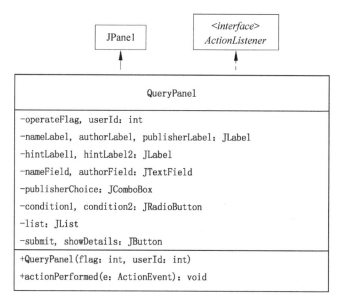

图 14-21　QueryPanel 类（完成"查看借书记录"操作）的 UML 图

对字符串进行处理，然后将 detailsOfBook 方法返回的提示字符串显示于弹出的标准对话框。

3) 源代码（QueryPanel.java，请参考 14.4.5 节）

## 2. Service 类的 seek(int，String，String，String，String) 方法和 detailsOfBook(int，String) 方法

seek 方法是用户在 QueryPanel 面板单击"查询"按钮后，程序执行 actionPerformed(ActionEvent) 方法时调用的，int 参数接收 QueryPanel 面板传递过来的操作标记（BooksManager.LEND_RECORD），4 个字符串参数接收 QueryPanel 面板传递的查询条件信息。在 seek 方法中，首先组织查询借书记录的字符串，然后调用 DataOperater 对象的 generalQuery 方法查询出符合条件的借书记录，存储于 Vector＜String＞集合。该方法代码请参考 14.4.5 节。

detailsOfBook 方法是用户在 QueryPanel 面板单击"查看详细信息"按钮后，程序执行 actionPerformed(ActionEvent)方法时调用的，int 参数接收 QueryPanel 面板传递过来的操作标记（BooksManager.LEND_RECORD），String 参数接收用户在 QueryPanel 面板的列表中选择的一条借书记录信息。在该方法中，对借书记录信息进行字符串处理，生成提示字符串返回。具体代码如下：

```java
public static String detailsOfBook(int operateFlag, String book){
    String str[] = new String[9];
    int index = -1;
    for (int i = 0; i < 9; i++) {
        index = book.indexOf(',');
        str[i] = book.substring(0, index);
        book = book.substring(index + 1);
```

```
            }
            char c = (char) 0;
            if (operateFlag == BooksManager.LEND_RECORD)
                c = '借';
            if (operateFlag == BooksManager.RETURN_RECORD)
                c = '还';
            String details = "您要查看的一条" + c
                    + "书记录的详细信息如下：\n" + "记录序号：" + str[0] + "\n"
                    + "书号：" + str[1] + "\n" + "书名：" + str[2] + "\n"
                    + "作者：" + str[3] + "\n" + "出版社：" + str[4] + "\n"
                    + "出版时间：" + str[5] + "\n" + c + "书人姓名：" + str[6] + "\n"
                    + c + "书人单位：" + str[7] + "\n" + "操作人姓名：" + str[8] + "\n"
                    + c + "书时间：" + book;
            return details;
        }
```

**3. DataOperate 对象的 generalQuery(int, String, int, String, String, String)方法**

generalQuery 方法从数据库查询出符合条件的借书记录，两个 int 参数分别接收从 Service 类的 seek 方法传递过来的操作标记（BooksManager.LEND_RECORD）、查询字符串分类标记，4 个字符串参数分别接收从 Service 类的 seek 方法传递过来的查询语句字符串、书名、作者、出版社，其中后面 3 个字符串用于向预处理的查询语句传递参数。generalQuery 方法查询 myBooks 数据库的 lendRecord,books,users 三个表,得到包含图书信息、借书记录信息、用户信息的复合信息，存储于 Vector<String>集合中，返回。具体源代码请参考 14.4.5 节。

### 14.4.10 还书功能的实现

**1. 界面类（QueryPanel 面板类）**

在主界面，当用户选择"还书"菜单下的"还书"菜单项时，创建 QueryPanel 面板类对象，将其放在主界面新建的内部窗口框架中。在该面板上，用户首先查询符合条件的借书记录信息显示于列表，然后从列表选择一条借书记录，可以通过查看图片的方式确认要还的图书，最后单击"还书"按钮执行还书操作。

1）效果图（图 14-22）

图 14-22 "还书"功能运行效果——QueryPanel 面板

2）UML 图（图 14-23）

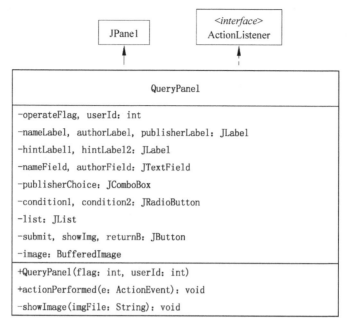

图 14-23　QueryPanel 类（完成"还书"操作）的 UML 图

图 14-23 的 UML 图标识出 QueryPanel 类中完成"还书"操作的主要成员。与图 14-10 相比，多了 returnB（"还书"按钮）这一成员变量。

QueryPanel 面板对象从主界面 BooksManager 对象接收到操作标记（BooksManager.RETURN）、用户标识号两个参数后，初始化组件（组合框中的可选项需要调用 Service 类的 publishers 方法获取，具体见 14.4.5 节），并为按钮注册动作监听器；当用户单击"查询"按钮后，actionPerformed（ActionEvent）方法获取用户在界面给出的条件，调用 Service 类的 seek 方法（具体见 14.4.9 节），查询出符合条件的借书记录，显示于列表中；当用户在列表中选择一条借书记录，点击"查看图片"按钮，actionPerformed（ActionEvent）方法获取用户选择的借书记录信息，调用 Service 类的 getImgFile 方法，获取被借图书的图像文件名字符串，再调用 showImage 方法，在弹出的窗口中显示图片；当用户在列表中选择一条借书记录，单击"还书"按钮，actionPerformed（ActionEvent）方法获取用户选择的借书记录信息，调用 Service 类的 returnBook 方法，完成还书操作。

3）源代码（QueryPand.java，请参考 14.4.5 节）

2．Service 类的 getImgFile（int，string）方法和 returnBook（Component，int，String）方法

getImgFile 方法是用户在 QueryPanel 面板点击"查看图片"按钮后，程序执行 actionPerformed（ActionEvent）方法时调用，int 参数接收从 QueryPanel 面板对象传递的操作标记（operateFlag 为 BooksManager.RETURN），String 参数接收用户选择的要还书的借书记录信息。该方法对参数字符串进行处理，从中找出借书记录索引，然后调用 DateOperator 对象的 imgFileQuery（int）方法，得到图书的封面图像文件名，返回。该方法的源代码见 14.4.5 节。

returnBook 方法是用户在 QueryPanel 面板单击"还书"按钮后,程序执行 actionPerformed (ActionEvent)方法时调用的,Component 参数接收 QueryPanel 面板引用,用于在 returnBook 方法中用标准对话框请求用户输入信息,int 参数接收用户标识号,String 参数接收用户选择的要还书的借书记录信息。方法返回 0 表示还书成功,返回 -1 表示数据库操作失败,返回 1 表示未还书(检测出该书已还、或用户通过对话框取消了还书)。

returnBook 方法首先对接收的借书记录信息字符串进行处理,提取出该条记录的状态值,据此判断这本书是否已还,如已还,则显示提示信息后,返回 1;如未还,再处理借书记录信息字符串,提取出借书记录标识号,调用 DataOperate 对象的 bookIdQueryWithLendRecordId 方法,根据借书记录标识查询出图书标识,之后调用 DataOperate 对象的 bookInfoQueryWithBookId 方法,根据图书标识查询出图书信息,据此组织提示信息字符串,在标准对话框显示出来请用户确认还书;用户确认后,两次弹出标准对话框,收集用户输入的还书人姓名、还书人所在单位;最后调用 DateOperator 对象的 returnB 方法完成还书的数据库操作。具体代码如下:

```java
public static int returnBook(Component c, int userId, String book) {
    int index = book.indexOf(',');
    if (book.charAt(book.length() - 1) == '1') {
        //根据借书记录中的状态值,判断是否已还
        JOptionPane.showMessageDialog(c, "该书已还!");
        return 1;
    }
    //根据借书记录标识查询出图书标识
    int lendRecordId = Integer.parseInt(book.substring(0, index));
    int bookId = dataOperate.bookIdQueryWithLendRecordId(lendRecordId);
    //根据图书标识查询出图书信息
    String bookInfo = dataOperate.bookInfoQueryWithBookId(bookId);
    String hintMessage = "您决定要还的一本书的信息如下: \n" + bookInfo
                    + "\n 确定还书吗?";
    int confirm = JOptionPane.showConfirmDialog(c, hintMessage);
    if (confirm > 0)
        return 1;
    index = bookInfo.lastIndexOf(': ');
    int lentQuantity = Integer.parseInt(bookInfo.substring(index + 1).trim()) - 1;
    String s1 = JOptionPane.showInputDialog(c, "请输入还书人姓名");
    String s2 = JOptionPane.showInputDialog(c, "请输入还书人所在单位");
    if (dataOperate.returnB(lendRecordId, bookId, lentQuantity, s1, s2, userId) != 0)
        return -1;
    return 0;
}
```

3. DateOperator 对象的 imgFileQuery(int)方法、bookIdQueryWithLendRecordId(int)方法、bookInfoQueryWithBookId(int)方法、returnB(int,int,int,String,String,int)方法、updateStateOfLendRecord(int)方法、updateStock(int,int)方法和 insertReturnRecord(int,String,String,int)方法

imgFileQuery 方法根据借书记录标识号,从 lendRecord 表和 books 表中查询出所借图书的封面图像文件名,返回。int 参数接收从 Service 类的 getImgFile 方法传递的借书记录

标识号。

　　bookIdQueryWithLendRecordId 方法根据借书记录标识查询出图书标识,查询成功返回图书标识,数据库操作出错返回 −1,未查出返回 0。参数接收从 Service 类的 returnBook 方法传递过来的借书记录标识号。

　　bookInfoQueryWithBookId 方法根据图书标识查询出图书信息,查询成功返回图书信息字符串,数据库操作出错返回 null。参数接收从 Service 类传递过来的图书标识号。

　　returnB 完成还书的数据库操作,还书成功返回 0,还书失败返回 −1。4 个 int 参数分别接收从 Service 类的 returnBook 方法传递的借书记录标识号、图书标识号、修改库存时应存储的借出数量、用户标识号,两个 String 参数分别接收从 Service 类的 returnBook 方法传递的还书人姓名、还书人所在单位。

　　在 returnB 方法中,首先关闭数据库事务自动更新模式,然后调用 updateStateOfLendRecord 方法修改借书记录的状态值,调用 updateStock 方法修改库存,调用 insertReturnRecord 方法保存还书记录,最后提交事务,恢复数据库事务自动更新模式。如在数据库操作中发生异常,则调用 rollback 方法(代码参见 14.4.8 节)撤销事务。

　　updateStateOfLendRecord(int)方法将 lendRecord 表中借书记录的状态值修改为 true(表示已还),修改成功返回 0,数据库操作出错返回 −1; int 参数接收从 retureB 方法传递的借书记录号。

　　updateStock(int, int)方法完成 books 表中借阅图书的借出数量(减 1 后)字段的存储。两个 int 参数分别接收借阅图书的标识号、借出数量(减 1 后)。其代码可参考 14.4.8 节。

　　insertReturnRecord(int, String, String, int)方法将还书信息存储于还书记录表(returnRecord)中。两个 int 参数分别接收借阅图书的标识号、用户标识号,两个字符串参数分别接收还书人姓名、还书人所在单位。

　　具体代码如下:

```java
public String imgFileQuery(int lendRecordId){
    //根据借书记录索引,查询出所借图书的封面图像文件名
    String imgFile = null;
    try {
        sql = "SELECT imgFile FROM lendRecord "
            + "INNER JOIN books ON lendRecord.bookId = books.id "
            + "WHERE lendRecord.id = ?";
        pstmt = con.prepareStatement(sql);
        pstmt.setInt(1, lendRecordId);
        ResultSet rs = pstmt.executeQuery();
        while (rs.next())
            imgFile = rs.getString(1);
    } catch (SQLException se) {
        System.err.println("数据库查询出错!");
        System.err.println(se);
    }
    return imgFile;
}
public int bookIdQueryWithLendRecordId(int lendRecordId) {
    //还书时,根据借书记录标识查询出图书标识
```

```java
            try {
                sql = "SELECT bookId FROM lendRecord WHERE id = ?";
                pstmt = con.prepareStatement(sql);
                pstmt.setInt(1, lendRecordId);
                ResultSet rs = pstmt.executeQuery();
                if (rs.next())
                    return rs.getInt(1);
            } catch (SQLException se) {
                System.err.println("根据借书记录标识" + lendRecordId + "查询图书标识出错!");
                System.err.println(se);
                return -1;
            }
            return 0;
        }
        public String bookInfoQueryWithBookId(int bookId) {
            //还书时,根据图书标识查询出图书信息
            String bookInfo = "";
            String str[] = new String[10];
            try {
                sql = "SELECT * FROM books WHERE id = ?";
                pstmt = con.prepareStatement(sql);
                pstmt.setInt(1, bookId);
                ResultSet rs = pstmt.executeQuery();
                if (rs.next()) {
                    for(int i = 0; i < 10; i++)
                        str[i] = rs.getString(i+1);
                }
            } catch (SQLException se) {
                System.err.println("根据图书标识" + bookId + "查询图书信息出错!");
                System.err.println(se);
                return null;
            }
            bookInfo = "书号:" + str[1] + "\n" + "书名:" + str[2] + "\n"
                + "作者:" + str[3] + "\n" + "出版社:" + str[4] + "\n"
                + "价格:" + str[5] + "\n" + "出版时间:" + str[6] + "\n"
                + "存放位置:" + str[7] + "\n" + "数量:" + str[8] + "\n"
                + "借出数量:" + str[9];
            return bookInfo;
        }
        public int returnB(int lendRecordId, int bookId, int lentQuantity, String s1,
        String s2, int userId){
            try {
                con.setAutoCommit(false);                    // 关闭数据库事务自动更新模式
                if (updateStateOfLendRecord(lendRecordId) != 0) // 修改借书记录的状态值
                    return -1;
                if (updateStock(bookId, lentQuantity) == -1) // 修改库存
                    return -1;
                // 存储还书记录
```

```java
            if (insertReturnRecord(bookId, s1, s2, userId) == -1)
                return -1;
            con.commit();
            con.setAutoCommit(true);
            return 0;
        } catch (SQLException e){
            System.err.println("事务提交或设置事务自动提交出错!");
            System.err.println(e);
            rollback();
            return -1;
        }
    }
    public int updateStateOfLendRecord(int lendRecordId) {
        // 还书时,修改借书记录的状态值
        try {
            sql = "UPDATE lendRecord SET state = ? WHERE id = ?";
            pstmt = con.prepareStatement(sql);
            pstmt.setBoolean(1, true);
            pstmt.setInt(2, lendRecordId);
            pstmt.executeUpdate();
        } catch (SQLException se) {
            System.err.println("修改借书记录的状态值出错!");
            System.err.println(se);
            rollback();
            return -1;
        }
        return 0;
    }
    public int insertReturnRecord(int bookId, String s1, String s2, int userId) {
        // 还书时, 存储还书记录
        try {
            sql = "INSERT into returnRecord Values (?,?,?,?,?)";              // 存储还书记录
            pstmt = con.prepareStatement(sql);
            pstmt.setInt(1, bookId);
            pstmt.setString(2, s1);
            pstmt.setString(3, s2);
            pstmt.setInt(4, userId);
            java.util.Date d = new java.util.Date();
            pstmt.setString(5, d.toString());
            pstmt.executeUpdate();
        } catch (SQLException se) {
            System.err.println("存储还书记录出错!");
            System.err.println(se);
            rollback();
            return -1;
        }
        return 0;
    }
```

## 14.4.11 查看还书记录功能的实现

**1. 界面类（QueryPanel 面板类）**

在主界面，当用户选择"还书"菜单下的"还书记录"菜单项时，创建 QueryPanel 面板类对象，将其放在主界面新建的内部窗口框架中。在该面板上，用户查询符合条件的还书记录显示于列表，然后从列表选择一条记录，可以查看该条还书记录的详细信息。

1）效果图（图 14-24）

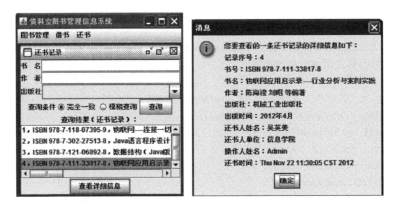

图 14-24 "查看还书记录"功能运行效果——QueryPanel 面板及弹出的信息框

2）UML 图（图 14-25）

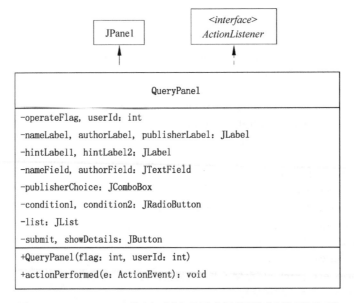

图 14-25 QueryPanel 类（完成"查看还书记录"操作）的 UML 图

图 14-25 的 UML 图标识出 QueryPanel 类中完成"查看还书记录"操作的主要成员。与图 14-10 相比，成员变量 showImg（"查看图片"按钮）换成了 showDetails（"查看详细信

息"按钮)。

QueryPanel 面板对象从主界面 BooksManager 对象接收到操作标记(BooksManager.RETURN_RECORD)、用户标识号两个参数后,初始化组件(组合框中的可选项需要调用 Service 类的 publishers 方法获取,具体见 14.4.5 节);当用户单击"查询"按钮后,actionPerformed(ActionEvent)方法获取用户在界面给出的条件,调用 Service 类的 seek 方法查询出符合条件的还书记录,显示于列表中;用户在列表中选择一条记录,单击"查看详细信息"按钮后,actionPerformed(ActionEvent)方法获取用户在列表中选择的一条还书记录字符串,调用 Service 类的 detailsOfBook 方法对字符串进行处理,然后将 detailsOfBook 方法返回的提示字符串显示于弹出的标准对话框。

3) 源代码(QueryPanel.java,请参考 14.4.5 节)

### 2. Service 类的 seek(int, String, String, String, String) 方法和 detailsOfBook(int, String) 方法

seek 方法是用户在 QueryPanel 面板单击"查询"按钮后,程序执行 actionPerformed(ActionEvent) 方法时调用的,int 参数接收 QueryPanel 面板传递过来的操作标记(BooksManager.RETURN_RECORD),4 个字符串参数接收 QueryPanel 面板传递的查询条件信息。在 seek 方法中,首先组织查询还书记录的字符串,然后调用 DataOperater 对象的 generalQuery 方法查询出符合条件的还书记录,存储于 Vector<String>集合中。该方法代码可参考 14.4.5 节。

detailsOfBook 方法是用户在 QueryPanel 面板单击"查看详细信息"按钮后,程序执行 actionPerformed(ActionEvent)方法时调用的,int 参数接收 QueryPanel 面板传递过来的操作标记(BooksManager.RETURN_RECORD),String 参数接收用户在 QueryPanel 面板的列表中选择的一条还书记录信息。在该方法中,对还书记录信息进行字符串处理,生成提示字符串返回。该方法代码可参考 14.4.9 节。

### 3. DataOperate 对象的 generalQuery(int, String, int, String, String, String)方法

generalQuery 方法从数据库查询出符合条件的还书记录,两个 int 参数分别接收从 Service 类的 seek 方法传递过来的操作标记(BooksManager.RETURN_RECORD)、查询字符串分类标记,4 个字符串参数分别接收从 Service 类的 seek 方法传递过来的查询语句字符串、书名、作者、出版社,其中后面 3 个字符串用于向预处理的查询语句传递参数。generalQuery 方法查询 myBooks 数据库的 returnRecord、books、users 三个表,得到包含图书信息、还书记录信息、用户信息的复合信息,存储于 Vector<String>集合中,返回。具体代码可参考 14.4.5 节。

## 14.5 资料室图书管理系统调试与软件发布

### 14.5.1 系统调试

在 Eclipse 环境下,创建一个新的工程,在工程中加载外部包 sqljdbc.jar(加载方法:将

光标移至工程名,右击,在弹出的菜单中选择 Build Path→Add External Archives,在弹出的打开文件对话框中选择打开 sqljdbc.jar 文件即可)。在该工程中用来存放源代码的路径 src 的缺省下,创建 14.4 节中讲到的 8 个类。然后运行主类 LoginManager 即可。

MD5 类(用于对密码进行加密)的参考代码如下:

```java
import java.security.MessageDigest;
import java.security.NoSuchAlgorithmException;
/*
 * MD5 算法
 */
public class MD5 {
    private final static String[] strDigits = { "0", "1", "2", "3", "4", "5",
            "6", "7", "8", "9", "a", "b", "c", "d", "e", "f" };          // 全局数组
    public MD5() {
    }
    // 返回形式为数字与字符串
    private static String byteToArrayString(byte bByte) {
        int iRet = bByte;
        if (iRet < 0) {
            iRet += 256;
        }
        int iD1 = iRet / 16;
        int iD2 = iRet % 16;
        return strDigits[iD1] + strDigits[iD2];
    }
    // 返回形式只为数字
    private static String byteToNum(byte bByte) {
        int iRet = bByte;
        System.out.println("iRet1 = " + iRet);
        if (iRet < 0) {
            iRet += 256;
        }
        return String.valueOf(iRet);
    }
    // 转换字节数组为十六进制字串
    private static String byteToString(byte[] bByte) {
        StringBuffer sBuffer = new StringBuffer();
        for (int i = 0; i < bByte.length; i++) {
            sBuffer.append(byteToArrayString(bByte[i]));
        }
        return sBuffer.toString();
    }
    public static String GetMD5Code(String strObj) {
        String resultString = null;
        try {
            resultString = new String(strObj);
            MessageDigest md = MessageDigest.getInstance("MD5");
            // md.digest() 该函数返回值为存放哈希值结果的 byte 数组
            resultString = byteToString(md.digest(strObj.getBytes()));
        } catch (NoSuchAlgorithmException ex) {
            ex.printStackTrace();
        }
        return resultString;
```

```
    }
    public static void main(String[] args) {
        MD5 getMD5 = new MD5();
        System.out.println(MD5.GetMD5Code("yu"));
    }
}
```

### 14.5.2 软件发布

在 Eclipse 环境下,在菜单中单击"文件"→Export,会弹出一个界面;在弹出的界面中,展开 Java 后,选择 Runnable JAR file,然后单击 Next 按钮,会弹出另外一个界面;在该界面中,有两个组合框,在提示有 Launch configuration 的组合框中选择工程的主类 LoginManager,在提示有 Export destination 的组合框中给出生成的 JAR 文件的文件名(这里给出的文件名为 BooksManager.jar)及其存放路径,然后单击 Finish 按钮。

现在,可以将生成的文件 BooksManager.jar 复制到任何一台安装了 Java 运行环境的计算机上,读者用鼠标双击该文件的图标,即可运行软件。

## 14.6 综合课程设计作业

### 14.6.1 资料室图书管理信息系统扩展

**1. 对系统增设"用户管理"功能**

可添加、修改、删除及查询用户信息,并设计和实现对不同用户设置不同"权限",以实现系统的安全性管理。

**2. 完善人机交互时的"数据验证"机制**

可在添加、修改数据信息时,验证用户输入数据格式的正确性。如用户输入数据格式不正确,可弹出提示框给出提示信息,且不在相应的数据表中进行记录的添加或修改操作。

**3. 对"录入"操作,进一步实现"重复数据检索"功能**

在"录入"或"修改"图书时,检索"录入"或"修改"的图书信息是否与 books 表中已有图书的"ISBN 号"相同,如相同,则向用户发出提示,且不在 books 表中进行记录添加或记录修改操作。

在"添加"用户时,检索"添加"用户是否与 users 表中已有用户的"用户名"相同,如相同,则向用户发出提示,且不在 users 表中进行记录添加或记录修改操作。

### 14.6.2 综合课程设计题目

(1) 设计和实现班级学生管理信息系统。
(2) 设计和实现家庭收支管理信息系统。

# 第15章

# Java与网络：P2P聊天系统

## 15.1 P2P聊天系统需求分析

P2P聊天系统应满足P2P即时通信要求，即聊天信息不经服务器中转，由参加聊天的一方直接发送至另一方。

除上述基本需求外，实现以下功能。

(1) 多人聊天。

(2) 信息服务器。该服务器为上线的P2P端提供注册、选择其他聊天端的服务。

(3) P2P聊天端。具体说明如下：

① 启动后向信息服务器注册其用户名和地址信息，并从信息服务器查询在线的P2P端，从中选择聊天方（可选择多个P2P端聊天）。

② 不通过信息服务器，直接编辑并向选择的聊天方发送信息（第1次发送的是聊天请求），开始聊天。

③ 检测是否有其他P2P端发来的聊天请求，如有，选择同意或拒绝。如同意，开始聊天。

④ 在聊天过程中，接收并显示其他P2P端发来的聊天信息。

⑤ 在聊天过程中，检测其他P2P端是否都已经退出。

## 15.2 P2P聊天系统设计

### 15.2.1 信息服务器功能设计

根据需求分析，信息服务器启动后，主程序不断检测新上线的P2P端发来的连接请求。每当接收到新的连接请求后，信息服务器需并行完成以下多项工作。

(1) 继续检测新上线的P2P端发来的注册请求。

(2) 继续为已连接的P2P端提供服务。

(3) 与刚发送连接请求的P2P端建立TCP连接后，为其提供服务。为此，需循环执行以下操作：

① 接收请求；

② 解析请求；
③ 发送响应。

在②中，解析出接收到的请求类型，完成一定的操作，并且据此生成响应。

具体对应的请求和响应类型以及服务器完成的操作见表 15-1。

表 15-1 信息服务器收到的请求、需完成的操作以及对应发送的响应

| 请求 | 服务器完成的操作步骤 | | 响应 |
| --- | --- | --- | --- |
| 请求类型 1：P2P 端注册 | 步骤 1：从服务器保存的 P2P 端信息中查阅是否已存在此注册名 | 已存在此注册名 | 响应类型 1（见注释）："注册名已被他人使用" |
| | | 步骤 2：不存在此注册名，保存此 P2P 端信息（注册名、地址） | 响应类型 1："注册成功" |
| 请求类型 2：获取在线 P2P 端 | 步骤 1：从保存的 P2P 端信息中查阅是否已存在发送请求的 P2P 端 | 不存在此 P2P 端 | 响应类型 1："未注册" |
| | | 步骤 2：存在此 P2P 端。从保存的 P2P 端信息中提取生成"已注册的 P2P 端注册名列表" | 响应类型 2：所有在线 P2P 端注册名列表 |
| 请求类型 3：P2P 端获取"已选择聊天对象（P2P 端）的地址" | 步骤 1：从保存的 P2P 端信息中查阅是否已存在发送请求的 P2P 端 | 不存在此 P2P 端 | 响应类型 1："未注册" |
| | | 步骤 2：存在此 P2P 端。根据请求中"已选择 P2P 端（与之聊天）的注册名"，从保存的 P2P 端信息中查阅其地址信息 | 响应类型 3：已选择聊天对象（P2P 端）的地址 |
| 请求类型 4：P2P 端退出信息服务器 | 步骤 1：从保存的 P2P 端信息中查阅是否已存在发送请求的 P2P 端 | 不存在此 P2P 端 | 响应类型 1："未注册" |
| | | 步骤 2：存在此 P2P 端。从保存的 P2P 端信息中去除该 P2P 端 | 响应类型 1："已退出" |

注：响应类型 1 表示此响应中只包含应答字符串信息。

主程序不断检测新上线的 P2P 端发来的注册请求，每当接收到新的注册请求后，创建子线程。在子线程中，为当前 P2P 端提供服务。

信息服务器主程序的工作过程以及与 P2P 聊天端的通信参见图 15-3，信息服务器子线程的工作过程以及与 P2P 聊天端的通信参见图 15-10。

### 15.2.2 P2P 聊天端设计

根据需求分析，设计 P2P 聊天端具有主程序以及"与信息服务器通信"、"从其他 P2P 端接收聊天信息"、"聊天过程中，检测其他 P2P 端是否都已经退出"的 3 个子线程。主程序、各子线程设计如下。

#### 1. 主程序

启动后，进入主界面。启动工作过程及主界面功能选项分别如图 15-1 和图 15-2 所示。

图 15-1 P2P 聊天端(主程序)启动过程

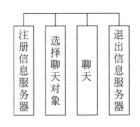
图 15-2 P2P 聊天端主界面功能选项单

在主界面中,选择"注册信息服务器"后,进入到"注册"界面。在界面中输入"注册名"、"服务器 IP",单击"提交"按钮后,将用户在界面输入注册名和建立的本地端口信息整理为"P2P 端注册"请求对象,随后启动已经建立的"与信息服务器通信"子线程,并将请求对象传递给该子线程,后者连接至指定 IP 和端口(本例使用 8000)的服务器,发送请求对象进行注册。当"与信息服务器通信"的子线程接收到信息服务器的"注册成功"响应后,通过建立的通信管道传递响应给主程序。主程序随即建立一个新的子线程,用于循环检测和接收从其他 P2P 端发来的聊天请求或聊天信息。P2P 聊天端主程序注册功能的具体工作过程以及与信息服务器(主程序)的通信如图 15-3 所示。

在主界面中,单击"选择聊天对象"后,进入到相应的界面。在界面中单击"获取在线 P2P 端"按钮后,将生成的请求对象传递给"与信息服务器通信"的子线程,后者发送请求给信息服务器,并从服务器接收到"在线 P2P 端的注册名列表"响应信息,随后通过建立的通信管道传递给主程序。主程序显示"在线 P2P 端的注册名"于界面的多行文本框中,以供选择。接着,用户在界面中选择与之聊天的 P2P 端(可多选),主程序逐一构建"获取聊天对象地址"的请求对象,传递给"与信息服务器通信"的子线程,后者发送请求给信息服务器,并从服务器接收到"P2P 端聊天对象地址"响应,用管道传递给主程序。主程序将该地址存入聊天对象地址列表。具体工作过程如图 15-4 所示。

在主界面中,选择"聊天"功能后,打开聊天窗口界面,并建立和启动一个新的子线程,用于在聊天过程中,检测其他 P2P 端是否都已经退出。具体工作过程如图 15-5 所示。

聊天窗口界面功能如图 15-6 所示。

在聊天窗口界面中,单击"发送"按钮后,主程序获取用户在窗口中输入的信息,发送给"地址列表"中所有的 P2P 端聊天对象(对方使用"从其他 P2P 端接收聊天信息"的子线程接收信息)。具体工作过程以及与 P2P 端聊天对象的子线程之间的通信如图 15-7 所示。

在聊天窗口界面中,单击"退出"按钮后,主程序提示用户正在聊天,询问是否退出。得到用户确认后,主程序向"地址列表"中所有的聊天对象(P2P 端)发送"再见",并结束"聊天过程中,检测其他 P2P 端是否都已经退出"的子线程。之后,判断主程序的主界面是否已关闭,如主界面已经关闭,则结束应用程序;否则,将聊天窗口界面设置为不可见,等待用户从主界面选择"聊天"后再次使用。具体工作过程如图 15-8 所示。

在主界面中,选择"退出信息服务器"功能后,将生成的请求对象传递给"与信息服务器通信"的子线程,后者发送请求给信息服务器,并从服务器接收响应,将响应通过建立的通信管道传递给主程序。随后,主程序结束"与信息服务器通信"的子线程,判断聊天窗口界面是否已关闭,如聊天窗口界面已经关闭,则结束应用程序;否则,将主界面设置为不可见。具体工作过程如图 15-9 所示。

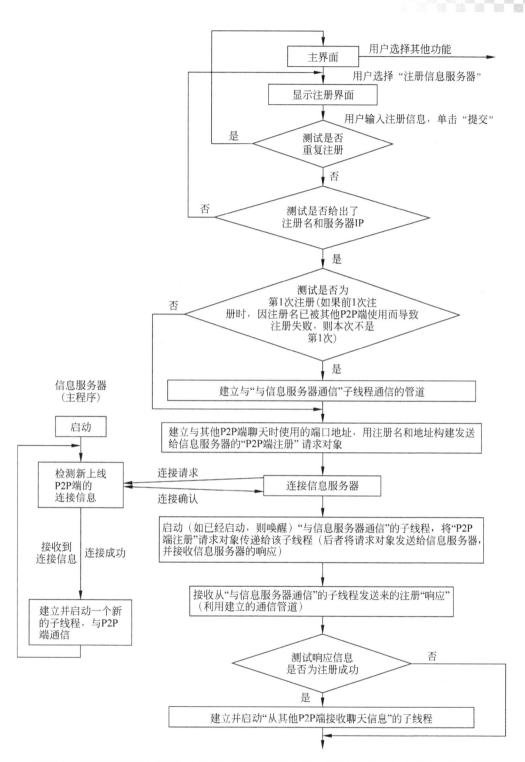

图 15-3　信息服务器（主程序）、P2P 聊天端"注册信息服务器"功能的工作过程及通信流程图

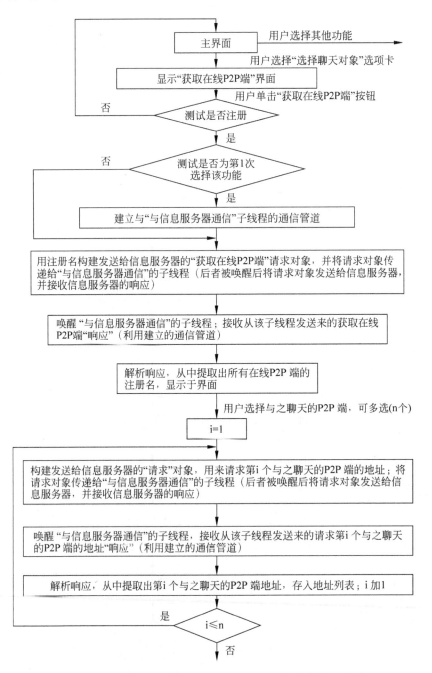

图 15-4  P2P 聊天端主界面——"选择聊天对象"功能的工作过程

## 2. "与信息服务器通信"的子线程

P2P 聊天端主程序启动时建立该子线程,注册时启动了该子线程的运行。P2P 聊天端"与信息服务器通信"的子线程具体的工作过程以及与信息服务器(子线程)的通信如图 15-10 所示。

# 第15章 Java与网络：P2P聊天系统

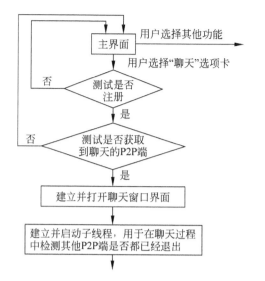

图 15-5　P2P 聊天端主界面——"聊天"功能的工作过程　　　图 15-6　P2P 聊天端（聊天窗口界面）

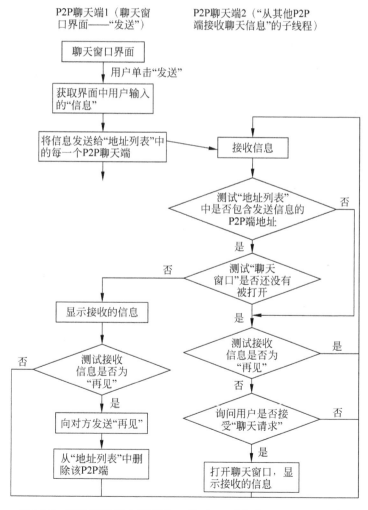

图 15-7　P2P 聊天端 1（聊天窗口界面——"发送"）、P2P 聊天端 2（"从其他 P2P 端接收聊天信息"的子线程）的工作过程及其通信

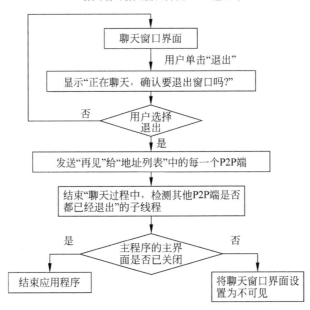

图 15-8　P2P 聊天端(聊天窗口界面——"退出")

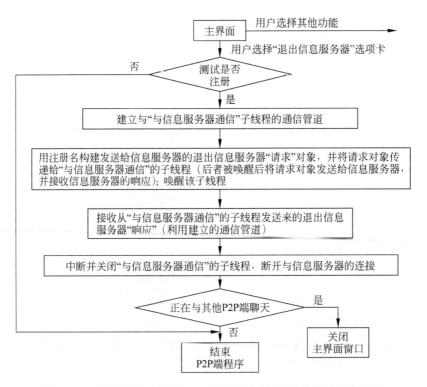

图 15-9　P2P 聊天端主界面——"退出信息服务器"功能的工作过程

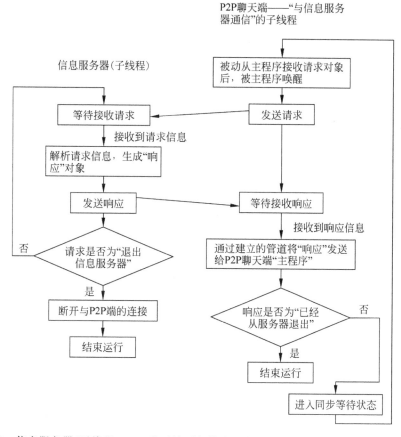

图 15-10　信息服务器(子线程)、P2P 聊天端("与信息服务器通信"的子线程)的工作过程与通信

### 3."从其他 P2P 端接收聊天信息"的子线程

P2P 聊天端主程序注册时建立并启动了该子线程(图 15-3)。P2P 聊天端"从其他 P2P 端接收聊天信息"的子线程具体的工作过程,以及与 P2P 端聊天对象主程序(聊天窗口界面——"发送"功能)的通信如图 15-7 所示。

### 4."聊天过程中,检测其他 P2P 端是否都已经退出"的子线程

P2P 聊天端主程序启动后,用户在主界面选择了"聊天"功能,这时需先建立并打开聊天窗口界面,然后建立并启动该子线程(图 15-5)。子线程具体的工作过程如图 15-11 所示。

综上所述,P2P 聊天端主程序在注册信息服务器时创建和启动了"与信息服务器通信"的子线程,在注册信息服务器、选择聊天对象、退出信息服务器时分别建立主程序与该子线程通信的管道。当主程序将创建的请求对象传递给子线程时,子线程被唤醒,发送请求对象给信息服务器,并接收信息服务器反馈的响应信息,将响应信息通过管道传递给主程序,随后进入到同步等待状态,等待唤醒。主程序选择退出信息服务器时,结束"与信息服务器通信"的子线程。

P2P 聊天端主程序在注册信息服务器时创建并启动了"从其他 P2P 端接收聊天信息"

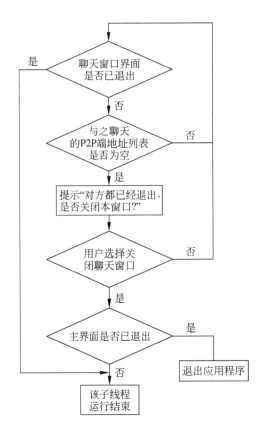

图 15-11 P2P 聊天端——"聊天过程中,检测其他 P2P 端是否都已经退出"子线程的工作过程

的子线程。该子线程不断接收其他 P2P 端发送的聊天信息,将其显示于聊天窗口界面中。当 P2P 聊天端应用程序结束运行时,该子线程也结束运行。

P2P 聊天端主程序在开始聊天时,创建并启动了"聊天过程中,检测其他 P2P 端是否都已经退出"的子线程。该子线程通过不断监测存放聊天对象的地址列表是否为空,来判断其他聊天对象是否都已经退出。当聊天窗口界面关闭,或应用程序结束运行时,该子线程也结束运行。

## 15.3 P2P 聊天系统实现思路

### 15.3.1 传输协议选择

一方面,P2P 聊天端需要与信息服务器通信,完成注册、选择聊天对象、退出信息服务器应用;另一方面,P2P 聊天端需要与其他 P2P 端聊天。

P2P 端与信息服务器的通信传递注册名、地址等重要信息,需要保证其可靠性,对实时性要求不迫切,所以选择使用 TCP 协议传输。

P2P 端与其他 P2P 端聊天进行即时通信,实时性要求高,选择使用 UDP 协议传输。

为此,P2P 端建立 DatagramSocket 对象,以便与其他 P2P 端之间交互聊天信息,并从

DatagramSocket 对象中获取 UDP 端口地址，和注册名一起封装为注册请求对象；P2P 端与信息服务器进行 TCP 连接，得到 Socket 对象，完成与信息服务器之间的通信。

### 15.3.2　P2P 端与信息服务器的应用协议

设计 Request 类和 Response 类，其中 Request 对象封装 P2P 聊天端发送给信息服务器的请求信息，而 Response 对象封装信息服务器反馈给 P2P 聊天端的响应信息。

P2P 端创建 Request 类的对象，发送给信息服务器；信息服务器收到请求后，生成相应的 Response 类的对象，反馈给 P2P 端。

各种类型的 Request 对象和 Response 对象封装的内容，二者的通信关系参见表 15-1。

## 15.4　P2P 聊天系统实现

### 15.4.1　Request 类和 Response 类

**1. Request 类**

Request 类封装了 P2P 端注册、获取在线 P2P 端、获取"已选择聊天对象（P2P 端）的地址"、退出信息服务器时，向服务器发送的请求信息。具体请求类型参见表 15-1。

1) UML 图

如图 15-12 所示。

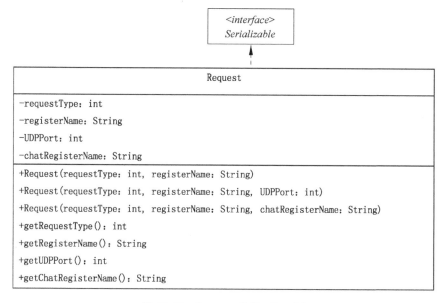

图 15-12　Request 类的 UML 图

Request 类创建的对象需在网络中传输，所以实现了 Serializable 接口。

成员变量：requestType 定义请求类型，registerName 定义注册名，UDPPort 定义聊天时使用的 UDP 端口地址；而 chatRegisterName 定义所选择聊天对象的注册名，发送包含此信息的请求对象的目的是获取聊天对象的地址。

2) 代码(Request.java)

```java
package appProtocol;
import java.io.Serializable;
public class Request implements Serializable {
    private int requestType;
    private String registerName;
    private int UDPPort;
    private String chatRegisterName;
    public Request(int requestType, String registerName) {
        this.requestType = requestType;
        this.registerName = registerName;
    }
    public Request(int requestType, String registerName, int UDPPort) {
        this(requestType, registerName);
        this.UDPPort = UDPPort;
    }
    public Request(int requestType, String registerName, String chatRegisterName){
        this(requestType, registerName);
        this.chatRegisterName = chatRegisterName;
    }
    public int getRequestType() {
        return requestType;
    }
    public String getRegisterName() {
        return registerName;
    }
    public int getUDPPort() {
        return UDPPort;
    }
    public String getChatRegisterName() {
        return chatRegisterName;
    }
}
```

## 2. Response 类

Response 类封装了当信息服务器收到 P2P 端注册、获取在线 P2P 端、获取"已选择聊天对象(P2P 端)的地址"、退出信息服务器的 Request 对象时，向 P2P 端发送的应答信息。

1) UML 图(图 15-13)

Response 类创建的对象需要在网络中传输，所以实现了 Serializable 接口。

成员变量 responseType 定义响应类型，message 定义应答字符串信息。

成员变量 allNameOfRegister 定义所有在线 P2P 端的注册名列表，包含该有效信息的响应类型为 2，是对获取在线 P2P 端请求(类型为 2)的应答。

成员变量 chatP2PEndAddress 定义发送请求的 P2P 端所选择聊天对象的地址，包含该

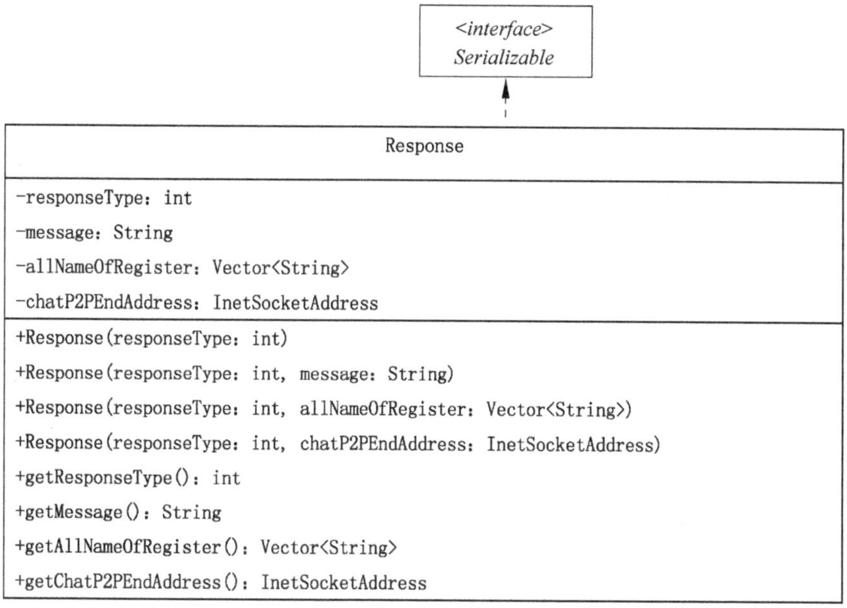

```
                    ┌─────────────────┐
                    │   <interface>   │
                    │   Serializable  │
                    └─────────────────┘
                              △
                              │
┌──────────────────────────────────────────────────────────┐
│                         Response                          │
├──────────────────────────────────────────────────────────┤
│ -responseType: int                                        │
│ -message: String                                          │
│ -allNameOfRegister: Vector<String>                        │
│ -chatP2PEndAddress: InetSocketAddress                     │
├──────────────────────────────────────────────────────────┤
│ +Response(responseType: int)                              │
│ +Response(responseType: int, message: String)             │
│ +Response(responseType: int, allNameOfRegister: Vector<String>) │
│ +Response(responseType: int, chatP2PEndAddress: InetSocketAddress) │
│ +getResponseType(): int                                   │
│ +getMessage(): String                                     │
│ +getAllNameOfRegister(): Vector<String>                   │
│ +getChatP2PEndAddress(): InetSocketAddress                │
└──────────────────────────────────────────────────────────┘
```

图 15-13 Response 类的 UML 图

有效信息的响应类型为 3，是对获取"已选择聊天对象的地址"请求（类型为 3）的应答。信息服务器需根据收到的请求对象所包含的"已选择聊天对象的注册名"，查阅到"已选择聊天对象的地址"。

具体请求和响应类型参见表 15-1。

2）代码（Response.java）

```java
package appProtocol;
import java.util.*;
import java.io.Serializable;
import java.net.InetSocketAddress;
public class Response implements Serializable {
    private int responseType;
    private String message;
    private Vector<String> allNameOfRegister;
    private InetSocketAddress chatP2PEndAddress;
    public Response(int responseType) {
        this.responseType = responseType;
    }
    public Response(int responseType, String message) {
        this(responseType);
        this.message = message;
    }
    public Response(int responseType, Vector<String> allNameOfRegister){
        this(responseType);
        this.allNameOfRegister = allNameOfRegister;
    }
    public Response(int responseType, InetSocketAddress chatP2PEndAddress) {
        this(responseType);
```

```java
        this.chatP2PEndAddress = chatP2PEndAddress;
    }
    public int getResponseType() {
        return responseType;
    }
    public String getMessage() {
        return message;
    }
    public Vector<String> getAllNameOfRegister() {
        return allNameOfRegister;
    }
    public InetSocketAddress getChatP2PEndAddress() {
        return chatP2PEndAddress;
    }
}
```

### 15.4.2 信息服务器的实现

根据以上设计和实现思路，信息服务器编写两个类：MessageServer 类和 MessageHandler 类。MessageServer 类实现主程序，MessageHandler 类是子线程的线程体类。

**1. MessageServer 类**

1）UML 图

如图 15-14 所示。

| MessageServer |
|---|
| +PORT{immutable}: int = 8000 |
| +MAX_QUEUE_LENGTH{immutable}: int = 100 |
| +start(): void |
| +main(args[]: String): void |

图 15-14 MessageServer 类的 UML 图

PORT 表示端口号，具有 public static final 属性，赋值 8000。信息服务器在该端口地址上等待 P2P 端的连接请求。

MAX_QUEUE_LENGTH 表示最大队列长度，也具有 public static final 属性，赋值 100。

在 start() 方法中，服务器创建 ServerSocket 对象，该对象通过调用其 accept() 方法在 8000 端口上监听。如监听到客户端，与之连接后返回 Socket 对象，start() 方法创建 MessageHandler 类型子线程，与该客户通信。start() 方法又回到循环开始，继续监听。

2）代码（MessageServer.java）

```java
package messageServer;
import java.net.ServerSocket;
```

```java
import java.net.Socket;
import java.io.IOException;
public class MessageServer {
    public static final int PORT = 8000;
    public static final int MAX_QUEUE_LENGTH = 100;
    public static void main(String args[]) {
        MessageServer messageServer = new MessageServer();
        messageServer.start();
    }
    public void start() {
        try {
            ServerSocket serverSocket = new ServerSocket(PORT, MAX_QUEUE_LENGTH);
            System.out.println("服务器已启动...");
            while (true) {
                Socket socket = serverSocket.accept();
                System.out.println("已接收到客户来自: " + socket.getInetAddress());
                MessageHandler handler = new MessageHandler(socket);
                handler.start();
            }
        } catch (IOException e) {
            e.printStackTrace();
        }
    }
}
```

3) 运行结果显示

服务器已启动...
已接收到客户来自:/127.0.0.1
"XiaoMei"注册成功...
已接收到客户来自:/127.0.0.1
"DaBaiTu"注册成功...
"XiaoMei"从服务器退出...

### 2. MessageHandler 类

1) UML 图(图 15-15)

MessageHandler 类是子线程的线程体类,所以实现了 Runnable 接口。

clientMessage 成员变量存储注册 P2P 端的注册名与地址信息,是所有子线程的线程体对象(每个子线程与一个 P2P 端通信)的公有哈希表变量,所以设置为 static 属性。其中注册名为键,地址为值,以便于查阅。当 P2P 端注册时,其注册名和地址信息被存入该哈希表;当 P2P 端请求退出信息服务器时,其注册名和地址信息被从哈希表变量中删除。

成员变量 listener 引用子线程对象,dataIn 与 dataOut 分别为子线程对象与 P2P 端的 CommWithServer 子线程通信时使用的输入输出流变量,request 和 response 分别为双方子线程交互时使用的协议数据单元。

构造方法 MessageHandler(Socket)创建子线程的线程体对象,利用参数接收从主程序传递的 Socket 对象。start()方法从 Socket 对象获取输入输出流对象,并利用线程体对象

图 15-15 MessageHandler 类的 UML 图

做参数构造出子线程对象 listener，启动 listener 子线程。stop()方法执行与 start()方法相反的操作。receiveRequest()方法和 sendResponse()方法分别用来接收请求和发送响应，且均为私有方法。parseRequest()方法解析接收到的请求对象类型，执行相应的操作后，生成响应对象。registerNameHasBeenUsed(registerName：String)方法则测试 P2P 端是否使用了其他 P2P 端已经用过的注册名。

在线程体的 run()方法中，使用布尔变量 keepListening 控制，循环执行接收请求、解析请求、生成响应、发送响应的操作。当 P2P 端请求退出信息服务器时，parseRequest()方法将 keepListening 变量设置为 false，循环被中断；此时调用 stop()方法断开与该 P2P 端的 TCP 连接后，线程体结束执行。执行过程参见图 15-10。

2) 代码(MessageHandler.java)

```java
package messageServer;
import java.net.*;
import java.io.*;
import java.util.*;
import appProtocol.Request;
import appProtocol.Response;
public class MessageHandler implements Runnable {
    private Socket socket;
    private ObjectInputStream dataIn;
    private ObjectOutputStream dataOut;
    private Thread listener;
```

```java
        private static Hashtable<String, InetSocketAddress> clientMessage
            = new Hashtable<String, InetSocketAddress>();
    private Request request;
    private Response response;
    private boolean keepListening = true;
    public MessageHandler(Socket socket) {
        this.socket = socket;
    }
    public synchronized void start() {
        if (listener == null) {
            try {
                dataIn = new ObjectInputStream(socket.getInputStream());
                dataOut = new ObjectOutputStream(socket.getOutputStream());
                listener = new Thread(this);
                listener.start();
            } catch (IOException e) {
                e.printStackTrace();
            }
        }
    }
    public synchronized void stop() {
        if (listener != null) {
            try {
                listener.interrupt();
                listener = null;
                dataIn.close();
                dataOut.close();
                socket.close();
            } catch (IOException e) {
                e.printStackTrace();
            }
        }
    }
    public void run() {
        try {
            while (keepListening) {          // 监听该 P2P 端
                receiveRequest();            // 接收请求
                parseRequest();              // 解析请求
                sendResponse();              // 发送响应
                request = null;              // 清除请求变量
            }
            stop();                          //结束线程前,断开与 P2P 端的 TCP 连接
        } catch (ClassNotFoundException e) {
            e.printStackTrace();
        } catch (IOException e) {
            stop();
            System.err.println("与客户端通信出现错误...");
        }
    }
    private void receiveRequest() throws IOException, ClassNotFoundException {
        request = (Request) dataIn.readObject();
```

```java
        }
        private void parseRequest() {
            if (request == null)
                return;
            response = null;
            int requestType = request.getRequestType();
            String registerName = request.getRegisterName();
            if (requestType != 1 && !registerNameHasBeenUsed(registerName)) {
                // 请求类型不为1,不是注册请求,且该 P2P 端还未注册过
                response = new Response(1, registerName + ",您还未注册!");
                return;
            }
            switch (requestType) {                      // 测试请求类型
            case 1: // P2P 端注册
                if (registerNameHasBeenUsed(registerName)) {
                    // 注册名已被他人使用,生成响应
                    response = new Response(1, "\"" + registerName + "\""
                            + "已被其他人使用,请您使用其他注册名注册!");
                    break;
                }
                clientMessage.put(registerName,
                    new InetSocketAddress(socket.getInetAddress(),
                        request.getUDPPort()));      // 注册成功,保存 P2P 端信息,生成响应
                response = new Response(1, registerName + ",您已经注册成功!");
                System.out.println("\"" + registerName + "\"注册成功...");
                break;
            case 2:// P2P 端请求获取已注册的 P2P 端注册名列表
                Vector<String> allNameOfRegister = new Vector<String>();
                for (Enumeration<String> e = clientMessage.keys(); e.hasMoreElements();)
                    // 生成已注册的 P2P 端注册名列表
                    allNameOfRegister.addElement(e.nextElement());
                response = new Response(2, allNameOfRegister);                  // 生成响应
                break;
            case 3:// P2P 端请求获取已选择 P2P 端(与之聊天)的地址
                String chatRegisterName = request.getChatRegisterName();
                InetSocketAddress chatP2PEndAddress = clientMessage.get(chatRegisterName);
                response = new Response(3, chatP2PEndAddress);                  // 生成响应
                break;
            case 4:// P2P 端请求退出信息服务器
                clientMessage.remove(registerName);// 从保存的信息中去除该 P2P 端
                response = new Response(1, registerName + ",您已经从服务器退出!");
                keepListening = false;              // 该 P2P 端已经退出,不再监听,结束本线程
                System.out.println("\"" + registerName + "\"从服务器退出...");
            }
        }
        private boolean registerNameHasBeenUsed(String registerName) {
            if (registerName != null && clientMessage.get(registerName) != null)
                return true;
            return false;
```

```
        }
        private void sendResponse() throws IOException {
            if (response != null) {
                dataOut.writeObject(response);
            }
        }
    }
```

### 15.4.3　P2P 聊天端的实现

根据以上设计和实现思路，P2P 聊天端程序编写 7 个类。其中 P2PChatEnd 类实现主程序的主界面，Register 类实现主程序的注册功能界面，GetOnlineP2PEnds 类实现主程序的选择聊天对象功能界面，Exit 类实现主程序的退出信息服务器功能界面；CommWithServer 类是与信息服务器通信的子线程类；Chat 类实现主程序的聊天功能界面，同时也是"从其他 P2P 端接收聊天信息"的子线程的线程体类；ChatWindow 类是聊天窗口界面类，同时也是"聊天过程中，检测其他 P2P 端是否都已经退出"的子线程的线程体类。

**1. P2PChatEnd 类**

1) 效果图（图 15-16）

图 15-16　P2P 聊天端主界面

2) UML 图（图 15-17）

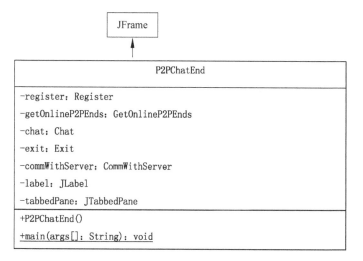

图 15-17　P2PChatEnd 类的 UML 图

图 15-17 所示的 UML 图标识出 P2PChatEnd 类的主要成员。

主程序的主界面 P2PChatEnd 类是 JFrame 类的子类。P2PChatEnd 类的标签成员变量 label 用来在窗口中显示系统封面；成员变量 tabbedPane 为选项卡窗格，该选项卡窗格中添加标题为"系统封面"、"注册信息服务器"、"选择聊天对象"、"聊天"、"退出信息服务器" 5 个选项卡。

成员变量 register、getOnlineP2PEnds、chat、exit 分别引用 Register 类、GetOnlineP2PEnds 类、Chat 类和 Exit 类创建的面板容器对象，各自负责提供注册功能界面、选择聊天对象功能界面、聊天功能界面和退出信息服务器功能界面，均是 P2PChatEnd 类的主要成员。

成员变量 commWithServer 是 CommWithServer 类创建的对象。

P2PChatEnd() 是构造方法，负责完成主程序主界面窗口的初始化，并创建 CommWithServer 类的对象，将"与信息服务器通信"的子线程对象传递给随后建立的 Register、GetOnlineP2PEnds、Exit 对象，以便在完成注册、选择聊天对象、退出信息服务器功能时，利用 CommWithServer 子线程与信息服务器通信；创建 Chat 类的对象，并将 Chat 对象传递给 Register 对象，以便在注册时将创建的聊天地址（UDP 套接字）传递给 Chat 对象。

3）代码（P2PChatEnd.java）

```java
package p2pChatEnd;
import javax.swing.*;
import java.awt.*;
public class P2PChatEnd extends JFrame {
    private Register register;
    private GetOnlineP2PEnds getOnlineP2PEnds;
    private Chat chat;
    private JLabel label;
    private JTabbedPane tabbedPane;
    private Exit exit;
    private CommWithServer commWithServer;
    public P2PChatEnd() {
        setTitle("P2P 聊天端");
        label = new JLabel();
        label.setText(" P2P 聊天端");
        label.setForeground(Color.blue);
        label.setFont(new Font("隶书", Font.BOLD, 22));
        label.setIcon(new ImageIcon("welcome.jpg"));
        label.setHorizontalTextPosition(SwingConstants.RIGHT);
        label.setBackground(Color.green);
        commWithServer = new CommWithServer();
        register = new Register(commWithServer);
        getOnlineP2PEnds = new GetOnlineP2PEnds(commWithServer);
        chat = new Chat(this);
        register.setChat(chat);
        exit = new Exit(commWithServer, this);
        tabbedPane = new JTabbedPane(JTabbedPane.LEFT);
        tabbedPane.add("系统封面", label);
        tabbedPane.add("注册信息服务器", register);
```

```
            tabbedPane.add("选择聊天对象", getOnlineP2PEnds);
            tabbedPane.add("聊 天", chat);
            tabbedPane.add("退出信息服务器", exit);
            add(tabbedPane, BorderLayout.CENTER);
            setBounds(120, 60, 400, 147);
            setVisible(true);
            setDefaultCloseOperation(JFrame.DO_NOTHING_ON_CLOSE);
     }
     public static void main(String args[]) {
            new P2PChatEnd();
     }
}
```

### 2. Register 类

1) 效果图（图 15-18）

图 15-18　P2P 聊天端注册界面

2) UML 图（图 15-19）

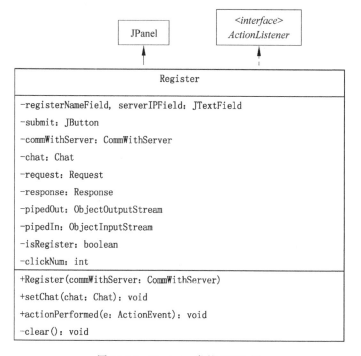

图 15-19　Register 类的 UML 图

图 15-19 所示的 UML 图标识出 Register 类的主要成员。

主程序的注册功能界面 Register 类是 JPanel 类的子类，并实现了 ActionListener 接口。

Register 类的成员变量 request 和 response 分别用来引用发送给信息服务器的注册请求对象，以及从信息服务器接收到的响应对象。

成员变量 pipedOut 和 pipedIn 分别引用主程序的 Register 类对象与 CommWithServer 子线程通信使用的管道输出流和管道输入流对象。该管道只在第一次注册时创建（如果前一次注册时，因注册名已被其他 P2P 端使用而导致注册失败，则本次注册不是第 1 次），成员变量 clickNum 用来控制注册次数。

布尔型成员变量 isRegister 控制不能重复注册。如果 isRegister 值为 true，则已经注册过，此次为重复注册。

成员变量 commWithServer 引用从主界面 P2PChatEnd 对象传递过来的"与信息服务器通信"的子线程对象。

Register(CommWithServer)是构造方法，负责接收从主界面 P2PChatEnd 对象传递过来的"与信息服务器通信"的子线程对象，并完成注册界面的初始化，以及为"提交"按钮 submit 注册动作监听器。

setChat(Chat)方法接收从主界面 P2PChatEnd 对象传递过来的 Chat 对象（赋给成员变量 chat），以便注册时将创建的聊天地址（UDP 套接字）传递给 Chat 对象，以及用该 Chat 对象作为线程体，建立并启动"从其他 P2P 端接收聊天信息"子线程。

actionPerformed(ActionEvent) 方法是 ActionListener 接口定义的方法。当用户在单行文本域 registerNameField 和 serverIPField 中输入注册名和服务器 IP 地址后，单击"提交"按钮时，执行该方法。该方法的工作过程参见图 15-3。

clear()为私有成员方法，用来在单击"提交"按钮后清除单行文本域 registerNameField 和 serverIPField 中的内容。

3）代码（Register.java）

```java
package p2pChatEnd;
import javax.swing.*;
import java.awt.event.*;
import java.awt.*;
import java.net.*;
import java.io.*;
import appProtocol.Response;
import appProtocol.Request;
public class Register extends JPanel implements ActionListener {
    private JLabel hintLabel;
    private JTextField registerNameField, serverIPField;
    private JButton submit;
    private CommWithServer commWithServer;
    private Chat chat;
    private Request request;
    private Response response;
    private ObjectOutputStream pipedOut;
```

```java
    private ObjectInputStream pipedIn;
    private int clickNum = 0;
    private boolean isRegister = false;
    public Register(CommWithServer commWithServer) {
        this.commWithServer = commWithServer;
        setLayout(new BorderLayout());
        hintLabel = new JLabel("注册", JLabel.CENTER);
        hintLabel.setFont(new Font("隶书", Font.BOLD, 18));
        registerNameField = new JTextField(10);
        serverIPField = new JTextField(10);
        submit = new JButton("提交");
        submit.addActionListener(this);
        Box box1 = Box.createHorizontalBox();
        box1.add(new JLabel("注 册 名：", JLabel.CENTER));
        box1.add(registerNameField);
        Box box2 = Box.createHorizontalBox();
        box2.add(new JLabel("服务器 IP：", JLabel.CENTER));
        box2.add(serverIPField);
        Box boxH = Box.createVerticalBox();
        boxH.add(box1);
        boxH.add(box2);
        boxH.add(submit);
        JPanel panelC = new JPanel();
        panelC.setBackground(new Color(210, 210, 110));
        panelC.add(boxH);
        add(panelC, BorderLayout.CENTER);
        JPanel panelN = new JPanel();
        panelN.setBackground(Color.green);
        panelN.add(hintLabel);
        add(panelN, BorderLayout.NORTH);
    }
    public void setChat(Chat chat) {
        this.chat = chat;
    }
    public void actionPerformed(ActionEvent e) {
        if (isRegister) {
            String hint = "不能重复注册!";
            JOptionPane.showMessageDialog(this, hint, "警告",
                    JOptionPane.WARNING_MESSAGE);
            clear();
            return;
        }
        clickNum++;
        String registerName = registerNameField.getText().trim();
        String serverIP = serverIPField.getText().trim();
        if (registerName.length() == 0 || serverIP.length() == 0) {
            String hint = "必须输入注册名和服务器 IP!";
            JOptionPane.showMessageDialog(this, hint, "警告",
                    JOptionPane.WARNING_MESSAGE);
            clear();
            return;
```

```java
            }
            try {
                if (clickNum == 1) {
                    PipedInputStream pipedI = new PipedInputStream();
                    PipedOutputStream pipedO = new PipedOutputStream(pipedI);
                    pipedOut = new ObjectOutputStream(pipedO);
                    pipedIn = new ObjectInputStream(pipedI);
                }
                DatagramSocket datagramSocket = new DatagramSocket();
                Chat.setSocket(datagramSocket);        //将创建的聊天地址传递给 Chat 对象
                int UDPPort = datagramSocket.getLocalPort();
                request = new Request(1, registerName, UDPPort);
                if (commWithServer != null) {
                    if (commWithServer.isAlive()){ //线程已经启动,已与信息服务器连接
                        commWithServer.close();       //断开与信息服务器的连接
                        commWithServer.connect(serverIP, request, pipedOut);
                        // 连接信息服务器
                        commWithServer.notifyCommWithServer();                //将线程唤醒
                    } else {
                        commWithServer.connect(serverIP, request, pipedOut);
                        // 连接信息服务器
                        commWithServer.start();      // 启动线程,与信息服务器通信
                    }
                }
                response = (Response) pipedIn.readObject();
            } catch (Exception ex) {
                JOptionPane.showMessageDialog(this, "无法连接或与服务器通信出错",
                    "警告",JOptionPane.WARNING_MESSAGE);
                clear();
                return;
            }
            String message = response.getMessage();
            boolean flag = true;
            if (message != null
                && message.equals(request.getRegisterName() + ",您已经注册成功!")){
                message += "请单击左侧的\"获取在线 P2P 端\"";
                flag = false;
            }
            JOptionPane.showMessageDialog(null, message, "信息提示",
                    JOptionPane.PLAIN_MESSAGE);
            if (flag) { //注册没有成功,清除单行文本域,返回重新注册
                clear();
                return;
            }
    /*注册成功,将注册名传递给 GetOnlineP2PEnds 类对象、Chat 类对象和 Exit 类对象 */
            GetOnlineP2PEnds.setRegisterName(registerName);
            Chat.setRegisterName(registerName);
            Exit.setRegisterName(registerName);
            isRegister = true;                        //设置注册成功标志,控制不能重复注册
                //建立并启动"从其他 P2P 端接收聊天信息"的子线程,等待接收信息
            new Thread(chat).start();
```

```
        clear();
    }
    private void clear(){
        registerNameField.setText("");
        serverIPField.setText("");
    }
}
```

### 3．GetOnlineP2PEnds 类

1）效果图（图 15-20）

图 15-20　P2P 聊天端选择聊天对象界面

2）UML 图（图 15-21）

图 15-21 所示的 UML 图标识出主程序的选择聊天对象功能界面 GetOnlineP2PEnds 类的主要成员。

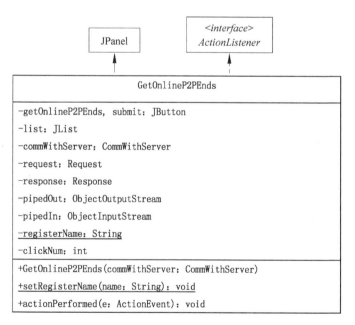

图 15-21　GetOnlineP2PEnds 类的 UML 图

GetOnlineP2PEnds 类是 JPanel 类的子类，并实现了 ActionListener 接口。

GetOnlineP2PEnds 类的成员变量 getOnlineP2PEnds 和 submit 分别为窗口中的"获取在线 P2P 端"按钮和"提交"按钮。成员变量 list 是多行文本框，用来显示获取到的在线 P2P 端注册名列表。

成员变量 request 和 response 分别用来引用发送给信息服务器的"获取在线 P2P 端"和"获取已选择聊天对象（P2P 端）的地址"请求对象，以及相应从信息服务器接收到的响应对象。

成员变量 pipedOut 和 pipedIn 分别引用主程序的 GetOnlineP2PEnds 类对象与 CommWithServer 子线程通信使用的管道输出流和管道输入流对象。该管道只在用户第一次单击"选择聊天对象"选项卡时创建，成员变量 clickNum 用来控制用户单击"选择聊天对象"选项卡的次数。

成员变量 commWithServer 引用从主界面 P2PChatEnd 对象传递过来的"与信息服务器通信"的子线程对象。

GetOnlineP2PEnds（CommWithServer）是构造方法，负责接收从界面 P2PChatEnd 对象传递过来的"与信息服务器通信"的子线程对象，并完成选择聊天对象界面的初始化，为按钮注册动作监听器。

setRegisterName(String)方法是静态方法，接收从 Register 对象传递过来的注册名（赋给静态成员变量 registerName）。

actionPerformed(ActionEvent) 方法是 ActionListener 接口定义的方法。当用户单击"获取在线 P2P 端"或"提交"按钮时，执行该方法。该方法的工作过程参见图 15-4。

3）代码（GetOnlineP2PEnds.java）

```java
package p2pChatEnd;
import javax.swing.*;
import java.awt.event.*;
import java.awt.*;
import java.net.*;
import java.io.*;
import java.util.*;
import appProtocol.Response;
import appProtocol.Request;
public class GetOnlineP2PEnds extends JPanel implements ActionListener{
    private JButton getOnlineP2PEnds, submit;
    private JList list;
    private CommWithServer commWithServer;
    private Request request;
    private Response response;
    private ObjectOutputStream pipedOut;
    private ObjectInputStream pipedIn;
    private static String registerName;
    private int clickNum = 0;
    public GetOnlineP2PEnds(CommWithServer commWithServer) {
        this.commWithServer = commWithServer;
        setLayout(new BorderLayout());
        getOnlineP2PEnds = new JButton("获取在线 P2P 端");
        getOnlineP2PEnds.setBackground(Color.green);
        submit = new JButton("提 交");
        submit.setBackground(Color.green);
```

```java
            getOnlineP2PEnds.addActionListener(this);
            submit.addActionListener(this);
            list = new JList();
            list.setFont(new Font("楷体", Font.BOLD, 15));
            JScrollPane scroll = new JScrollPane();
            scroll.getViewport().setView(list);
            Box box = Box.createHorizontalBox();
            box.add(new JLabel("单击"获取" : ", JLabel.CENTER));
            box.add(getOnlineP2PEnds);
            JPanel panelR = new JPanel(new BorderLayout());
            panelR.setBackground(new Color(210, 210, 110));
            panelR.add(submit, BorderLayout.SOUTH);
            JPanel panel = new JPanel(new BorderLayout());
            panel.setBackground(new Color(210, 210, 110));
            panel.add(box, BorderLayout.NORTH);
            panel.add(new JLabel("选择聊天 P2P 端: "), BorderLayout.WEST);
            panel.add(scroll, BorderLayout.CENTER);
            panel.add(panelR, BorderLayout.EAST);
            add(panel, BorderLayout.CENTER);
            submit.setEnabled(false);
            validate();
    }
    public static void setRegisterName(String name) {
            registerName = name;
    }
    public void actionPerformed(ActionEvent e) {
            if (registerName == null || commWithServer == null
                    || !commWithServer.isAlive()) {
                JOptionPane.showMessageDialog(null, "您还没有注册!", "信息提示",
                        JOptionPane.PLAIN_MESSAGE);
                return;
            }
            try {
                if (e.getSource() == getOnlineP2PEnds) {
                    clickNum++;
                    if (clickNum == 1) {
                        PipedInputStream pipedI = new PipedInputStream();
                        PipedOutputStream pipedO = new PipedOutputStream(pipedI);
                        pipedOut = new ObjectOutputStream(pipedO);
                        pipedIn = new ObjectInputStream(pipedI);
                    }
                    request = new Request(2, registerName);
                    //以下两行代码将请求和管道输出流对象传递给 CommWithServer 对象
                    commWithServer.setRequest(request);
                    commWithServer.setPipedOut(pipedOut);
                    //以下一行代码唤醒 CommWithServer 子线程
                    commWithServer.notifyCommWithServer();
                    response = (Response) pipedIn.readObject();
                    //以下一行代码从响应中得到在线的 P2P 端注册名列表
                    Vector<String> onLineP2PEnds = response.getAllNameOfRegister();
                    list.setListData(onLineP2PEnds);           //注册名列表显示于 list 中
```

```java
                    submit.setEnabled(true);
                }
                if (e.getSource() == submit) {
                    Object[] object = list.getSelectedValues();
                    int len = object.length;
                    if (len == 0) {
                        JOptionPane.showMessageDialog(this, "您还未选择聊天 P2P 端!",
                                "信息提示",JOptionPane.PLAIN_MESSAGE);
                        return;
                    }
                    String register[] = new String[object.length];
                    for (int i = 0; i < object.length; i++)
                        register[i] = (String) object[i];
                    Vector< InetSocketAddress > P2PEndAddress =
                        new Vector< InetSocketAddress >();              //创建聊天对象地址列表
                    int chatP2PEnds = 0;
                    for (int i = 0; i < len; i++) {
                        if (register[i].equals(registerName))
                            continue;
                        request = new Request(3, registerName, register[i]);
                        //以下两行代码将请求和管道输出流对象传递给 CommWithServer 对象
                        commWithServer.setRequest(request);
                        commWithServer.setPipedOut(pipedOut);
                        //以下一行代码唤醒 CommWithServer 子线程
                        commWithServer.notifyCommWithServer();
                        response = (Response) pipedIn.readObject();
                        //以下一行代码将从响应中得到的聊天对象地址加入列表中
                        P2PEndAddress.add(response.getChatP2PEndAddress());
                        chatP2PEnds++;
                    }
                    String message = null;
                    if (chatP2PEnds == 0) {
                        message = "您只选择了与自己聊天,请重新选择聊天端!";
                    } else { //将聊天对象地址列表传递给 Chat 对象
                        Chat.setChatP2PEndAddress(P2PEndAddress);
                        message = "已获取到您选择 P2P 端的地址,请单击左侧的\"聊天\"按钮";
                    }
                    JOptionPane.showMessageDialog(this, message, "信息提示",
                            JOptionPane.PLAIN_MESSAGE);
                    P2PEndAddress.clear();          //清空地址列表
                    list.setListData(P2PEndAddress);
                }
            } catch (Exception ex) {
                JOptionPane.showMessageDialog(this, "与服务器通信出错", "警告",
                        JOptionPane.WARNING_MESSAGE);
            }
        }
    }
```

### 4. Exit 类

1）效果图（图 15-22）

图 15-22　P2P 聊天端退出信息服务器界面

2）UML 图（图 15-23）

图 15-23 所示的 UML 图标识出主程序的退出信息服务器功能界面 Exit 类的主要成员。

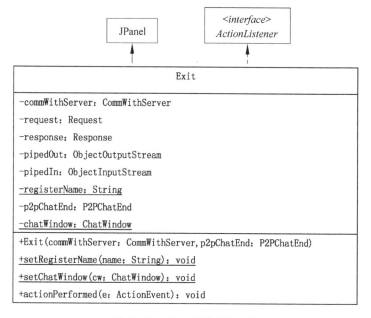

图 15-23　Exit 类的 UML 图

Exit 类是 JPanel 类的子类，并实现了 ActionListener 接口。

成员变量 request 和 response 分别用来引用发送给信息服务器的"退出信息服务器"请求对象，以及从信息服务器接收到的响应对象。只有当用户已经注册至信息服务器时，会使用这两个对象。否则，只是退出主界面。

成员变量 pipedOut 和 pipedIn 分别引用主程序的 Exit 类对象与 CommWithServer 子线程通信使用的管道输出流和管道输入流对象。该管道只有当用户已经注册至信息服务器时创建。

成员变量 commWithServer 引用从主界面 P2PChatEnd 对象传递过来的"与信息服务器通信"的子线程对象。

Exit(CommWithServer，P2PChatEnd)是构造方法，负责接收从界面 P2PChatEnd 对

象传递过来的"与信息服务器通信"的子线程对象,以及接收 P2PChatEnd 对象本身,并完成退出信息服务器界面的初始化。

setRegisterName(String)方法是静态方法,接收从 Register 对象传递过来的注册名,将其赋值于静态成员变量 registerName。

setChatWindow(ChatWindow)方法是静态方法,接收 ChatWindow 聊天窗口对象,将其赋值于静态成员变量 chatWindow。该方法在 ChatWindow 类中调用。

actionPerformed(ActionEvent) 方法是 ActionListener 接口定义的方法。当用户单击"退出"按钮时,执行该方法。该方法的工作过程参见图 15-9。

3) 代码(Exit.java)

```java
package p2pChatEnd;
import javax.swing.*;
import java.awt.event.*;
import java.awt.*;
import java.io.*;
import appProtocol.Response;
import appProtocol.Request;
public class Exit extends JPanel implements ActionListener {
    private CommWithServer commWithServer;
    private Request request;
    private Response response;
    private ObjectOutputStream pipedOut;
    private ObjectInputStream pipedIn;
    private static String registerName;
    private P2PChatEnd P2PChatEnd;
    private static ChatWindow chatWindow;
    public Exit(CommWithServer commWithServer, P2PChatEnd P2PChatEnd) {
        this.commWithServer = commWithServer;
        this.P2PChatEnd = P2PChatEnd;
        JLabel hint = new JLabel("确认要退出信息服务器吗?");
        JButton quit = new JButton("退出");
        quit.addActionListener(this);
        setLayout(new BorderLayout());
        add(hint, BorderLayout.CENTER);
        add(quit, BorderLayout.SOUTH);
    }
    public static void setRegisterName(String str) {
        registerName = str;
    }
    public static void setChatWindow(ChatWindow cw) {
        chatWindow = cw;
    }
    public void actionPerformed(ActionEvent e) {
        if (registerName != null && commWithServer.isAlive()) {//已注册
            try {
                PipedInputStream pipedI = new PipedInputStream();
                PipedOutputStream pipedO = new PipedOutputStream(pipedI);
                pipedOut = new ObjectOutputStream(pipedO);
```

```java
                pipedIn = new ObjectInputStream(pipedI);
                request = new Request(4, registerName);
                //以下两行代码将请求和管道输出流对象传递给 CommWithServer 对象
                commWithServer.setRequest(request);
                commWithServer.setPipedOut(pipedOut);
                //以下一行代码唤醒 CommWithServer 子线程
                commWithServer.notifyCommWithServer();
                //以下一行代码利用管道从 CommWithServer 子线程接收"响应"
                response = (Response) pipedIn.readObject();
            } catch (Exception ex) {
                JOptionPane.showMessageDialog(this,"与服务器通信出错","警告",
                        JOptionPane.WARNING_MESSAGE);
            }
            String message = response.getMessage() + "单击\"确定\"退出";
            JOptionPane.showMessageDialog(null, message, "信息提示",
                    JOptionPane.PLAIN_MESSAGE);
            //以下两行代码结束并关闭"与信息服务器通信"的子线程
            commWithServer.keepCommunicating = false;
            commWithServer.interrupt();
            commWithServer.close();              //断开与信息服务器的 TCP 连接
        }
        commWithServer = null;
        if (chatWindow == null || !chatWindow.isVisible())
            System.exit(0);
        else
            P2PChatEnd.setVisible(false);
    }
}
```

### 5．CommWithServer 类

1) UML 图(图 15-24)

图 15-24 所示的 UML 图标识出 CommWithServer 类的主要成员。

CommWithServer 类是 Thread 类的子类，实现了 P2P 端"与信息服务器通信"的子线程。

构造方法 CommWithServer()被在 P2P 端主界面类 P2PChatEnd 的构造方法中调用，创建的子线程对象被传递给 Register 对象、GetOnlineP2PEnds 对象和 Exit 对象。

connect(String,Request,ObjectOutputStream)成员方法被在注册界面类 Register 的 actionPerformed 方法中调用，connect 方法从 Register 类接收信息服务器的 IP 地址或主机名(赋值于成员变量 serverIP)、注册请求对象(赋值于成员变量 request)和管道输出流对象(赋值于成员变量 pipedOut，用来与注册功能界面通信，将从信息服务器返回的 response 响应对象发送给注册功能界面)，与信息服务器建立 TCP 套接字连接(赋值于成员变量 Socket，连接时服务器端口号使用静态常量 PORT 指定的值 8000)，并从 Socket 中获取到输入输出流(赋值于成员变量 in 和 out)，以便与信息服务器通信。

成员方法 setRequest(Request)和 setPipedOut(ObjectOutputStream)被在 GetOnlineP2PEnds 类和 Exit 类的 actionPerformed 方法中调用，分别用来被动地从 GetOnlineP2PEnds 对象和

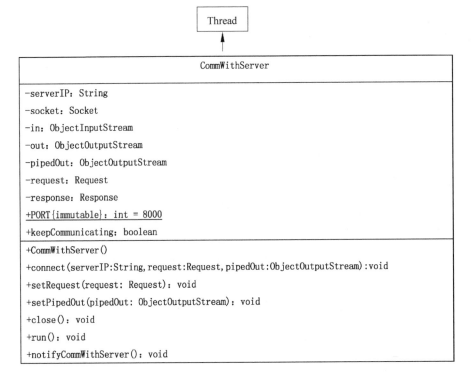

图 15-24 CommWithServer 类的 UML 图

Exit 对象中接收请求对象和管道输出流对象。

run()成员方法是线程体方法,其运行流程参见图 15-10。在 run()方法中,keepCommunicating(成员变量)是循环控制变量,该变量初始值为 true,在 Exit 类的 actionPerformed 方法中会被修改为 false。

close()成员方法关闭与信息服务器通信的输入输出流以及 TCP 连接。当发生异常时,close 方法在 run 方法中被调用。另外,close 方法会在 Exit 类的 actionPerformed 方法中被调用。

notifyCommWithServer()成员方法分别在 register 类、GetOnlineP2PEnds 类和 Exit 类的 actionPerformed 方法中被调用,该方法唤醒 CommWithServer 子线程。

2) 代码(CommWithServer.java)

```
package p2pChatEnd;
import java.net.*;
import java.io.*;
import appProtocol.Response;
import appProtocol.Request;
public class CommWithServer extends Thread {
    private String serverIP;
    private Socket socket;
    private ObjectInputStream in;
    private ObjectOutputStream out;
    private ObjectOutputStream pipedOut;
```

```java
    private Request request;
    private Response response;
    public static final int PORT = 8000;
    public boolean keepCommunicating = true;
    public CommWithServer() {
    }
    public void connect(String serverIP, Request request,
            ObjectOutputStream pipedOut) throws IOException {
        this.serverIP = serverIP;
        this.request = request;
        this.pipedOut = pipedOut;
        InetAddress address = InetAddress.getByName(serverIP);
        InetSocketAddress serverSocketA = new InetSocketAddress(address, PORT);
        socket = new Socket();
        socket.connect(serverSocketA);
        out = new ObjectOutputStream(socket.getOutputStream());
        in = new ObjectInputStream(socket.getInputStream());
    }
    public void setRequest(Request request) {
        this.request = request;
    }
    public void setPipedOut(ObjectOutputStream pipedOut) {
        this.pipedOut = pipedOut;
    }
    public synchronized void close() {
        try {
            in.close();
            out.close();
            socket.close();
        } catch (IOException e) {
            e.printStackTrace();
        }
    }
    public void run() {
        while (keepCommunicating) {
            synchronized (this) {
                try {
                    out.writeObject(request);       //被唤醒后向信息服务器发送请求
                    response = (Response) in.readObject();        //从信息服务器接收响应
                } catch (ClassNotFoundException e) {
                    e.printStackTrace();
                } catch (IOException e) {
                    close();
                    System.err.println("与服务器通信出现错误...");
                    return;
                }
                try {
                    pipedOut.writeObject(response);       //利用管道将响应发送给主程序
                    if (response.getResponseType() == 1) {
                        String message = response.getMessage();
                        if (message != null
```

```
                        && message.equals(request.getRegisterName()
                            + ",您已经从服务器退出!"))
                    return;
                }
                request = null;
                response = null;
                wait();                          //使子线程进入同步等待状态
            } catch (IOException e) {
                System.err.println("管道通信出现错误...");
            } catch (InterruptedException e) {
                System.err.println("线程同步出现错误...");
            }
        }
    }
    public synchronized void notifyCommWithServer() {
        notify();
    }
}
```

### 6. Chat 类

1) 效果图(图 15-25)

图 15-25　P2P 聊天端主程序的聊天功能界面

2) UML 图(图 15-26)

图 15-26 所示的 UML 图标识出主程序的聊天功能界面 Chat 类的主要成员。

Chat 类是 JPanel 类的子类,并实现了 ActionListener 接口和 Runnable 接口,是"从其他 P2P 端接收聊天信息"了线程的线程休类。

静态成员变量 registerName 和 socket 定义当前 P2P 端的注册名和用于聊天的 UDP 套接字地址。静态成员方法 setRegisterName(String) 和 setSocket(DatagramSocket) 在 Register 类的 actionPerformed 方法中被调用,用来从 Register 对象传递当前 P2P 端的注册名和用于聊天的 UDP 套接字地址给 Chat 对象。

静态成员变量 chatP2PEndAddress 定义存放各个聊天对象地址的数据结构。静态成员方法 setChatP2PEndAddress(Vector<InetSocketAddress>) 在 GetOnlineP2PEnds 类的 actionPerformed 方法中被调用,用来从 GetOnlineP2PEnds 对象传递聊天对象地址给 Chat 对象。

成员变量 p2pChatEnd 定义 P2P 端主程序的主界面。构造方法 Chat(P2PChatEnd) 在 P2PChatEnd 类的构造方法中被调用,接收从 P2PChatEnd 对象传递的主程序主界面对象,

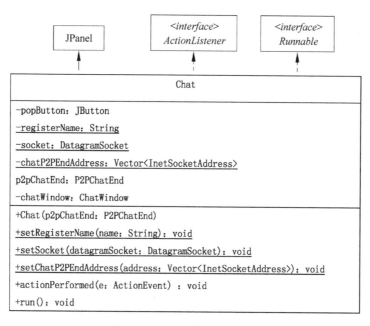

图 15-26　Chat 类的 UML 图

并完成如图 15-25 所示界面的初始化。

成员变量 chatWindow 定义弹出的聊天窗口界面。

成员变量 popButton 定义了图 15-25 所示的聊天功能界面中的"弹出聊天窗口"按钮，成员方法 actionPerformed(ActionEvent)是 ActionListener 接口定义的方法，当用户单击 popButton 按钮时，执行该方法。actionPerformed 方法的执行过程参见图 15-5。

run()方法是"从其他 P2P 端接收聊天信息"子线程的线程体。该方法的执行过程参见图 15-7。

3）代码（Chat.java）

```
package p2pChatEnd;
import javax.swing.*;
import java.awt.event.*;
import java.awt.*;
import java.net.*;
import java.io.IOException;
import java.util.Vector;
public class Chat extends JPanel implements ActionListener, Runnable {
    private JButton popButton;
    private static String registerName;
    private static DatagramSocket socket;
    private static Vector<InetSocketAddress> chatP2PEndAddress;
    P2PChatEnd p2pChatEnd;
    private ChatWindow chatWindow;
    public Chat(P2PChatEnd p2pChatEnd) {
        this.p2pChatEnd = p2pChatEnd;
        setLayout(new BorderLayout());
```

```java
            popButton = new JButton("弹出聊天窗口");
            popButton.addActionListener(this);
            add(popButton, BorderLayout.CENTER);
            chatP2PEndAddress = new Vector<InetSocketAddress>();
    }
    public static void setRegisterName(String name) {
            registerName = name;
    }
    public static void setSocket(DatagramSocket datagramSocket) {
            socket = datagramSocket;
    }
    public static void setChatP2PEndAddress(Vector<InetSocketAddress> address) {
            chatP2PEndAddress.addAll(address);
    }
    public void actionPerformed(ActionEvent e) {
        if (registerName == null) {
            JOptionPane.showMessageDialog(null, "您还没有注册!", "信息提示",
                    JOptionPane.PLAIN_MESSAGE);
            return;
        }
        if (chatP2PEndAddress.isEmpty()) {         //聊天对象地址列表为空
            JOptionPane.showMessageDialog(null, "您还没有获取到聊天的P2P端!",
"信息提示",JOptionPane.PLAIN_MESSAGE);
            return;
        }
        if (chatWindow == null)                    //如没有聊天窗口界面,则创建
            chatWindow = new ChatWindow(registerName, socket, p2pChatEnd);
        //如下一行代码传递"聊天对象地址列表"给聊天窗口界面chatWindow
        chatWindow.setChatP2PEndAddress(chatP2PEndAddress);
        if (!chatWindow.isVisible()) {
            //建立并启动子线程,用于在聊天过程中,检测其他P2P端是否都已经退出
            chatWindow.beginMonitor(true);
            chatWindow.setVisible(true);
        }
    }
    public void run() {
        byte[] buffer = new byte[256];
        DatagramPacket packet = null;
        try {
            while (true) {
                for (int i = 0; i < buffer.length; i++)
                    buffer[i] = (byte) 0;
                packet = new DatagramPacket(buffer, buffer.length);      //构建数据报
                socket.receive(packet);        // 从其他P2P端接收聊天信息
                InetAddress ip = packet.getAddress();
                int port = packet.getPort();    //从数据报提取发送方的IP地址与端口号
                InetSocketAddress socketAddress = new InetSocketAddress(ip,port);
                                    //上一行用IP地址与端口号构建发送方套接字地址
```

```java
            //以下从接收数据报中提取聊天信息
            String received = new String(packet.getData()).trim();
            int index = received.indexOf('>');
            boolean receiveGoodbye =
                received.indexOf("再见", index + 1) == index + 1;
            //以下测试聊天对象地址列表中是否包含该发送方地址
            boolean contain = chatP2PEndAddress.contains(socketAddress);
            if (!contain || chatWindow == null || !chatWindow.isVisible()) {
                if (receiveGoodbye) //如接收到"再见",回到循环起始
                    continue;
                String chatP2PEnd = received.substring(0, index);
                int option = JOptionPane.showConfirmDialog(this,
                    "收到\"" + chatP2PEnd + "\"聊天请求,是否接受?");
                if (option == 0) {
                    //接受聊天请求,将发送方地址加入聊天对象地址列表
                    chatP2PEndAddress.add(socketAddress);
                    if (chatWindow == null) { //创建聊天窗口界面
                        chatWindow = new ChatWindow(registerName, socket,
                            p2pChatEnd);
                        chatWindow.validate();
                    }
                    //以下将聊天对象地址列表传递给聊天界面窗口
                    chatWindow.setChatP2PEndAddress(chatP2PEndAddress);
                    if (!chatWindow.isVisible()) {
            //建立并启动子线程,用于在聊天过程中检测其他 P2P 端是否都已经退出
                        chatWindow.beginMonitor(true);
                        chatWindow.setVisible(true);
                    }
                    //以下将接收到的聊天信息显示于聊天窗口界面
                    chatWindow.setReceived(received);
                }
                continue; //回到循环起始
            }
            chatWindow.setReceived(received);
            if (receiveGoodbye) {                    //如收到"再见",则向发送方回送"再见"
                chatWindow.endChat(socketAddress);
                //以下从聊天对象地址列表中去除该发送方地址
                chatP2PEndAddress.remove(socketAddress);
            }                                        //if 块结束
        }                                            //while 循环块结束
    } catch (IOException e) {
        JOptionPane.showMessageDialog(this, "接收信息时,网络连接出现问题!");
    }                                                //try-catch 块结束
}                                                    //run 方法结束
}                                                    //chat 类结束
```

### 7. ChatWindow 类

1)效果图(图 15-27)

图 15-27　P2P 聊天端的聊天窗口界面

2)UML 图(图 15-28)

图 15-28 所示的 UML 图标识出主程序的聊天界面窗口 ChatWindow 类的主要成员。

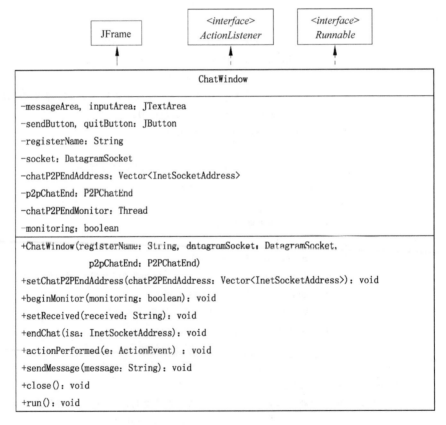

图 15-28　ChatWindow 类的 UML 图

ChatWindow 类是 JFrame 类的子类，并实现了 ActionListener 接口和 Runnable 接口，是"聊天过程中，检测其他 P2P 端是否都已经退出"子线程的线程体类。

构造方法 ChatWindow（String，DatagramSocket，P2PChatEnd）在 Chat 类的 actionPerformed 方法和 run 方法中调用，第一个参数接收从 Chat 对象传递的本地 P2P 端注册名，赋值于成员变量 registerName；第二个参数接收用于聊天的本地 P2P 端 UDP 套接字地址，赋值于成员变量 socket；第三个参数接收本地 P2P 端对象本身，赋值于成员变量 p2pChatEnd。构造方法完成如图 15-27 所示聊天界面的初始化，并将"聊天窗口对象"传递给"P2P 聊天端退出信息服务口界面，窗口对象（Exit）"。

成员变量 messageArea 和 inputArea 定义聊天界面中的聊天信息显示窗口和聊天信息输入窗口，成员变量 sendButton 和 quitButton 定义聊天界面中的发送信息按钮和退出聊天窗口按钮。

成员变量 chatP2PEndAddress 定义存放各个聊天对象地址的数据结构。成员方法 setChatP2PEndAddress(Vector＜InetSocketAddress＞) 在 Chat 类的 actionPerformed 方法和 run 方法中被调用，用来从 Chat 对象传递聊天对象地址给 ChatWindow 对象。

成员方法 setReceived(String) 在 Chat 类的线程体 run 方法中被调用，作用是将从其他 P2P 端接收到的聊天信息显示于 messageArea 聊天信息显示窗口。

成员变量 chatP2PEndMonitor 定义"聊天过程中，检测其他 P2P 端是否都已经退出"子线程对象，而成员变量 monitoring 定义子线程体中的循环控制变量。成员方法 beginMonitor(boolean) 在 Chat 类的 actionPerformed 方法和 run 方法中被调用，布尔型参数接收从 Chat 对象设置的循环控制变量值，赋值于成员变量 monitoring；在 beginMonitor 方法中，创建"聊天过程中，检测其他 P2P 端是否都已经退出"子线程对象，并启动子线程运行。

成员方法 endChat(InetSocketAddress) 在 Chat 类的线程体 run 方法中被调用，该方法的作用是将本地 P2P 端发出的"再见"信息显示于本地 messageArea 聊天信息显示窗口，并发送至方法参数指定地址的 P2P 聊天端；之后，将方法参数指定的 P2P 聊天端地址从"与之聊天的 P2P 端地址列表"中删除。

成员方法 sendMessage(String) 在 actionPerformed 方法和 close 方法中调用，作用是将参数指定的字符串信息发送至"与之聊天的 P2P 端地址列表"中的每一个 P2P 端。

成员方法 close() 执行关闭聊天窗口的操作，在关闭窗口前，结束（chatP2PEndMonitor 成员变量定义的）"聊天过程中，检测其他 P2P 端是否都已经退出"子线程的运行。该方法在本类的线程体 run 方法（该线程体属于"聊天过程中，检测其他 P2P 端是否都已经退出"子线程）、quitButton 按钮事件处理 actionPerformed 方法、关闭聊天窗口事件处理 windowClosing 中调用。

成员方法 actionPerformed(ActionEvent) 是 ActionListener 接口定义的方法，当用户单击 sendButton 和 quitButton 按钮时，执行该方法。actionPerformed 方法的执行过程参见图 15-7 和图 15-8。

run() 方法是"聊天过程中，检测其他 P2P 端是否都已经退出"的子线程的线程体。该方法的执行过程参见图 15-11。

3）代码（ChatWindow.java）

```java
package p2pChatEnd;
import javax.swing.*;
import java.awt.event.*;
import java.awt.*;
import java.net.*;
import java.io.IOException;
import java.util.Vector;
import javax.swing.border.BevelBorder;
public class ChatWindow extends JFrame implements ActionListener, Runnable {
    private JTextArea messageArea, inputArea;
    private JButton sendButton, quitButton;
    private JLabel hintMessage1, hintMessage2, statusBar;
    private String registerName;
    private DatagramSocket socket;
    private Vector<InetSocketAddress> chatP2PEndAddress;
    private P2PChatEnd p2pChatEnd;
    private Thread chatP2PEndMonitor;
    //"聊天过程中,检测其他 P2P 端是否都已经退出"子线程
    private boolean monitoring;                    //上述子线程体中的循环控制变量
    private String newline;
    public ChatWindow(String registerName, DatagramSocket datagramSocket,
            P2PChatEnd p2pChatEnd) {
        super("聊天窗口");
        this.registerName = registerName;
        socket = datagramSocket;
        this.p2pChatEnd = p2pChatEnd;
        monitoring = true;
        newline = System.getProperty("line.separator");
        hintMessage1 = new JLabel("显示聊天记录");
        hintMessage2 = new JLabel("编辑信息");
        messageArea = new JTextArea(4, 20);
        messageArea.setEditable(false);
        messageArea.setWrapStyleWord(true);
        messageArea.setLineWrap(true);
        inputArea = new JTextArea(4, 20);
        inputArea.setWrapStyleWord(true);
        inputArea.setLineWrap(true);
        sendButton = new JButton("发送");
        sendButton.addActionListener(this);
        quitButton = new JButton("退出");
        quitButton.addActionListener(this);
        statusBar = new JLabel("在线: " + registerName);
        statusBar.setBorder(new BevelBorder(BevelBorder.LOWERED));
        JPanel messagePanel = new JPanel();
        messagePanel.setLayout(new BorderLayout());
        messagePanel.add(hintMessage1, BorderLayout.NORTH);
        messagePanel.add(new JScrollPane(messageArea), BorderLayout.CENTER);
        JPanel buttonPanel = new JPanel();
        buttonPanel.setLayout(new GridLayout(2, 1));
```

```java
        buttonPanel.add(sendButton);
        buttonPanel.add(quitButton);
        Box box1 = new Box(BoxLayout.X_AXIS);
        box1.add(new JScrollPane(inputArea));
        box1.add(buttonPanel);
        Box box = new Box(BoxLayout.Y_AXIS);
        box.add(hintMessage2);
        box.add(box1);
        messagePanel.add(box, BorderLayout.SOUTH);
        Container contentPane = getContentPane();
        contentPane.add(messagePanel, BorderLayout.CENTER);
        contentPane.add(statusBar, BorderLayout.SOUTH);
        addWindowListener(new WindowAdapter() {
            public void windowClosing(WindowEvent e) {
                close();
            }
        });
        setSize(300, 400);
        Exit.setChatWindow(this);
    }
    public void setChatP2PEndAddress(Vector<InetSocketAddress> chatP2PEndAddress) {
        this.chatP2PEndAddress = chatP2PEndAddress;
    }
    public void beginMonitor(boolean monitoring) {
        this.monitoring = monitoring;
        chatP2PEndMonitor = new Thread(this);
        chatP2PEndMonitor.start();
    }
    public void setReceived(String received) {
        messageArea.append(received + newline);
    }
    public void endChat(InetSocketAddress isa) {
        String message = registerName + ">再见!";
        messageArea.append(message + newline);
        byte[] buf = message.getBytes();
        try {
            DatagramPacket packet = null;
            packet = new DatagramPacket(buf, buf.length, isa.getAddress(),
                    isa.getPort());
            socket.send(packet);
        } catch (IOException ee) {
            JOptionPane.showMessageDialog(this, "发送\"再见\"时,网络连接出错!");
        }
        chatP2PEndAddress.remove(isa);
    }
    public void actionPerformed(ActionEvent e) {
        if (e.getSource() == sendButton) {
            String message = registerName + ">" + inputArea.getText();
            sendMessage(message);
            inputArea.setText("");
        }
```

```java
            if (e.getSource() == quitButton) {
                close();
            }
        }
        public void sendMessage(String message) {
            messageArea.append(message + newline);
            byte[] buf = message.getBytes();
            DatagramPacket packet = null;
            try {
                for (InetSocketAddress isa : chatP2PEndAddress) {
                    packet = new DatagramPacket(buf, buf.length, isa.getAddress(),isa.getPort());
                    socket.send(packet);
                }
            } catch (IOException ee) {
                JOptionPane.showMessageDialog(this, "发送信息时,网络连接出错!");
            }
        }
        public void close() {
            if (!chatP2PEndAddress.isEmpty()) {
                int option = JOptionPane.showConfirmDialog(this, "正在聊天,确认要退出窗口吗?");
                if (option != 0)
                    return;
                String message = registerName + ">再见";
                sendMessage(message);
                monitoring = false;
                int i = 0;
                do {
                } while (!chatP2PEndAddress.isEmpty() && ++i <= 30);
            }
            monitoring = false;
            chatP2PEndMonitor = null;
            if (p2pChatEnd == null || !p2pChatEnd.isVisible())
                System.exit(0);
            else
                setVisible(false);
        }
        public void run() {
            while (monitoring) {
                if (!isVisible() || !chatP2PEndAddress.isEmpty())
                    continue;
                int option = JOptionPane.showConfirmDialog(this,
                        "对方都已经退出,是否关闭本窗口?");
                if (option != 0) {
                    try {
                        chatP2PEndMonitor.sleep(1000);
                    } catch (InterruptedException e) {
                        e.printStackTrace();
                    }
                    continue;
                }
                close();
```

                }
            }
        }

## 15.5 P2P聊天系统调试与软件发布

### 15.5.1 系统调试

#### 1. 信息服务器

在Eclipse环境下创建一个新的工程。在该工程中用来存放源代码的路径src下创建两个包：appProtocol包和messageServer包。在appProtocol包中创建两个类：Request类和Response类（详见15.4.1节）。在messageServer包中创建两个类：MessageServer类和MessageHandler类（详见15.4.2节）。然后运行主类MessageServer即可。

#### 2. P2P端

在Eclipse环境下创建一个新的工程。在该工程中用来存放源代码的路径src下创建两个包：appProtocol包和p2pChatEnd包。在appProtocol包中创建两个类：Request类和Response类（详见15.4.1节）。在p2pChatEnd包中创建7个类：P2PChatEnd类、Register类、GetOnlineP2PEnds类、Exit类、CommWithServer类、Chat类和ChatWindow类（详见15.4.3节）。然后运行主类P2PChatEnd即可。

### 15.5.2 软件发布

#### 1. 信息服务器

在Eclipse环境下，在菜单中单击"文件"→Export，会弹出一个界面；在弹出的界面中，展开Java后，选择Runnable JAR file，然后单击Next按钮，会弹出另外一个界面；在该界面中，有两个组合框，在提示有Launch configuration的组合框中选择工程的主类MessageServer，在提示有Export destination的组合框中给出生成的JAR文件的文件名（这里给出的文件名为MessageServer.jar）及其存放路径，然后单击Finish按钮。

现在，可以将生成的文件MessageServer.jar复制到任何一台安装了Java运行环境的计算机上，读者用鼠标双击该文件的图标，即可运行信息服务器软件。

#### 2. P2P端

在Eclipse环境下，在菜单中单击"文件"→Export，会弹出一个界面；在弹出的界面中，展开Java后，选择Runnable JAR file，然后单击Next按钮，会弹出另外一个界面；在该界面中，有两个组合框，在提示有Launch configuration的组合框中选择工程的主类P2PChatEnd，在提示有Export destination的组合框中给出生成的JAR文件的文件名（这里给出的文件名为P2PChatEnd.jar）及其存放路径，然后单击Finish按钮。

现在，可以将生成的文件 P2PChatEnd.jar 复制到任何一台安装了 Java 运行环境的计算机上，读者用鼠标双击该文件的图标，即可运行 P2P 端软件。

## 15.6 综合课程设计作业

### 15.6.1 P2P 聊天系统扩展

对系统增设"离线文件发送"功能。在 P2P 聊天过程中，可以向聊天对象发送离线文件。

### 15.6.2 综合课程设计题目

设计和实现 P2P 文件共享系统。

# 第16章

# Java与网络：Web服务器与浏览器

## 16.1 Web 服务器与浏览器需求分析

### 16.1.1 Web 服务器需求分析

Web 服务器应实现以下功能：
(1) 等待接收浏览器主动发起的连接请求，与之建立 TCP 连接。
(2) 接收浏览器发送的 Http 请求，解析请求，向浏览器发送 Http 响应。
(3) 断开与浏览器的连接。
此外，Web 服务器应能同时为多个浏览器客户提供上述服务。

### 16.1.2 浏览器需求分析

浏览器应实现以下功能：
(1) 接收用户输入的 URL 或用户单击的超链接。
(2) 主动向 Web 服务器发送连接请求，与之建立 TCP 连接。
(3) 向 Web 服务器发送 Http 请求，接收 Web 服务器发送的 Http 响应。
(4) 断开与 Web 服务器的连接。
(5) 解析 Http 响应。
(6) 显示接收到的信息。

## 16.2 Web 服务器与浏览器系统设计

### 16.2.1 Web 服务器功能设计

根据需求分析，Web 服务器启动后，不断检测新上线的浏览器发来的连接请求。当接收到新的连接请求后，Web 服务器需并行完成以下多项工作：
(1) 继续检测新上线的浏览器发来的连接请求。
(2) 继续为已连接的浏览器提供信息服务。
(3) 与刚发送连接请求的浏览器建立 TCP 连接后，为其提供信息服务。为此，需执行以下操作：

① 接收 Http 请求；
② 解析 Http 请求；
③ 发送 Http 响应。

在②中，解析出接收到的 Http 请求，完成一定的操作，并且据此生成 Http 响应。

主程序不断检测新上线的浏览器发来的连接请求，当接收到新的连接请求后，创建子线程。在子线程中，为当前浏览器提供信息服务。

Web 服务器的工作过程以及与浏览器之间的通信参见图 16-1。

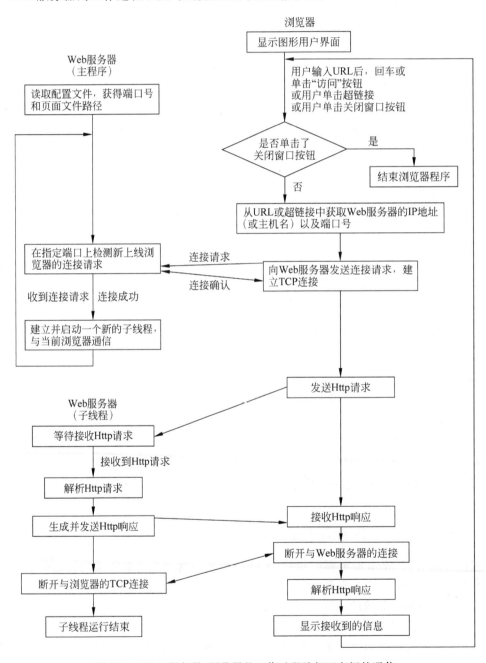

图 16-1　Web 服务器、浏览器的工作过程及相互之间的通信

Web 服务器不需要与用户交互和显示信息,所以不需提供图形用户界面。

此外,Web 服务器启动前应从配置文件中读出存取的端口号和页面文件路径的相关信息。

### 16.2.2 浏览器功能设计

根据需求分析,设计浏览器具有图形用户界面,能与用户交互和显示接收到的信息。浏览器启动后,进入图形用户界面。用户可以通过以下两种方式与其交互:

(1) 输入 URL 后,回车(或单击"访问"按钮)。

(2) 单击超链接。

在进行事件处理时,浏览器依次完成的工作参见图 16-1。

## 16.3 Web 服务器与浏览器系统实现思路

### 16.3.1 传输协议选择

浏览器与 Web 服务器的通信主要需保证其可靠性,对实时性要求不迫切,属于弹性应用,所以选择使用 TCP 协议传输。

### 16.3.2 浏览器与 Web 服务器的应用协议

设计 Request 类和 Response 类,在 Request 类中封装发送 Http 请求、接收 Http 请求和解析 Http 请求的方法(方法名分别为 sendHttpRequest、receiveHttpRequest 和 parseHttpRequest),在 Response 类中封装发送 Http 响应、接收 Http 响应和解析 Http 响应的方法(方法名为 sendHttpResponse、receiveHttpResponse 和 parseHttpResponse)。

Web 服务器与浏览器建立 TCP 连接后,创建子线程,用于与浏览器通信。在 Web 服务器的子线程体中,创建 Request 类的对象,通过对象先后调用 receiveHttpRequest()和 parseHttpRequest()方法;之后创建 Response 类的对象,通过对象调用 sendHttpResponse 方法。对于浏览器,主动与 Web 服务器建立 TCP 连接后,创建 Request 类的对象,通过对象调用 sendHttpRequest()方法;之后创建 Response 类的对象,通过对象调用 receiveHttpResponse()和 parseHttpResponse()方法。

### 16.3.3 增加"生成配置文件"功能

Web 服务器在启动前,必须从配置文件中获取到下列信息:

(1) 运行的端口号;

(2) 存放页面文件的路径。

该配置文件如何得到?设计一个应用程序,该程序运行后,自动生成一配置文件(文件名为 web.xml)。配置文件所存储的端口号,在该应用程序运行时,由用户从界面中给出;页面文件的存储路径,取应用程序运行时所在路径下的 webroot 目录。

## 16.4 Web 服务器与浏览器系统实现

### 16.4.1 应用协议的实现

**1. Request 类**

1) UML 图(图 16-2)

图 16-2 所示的 UML 图标识出 Request 类的主要成员。

| Request |
|---|
| -url: URL |
| -input: InputStream |
| -output: PrintStream |
| -httpRequestMessage: String |
| -uri: String |
| +Request(input: InputStream) |
| +Request(url: URL, output: OutputStream) |
| +sendHttpRequest(): void |
| +receiveHttpRequest(): void |
| +parseHttpRequest(): void |
| +parseUri(): void |
| +getUri(): String |

图 16-2 Request 类的 UML 图

成员变量：url 定义资源的 URL 地址，Request 对象以此生成 Http 请求报文，从浏览器发送至资源所在的 Web 服务器；output 定义浏览器向 Web 服务器发送 Http 请求报文时所需要的输出流；input 定义 Web 服务器从浏览器接收 Http 请求报文时所需要的输入流；httpRequestMessage 定义请求报文；uri 定义从请求报文中解析出的资源 URI 字符串。

构造方法：Request(input: InputStream)在 Web 服务器的子线程体中被调用，创建 Request 对象用于接收和解析 Http 请求报文，调用时，参数 input 被传递给成员变量 input；Request(url: URL, output: OutputStream)在浏览器 Browser 对象的 connectServer()方法中被调用，创建 Request 对象用于发送 Http 请求报文，调用时，参数 url(资源的 URL 地址)和 output 分别被传递给成员变量 url 和 output。

成员方法：sendHttpRequest()方法在浏览器界面的事件处理方法中被调用，用于发送 Http 请求报文；receiveHttpRequest()和 parseHttpRequest()方法均在 Web 服务器的子线程体中被调用，分别用来接收和解析 Http 请求；parseUri()由 parseHttpRequest()方法调用，用来从请求报文中解析出资源的 URI 字符串；getUri()在 Response 对象的 sendHttpResponse()方法中调用，用来获取解析出的资源 URI 字符串。

2）代码（Request.java）

```java
package appProtocol;
import java.io.IOException;
import java.io.*;
import java.net.URL;
public class Request {
    private URL url;
    private InputStream input;
    private PrintStream output;
    private String httpRequestMessage;
    private String uri;
    public Request(InputStream input) {
        this.input = input;
    }
    public Request(URL url,OutputStream output) {
        this.url = url;
        this.output = new PrintStream(output);
    }
    public void sendHttpRequest() throws IOException {
        String host = url.getHost();
        int port = url.getPort();
        if (port == -1)
            port = 80;
        String path = url.getPath();
        httpRequestMessage = "GET " + path + " HTTP/1.1\r\n" + "Host:"
            + host + ":" + port + "\r\n"
            + "Connection: Close\r\n" + "\r\n";
        output.println(httpRequestMessage);
    }
    public void receiveHttpRequest() {
        StringBuffer sb = new StringBuffer(2048);
        int i;
        byte[] buffer = new byte[2048];
        try {
            i = input.read(buffer);
        } catch (IOException e) {
            e.printStackTrace();
            i = -1;
        }
        for (int j = 0; j < i; j++) {
            sb.append((char) buffer[j]);
        }
        httpRequestMessage = sb.toString();
    }
    public void parseHttpRequest() {
        parseUri();
    }
    private void parseUri() {
        int index1, index2;
        index1 = httpRequestMessage.indexOf(' ');
```

```java
            if (index1 != -1) {
                index2 = httpRequestMessage.indexOf(' ', index1 + 1);
                if (index2 > index1)
                    uri = httpRequestMessage.substring(index1 + 1, index2);
            }
        }
        public String getUri() {
            return uri;
        }
    }
```

**2. Response 类**

1) UML 图(图 16-3)

图 16-3 所示的 UML 图标识出 Response 类的主要成员。

| Response |
|---|
| request: Request<br>-input: Reader<br>-output: PrintWriter<br>-httpResponseMessage: String |
| +Response(input: InputStream)<br>+Response(output: OutputStream)<br>+setRequest(request: Request): void<br>+sendHttpResponse(): void<br>+receiveHttpResponse(): void<br>+parseHttpResponse(): String |

图 16-3 Response 类的 UML 图

成员变量：request 定义 Request 对象，以便在 sendHttpResponse()方法中通过 request 调用其 getUri()方法获取页面文件的 URI(请参考源代码 Response.java)，request 变量的值在 setRequest()方法被调用时传递进来；output 定义 Web 服务器向浏览器发送响应报文时所需要的输出流，input 定义浏览器从 Web 服务器接收响应报文时所需要的输入流；httpResponseMessage 定义响应报文或响应报文头部。

构造方法：Response(output：OutputStream)在 Web 服务器的子线程体中被调用，创建 Response 对象用于发送 Http 响应报文，调用时，参数 output 被传递给成员变量 output；Response(input：InputStream)在浏览器 Browser 对象的 connectServer()方法中被调用，创建 Response 对象用于接收 Http 响应报文，调用时，参数 input 被传递给成员变量 input。

成员方法：setRequest(request：Request)在 Web 服务器的子线程体中被调用，参数 request 被传递给 Response 对象的成员变量 request；sendHttpResponse()在 Web 服务器的子线程体中被调用，用来从 Web 服务器向浏览器发送 Http 响应；receiveHttpResponse()方法和 parseHttpResponse()方法在浏览器界面的事件处理方法中被调用，用于接收和解析 Http 响应报文，parseHttpResponse()方法返回接收到的页面文件内容，用字符串表示。

2）代码（Response.java）

```java
package appProtocol;
import java.io.*;
import java.util.*;
import webServer.WebServer;
public class Response {
    private static final int BUFFER_SIZE = 8192;
    Request request;
    private Reader input;
    private PrintWriter output;
    private String httpResponseMessage;
    public Response(OutputStream output) {
        this.output = new PrintWriter(new OutputStreamWriter(output));
    }
    public Response(InputStream input) {
        this.input = new InputStreamReader(input);
    }
    public void setRequest(Request request) {
        this.request = request;
    }
    public void sendHttpResponse() throws IOException {
        StringBuffer sb = new StringBuffer(BUFFER_SIZE);
        BufferedReader fbr = null;
        try {
            File file = new File(WebServer.WEB_ROOT, request.getUri());
            if (file.exists()) {
                httpResponseMessage = "HTTP/1.1 200 OK\r\n"
                        + "Connection: close\r\n" + "Date: "
                        + new GregorianCalendar().getTime() + "\r\n"
                        + "Content-Length: " + file.length() + "\r\n"
                        + "Content-Type: text/html\r\n" + "\r\n";
                sb.append(httpResponseMessage);
                fbr = new BufferedReader(new FileReader(file));
                String str = fbr.readLine();
                while (str != null) {
                    sb.append(str);
                    sb.append("\r\n");
                    str = fbr.readLine();
                }
                output.write(sb.toString());
                output.flush();
            } else {
                httpResponseMessage = "HTTP/1.1 404 File Not Found\r\n"
                        + "Content-Type:text/html\r\n"
                        + "Content-Length:23\r\n" + "\r\n"
                        + "<HTML>\r\n" + "<Body>\r\n"
                        + "<h1>File Not Found</h1>\r\n"
                        + "</Body>\r\n" + "</HTML>";
```

```java
                output.write(httpResponseMessage);
                output.flush();
            }
        } catch (IOException e) {
            System.out.println(e.getMessage());
        } finally {
            if (fbr != null)
                fbr.close();
        }
    }
    public void receiveHttpResponse() throws IOException {
        boolean loop = true;
        StringBuffer sb = new StringBuffer(BUFFER_SIZE);
        int i;
        char[] buffer = new char[BUFFER_SIZE];
        while (loop) {
            if (input.ready()) {
                i = input.read(buffer);
                while (i != -1) {
                    sb.append(buffer);
                    i = input.read(buffer);
                }
                loop = false;
            }
        }
        httpResponseMessage = sb.toString();
    }
    public String parseHttpResponse() {
        int index1, index2;
        String state;
        StringBuffer sb = new StringBuffer(BUFFER_SIZE);
        index1 = httpResponseMessage.indexOf("HTTP");
        if (index1 == -1 || index1 > 0)
            return httpResponseMessage;
        state = httpResponseMessage.substring(index1 + 9, index1 + 12);
        if (!state.equals("200")) {
            index2 = httpResponseMessage.indexOf("\r\n");
            sb.append("<h1>");
            sb.append(httpResponseMessage.substring(index1 + 9, index2));
            sb.append("</h1>");
        }
        index2 = httpResponseMessage.indexOf("\r\n\r\n");
        sb.append("\r\n\r\n");
        sb.append(httpResponseMessage.substring(index2 + 4));
        return sb.toString();
    }
}
```

## 16.4.2 Web 服务器的实现

根据以上设计和实现思路,Web 服务器编写两个类:WebServer 类和 SocketThread

类。WebServer 类实现主程序，SocketThread 类是子线程类。在此例中，考虑到 SocketThread 对象只在 WebServer 对象中创建，则将 SocketThread 类作为私有内部类，嵌套于 WebServer 类中。

1. WebServer 类

1）UML 图（图 16-4）

```
┌─────────────────────────────────────┐
│             WebServer               │
├─────────────────────────────────────┤
│ -port: int                          │
│ +WEB_ROOT: String                   │
│ +WEB_CONFIG: String                 │
├─────────────────────────────────────┤
│ +WebServer()                        │
│ -getConfig(fileName: String): void  │
│ -start(): void                      │
│ +main(args[]: String): void         │
└─────────────────────────────────────┘
```

图 16-4 WebServer 类的 UML 图

图 16-4 所示的 UML 图标识出 WebServer 类的主要成员。

port 表示端口号，在 getConfig() 方法中，从配置文件中获取到属性值，默认赋值 80。Web 服务器在 port 值表示的端口地址上等待浏览器的连接请求。

WEB_ROOT 表示 Web 服务器存放页面文件的根路径字符串，在 getConfig() 方法中，从配置文件中获取到属性值，默认赋值为当前路径下的 webroot 目录。WEB_ROOT 具有 public static 属性。

WEB_CONFIG 表示配置文件路径＋配置文件名字符串，赋值为当前路径＋"web. xml"，即配置文件名为 web. xml。WEB_CONFIG 具有 public static final 属性。

在入口方法 main() 中调用构造方法 WebServer() 创建 WebServer 类的对象。构造方法 WebServer() 中，先后调用 getConfig(String) 方法和 start() 方法。getConfig(String) 方法将成员变量 WEB_CONFIG 的值传给入口参数 fileName，从配置文件中获取 webroot 属性值和 port 属性值，分别赋予成员变量 WEB_ROOT 和 port；start() 方法中，服务器创建 ServerSocket 对象，该对象通过调用其 accept() 方法在 port 变量值指明的端口上监听，如监听到客户端，与之连接后返回 Socket 对象，然后 start() 方法创建其内部类 SocketThread 类型的子线程，将 Socket 对象作为构造方法参数传递给子线程，并将子线程启动，与该客户通信。start() 方法又回到循环开始，继续监听。其工作过程请参见图 16-1。

2）代码（WebServer.java）

```
package webServer;
import java.net.*;
import java.io.*;
import java.util.Iterator;
import org.dom4j.Document;
import org.dom4j.DocumentException;
import org.dom4j.Element;
```

```java
import org.dom4j.io.SAXReader;
import appProtocol.*;
public class WebServer {
    private int port = 80;
    public static String WEB_ROOT = System.getProperty("user.dir")
            + File.separator + "webroot";
    public static final String WEB_CONFIG = System.getProperty("user.dir")
            + File.separator + "web.xml";
    public static void main(String args[]) {
        new WebServer();
    }
    public WebServer() {
        getConfig(WEB_CONFIG);
        start();
    }
    private void getConfig(String fileName) {
        File inputXml = new File(fileName);
        if (!inputXml.exists())
            return;
        SAXReader saxReader = new SAXReader();
        try {
            Document document = saxReader.read(inputXml);
            Element root = document.getRootElement();
            boolean flag1 = false, flag2 = false;
            for (Iterator i = root.elementIterator(); i.hasNext();) {
                Element node = (Element) i.next();
                String nodeName = node.getName().trim();
                if (nodeName.equals("webroot")) {
                    WEB_ROOT = node.getText();
                    flag1 = true;
                } else if (nodeName.equals("port")) {
                    port = Integer.parseInt(node.getText());
                    flag2 = true;
                }
                if (flag1 && flag2)
                    break;
            }
        } catch (DocumentException e) {
            System.out.println(e.getMessage());
        }
    }
    private void start() {
        System.out.println("Web server starting ...");
        ServerSocket serverSocket = null;
        try {
            serverSocket = new ServerSocket(port, 100);
        } catch (IOException e) {
            e.printStackTrace();
            System.exit(0);
        }
        System.out.println("Web server started .");
        System.out.println("Port Number: " + port);
        while (true) {
            Socket socket = null;
```

```java
            int clientNumber = 0;
            try {
                socket = serverSocket.accept();
            } catch (IOException e) {
                e.printStackTrace();
                continue;
            }
            new SocketThread(socket, ++clientNumber).start();
        }
    }
    private class SocketThread extends Thread {
        Socket socket;
        int clientNumber;
        InputStream inputStream;
        OutputStream outputStream;
        public SocketThread(Socket socket, int clientNumber) {
            this.socket = socket;
            this.clientNumber = clientNumber;
        }
        public void run() {
            try {
                inputStream = socket.getInputStream();
                outputStream = socket.getOutputStream();
                Request request = new Request(inputStream);
                request.receiveHttpRequest();
                request.parseHttpRequest();
                Response response = new Response(outputStream);
                response.setRequest(request);
                response.sendHttpResponse();
                socket.close();
            } catch (IOException e) {
                e.printStackTrace();
            }
        }
    }
}
```

3) 运行结果显示

```
Web server starting ...
Web server started .
Port Number: 80
```

### 2. SocketThread 类

1) UML 图（图 16-5）

SocketThread 类是 Web 服务器的子线程类，是 Thread 类的子类。该类被嵌套在 WebServer 类中，是 WebServer 类的私有内部类。

成员变量 socket 表示套接字连接，clientNumber 表示与 Web 服务器建立连接的浏览器客户顺序号，输入流 inputStream 和输出流 outputStream 通过 socket 获取，分别用于从浏览器接收信息和向浏览器发送信息。

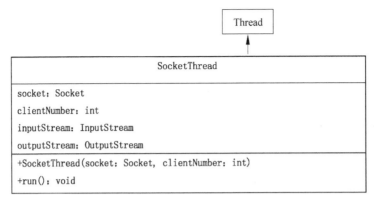

图 16-5　SocketThread 类的 UML 图

构造方法 SocketThread() 在 WebServer 对象的 start() 方法中被调用,从 WebServer 对象接收 Socket 对象和浏览器客户顺序号,赋值于成员变量 socket 和 clientNumber。

线程体的 run() 方法中,由 socket 获取到输入流 inputStream 和输出流 outputStream,创建 Request 对象,执行接收 Http 请求报文、解析 Http 请求报文操作,创建 Response 对象,执行发送 Http 响应报文的操作,关闭与浏览器的 TCP 连接。该方法的工作过程请参见图 16-1。

2) 代码

请参考 WebServer 类的源代码 WebServer.java 中——内部类 SocketThread 部分。

### 16.4.3　浏览器的实现

根据以上设计和实现思路,浏览器程序需编写三个类——一个 Browser 类和两个事件处理类:第一个事件处理类——URLHandler 处理单行文本框回车(输入 URL 地址后)或单击按钮事件,第二个事件处理类处理超链接事件,参见图 16-6。在此例中,考虑到两个事件处理对象只在 Browser 对象中创建,则将两个事件处理类作为私有内部类(其中超链接事件类为匿名内部类),嵌套于 Browser 类中。

1. Browser 类

该类主要实现浏览器的图形用户界面,以及供两个事件处理类对象调用的一些算法。

1) 效果图(图 16-6)

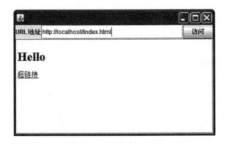

图 16-6　浏览器的运行效果

2）UML 图(图 16-7)

图 16-7 所示的 UML 图标识出 Browser 类的主要成员。

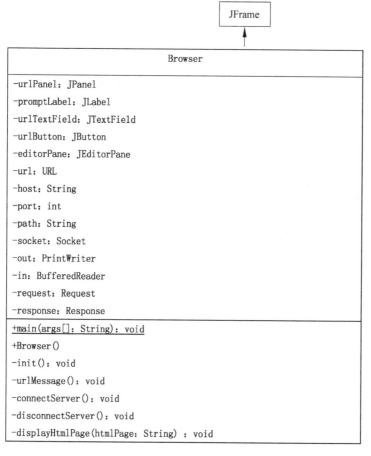

图 16-7　Browser 类的 UML 图

Browser 类是 JFrame 类的子类。

成员变量如下：

Browser 类的面板成员变量 urlPanel 上排列放置提示标签 promptLabel(提示用户输入 URL 地址)、单行文本框 urlTextField(用户在此输入 URL 地址)、"访问"按钮 urlButton；成员变量 editorPane 为编辑面板，用来显示从 Web 服务器获取到的页面，参见图 16-6。

成员变量 url 是 URL 类的对象，根据不同事件类型，或使用从 urlTextField 单行文本框中获取用户输入的 URL 地址字符串创建对象，或使用超链接事件对象 e 调用 getURL() 方法获取；成员变量 host、port、path 的值是调用 urlMessage() 成员方法从 URL 对象获取到的，分别存取 URL 对象中的主机名、端口号和路径；成员变量 socket、out、in 在 connectServer() 成员方法中创建，socket 是利用 host、port 与 Web 服务器建立 TCP 连接的套接字，out、in 从 socket 中获取，分别用于向 Web 服务器发送和从 Web 服务器接收；成员变量 request 和 response 分别是 Request 对象和 Response 对象，在 connectServer() 成员方法中，利用 url、out、in 创建，在事件处理的方法中，浏览器调用 request 的 sendHttpRequest() 方法向 Web 服务器发送 Http 请求,调用 response 的 receiveHttpResponse()方法和 parseHttpResponse()方法从 Web 服务器接收和解析 Http 响应。

公有方法:入口方法 main([]String)调用构造方法 Browser()创建浏览器对象,构造方法 Browser()则调用 init()私有方法。

私有方法:init()方法初始化图形界面的组件,并为 urlTextField 单行文本框、urlButton 按钮注册动作事件监听器,为 editorPane 编辑面板注册超链接事件监听器;urlMessage()方法从 URL 对象中获取到主机名、端口号和路径信息;connectServer()方法根据主机名和端口号连接 Web 服务器,得到套接字引用,从套接字中获取输入流和输出流,并创建 Request 对象和 Response 对象,将 URL 对象、输入流和输出流对象传递给请求对象和响应对象,用于发送 Http 请求、接收和解析 Http 响应;disconnectServer()方法断开与 Web 服务器的连接,关闭输入流、输出流以及套接字引用;displayHtmlPage(String)方法将接收和解析出的 Html 页面文本解释并显示于编辑面板上,请参见代码。

3) 代码(Browser.java)

```java
package browser;
import java.net.*;
import javax.swing.*;
import java.awt.Container;
import java.awt.BorderLayout;
import java.awt.Dimension;
import java.awt.event.*;
import javax.swing.event.*;
import java.io.*;
import appProtocol.Request;
import appProtocol.Response;
public class Browser extends JFrame {
    private Container mainPanel = getContentPane();
    private JPanel urlPanel = new JPanel();
    private JLabel promptLabel = new JLabel();
    private JTextField urlTextField = new JTextField();
    private JButton urlButton = new JButton();
    private JEditorPane editorPane = new JEditorPane();
    public static final int BUFFER_SIZE = 8192;
    private URL url;
    private String host;
    private int port;
    private String path;
    private Socket socket;
    private OutputStream out;
    private InputStream in;
    private Request request;
    private Response response;
    public static void main(String args[]) {
        new Browser();
    }
    public Browser() {
        try {
```

```java
                init();
        } catch (Exception e) {
            e.printStackTrace();
        }
    }
    private void init() throws Exception {
        mainPanel.setLayout(new BorderLayout());
        urlPanel.setLayout(new BoxLayout(urlPanel, BoxLayout.X_AXIS));
        promptLabel.setText("URL 地址");
        urlButton.setText("访问");
        URLHandler urlHandler = new URLHandler();
        urlTextField.addActionListener(urlHandler);
        urlButton.addActionListener(urlHandler);
        editorPane.setEditable(false);
        editorPane.addHyperlinkListener(new HyperlinkListener() {
            public void hyperlinkUpdate(HyperlinkEvent e) {
                editorPane_hyperlinkUpdate(e);
            }
            private void editorPane_hyperlinkUpdate(HyperlinkEvent e) {
                if (e.getEventType() == HyperlinkEvent.EventType.ACTIVATED) {
                    try {
                        url = e.getURL();
                        if (url != null)
                            urlMessage();
                        else {
                            String str = e.getDescription();
                            if (str.charAt(0) == '/')
                                path = str;
                            else
                                path = "/" + str;
                            url = new URL("http://" + host + ":" + port + path);
                        }
                        connectServer();
                        editorPane.setText("正在发送请求...");
                        request.sendHttpRequest();
                        editorPane.setText("正在接收响应...");
                        response.receiveHttpResponse();
                        disconnectServer();
                    } catch (Exception ee) {
                        new JOptionPane().showMessageDialog(Browser.this, "无法打开链接", "", 0);
                        return;
                    }
                    editorPane.setText("正在解析响应...");
                    String htmlPage = response.parseHttpResponse();
                    displayHtmlPage(htmlPage);
                    urlTextField.setText(url.toString());
                }
            }
        });
        JScrollPane scrollPane = new JScrollPane();
```

```java
            scrollPane.getViewport().add(editorPane);
            urlPanel.add(promptLabel);
            urlPanel.add(urlTextField);
            urlPanel.add(urlButton);
            mainPanel.add(urlPanel, BorderLayout.NORTH);
            mainPanel.add(scrollPane, BorderLayout.CENTER);
            setDefaultCloseOperation(JFrame.EXIT_ON_CLOSE);
            setSize(new Dimension(600, 500));
            setVisible(true);
        }
        private class URLHandler implements ActionListener {
            public void actionPerformed(ActionEvent e) {
                url_actionPerformed(e);
            }
            private void url_actionPerformed(ActionEvent e) {
                try {
                    url = new URL(urlTextField.getText());
                    urlMessage();
                    connectServer();
                    editorPane.setText("正在发送请求...");
                    request.sendHttpRequest();
                    editorPane.setText("正在接收响应...");
                    response.receiveHttpResponse();
                    disconnectServer();
                } catch (Exception ee) {
                    new JOptionPane().showMessageDialog(Browser.this, "错误的 URL 地址：" +
urlTextField.getText(), "", 0);
                    return;
                }
                editorPane.setText("正在解析响应...");
                String htmlPage = response.parseHttpResponse();
                displayHtmlPage(htmlPage);
            }
        }
        private void urlMessage() {
            host = url.getHost();
            port = url.getPort();
            if (port == -1)
                port = 80;
            path = url.getPath();
        }
        private void connectServer() throws IOException {
            editorPane.setText("正在连接服务器...");
            socket = new Socket(host, port);
            out = socket.getOutputStream();
            in = socket.getInputStream();
            request = new Request(url, out);
            response = new Response(in);
```

```
        }
        private void disconnectServer() throws IOException {
            editorPane.setText("正在与服务器断开连接...");
            in.close();
            out.close();
            socket.close();
        }
        private void displayHtmlPage(String htmlPage) {
            editorPane.setContentType("text/html");
            editorPane.setText(htmlPage);
        }
    }
```

#### 2. URLHandler 类

1）UML 图（图 16-8）

图 16-8 所示的 UML 图标识出 URLHandler 类的成员。该类被嵌套在 Browser 类中，是 Browser 类的私有内部类。

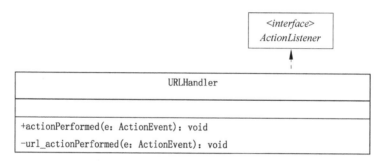

图 16-8　URLHandler 类的 UML 图

URLHandler 类实现了 ActionListener 接口。

actionPerformed(ActionEvent) 方法是 ActionListener 接口定义的方法。当用户在 Browser 对象的 urlTextField 单行文本框中输入 URL 地址后，回车或单击"访问"按钮时，执行该方法。在该方法中，调用了私有方法 url_actionPerformed(ActionEvent)。

私有方法 url_actionPerformed(ActionEvent) 的工作过程参见图 16-1。

2）源代码

请参考 Browser 类的源代码 Browser.java 中的私有内部类 URLHandler 部分。

#### 3. 私有匿名内部类

1）UML 图（图 16-9）

图 16-9 所示的 UML 图标识出 Browser 类的私有匿名内部类的成员。该类被嵌套在 Browser 类中。

该类实现了 HyperlinkListener 接口。

hyperlinkUpdate(HyperlinkEvent) 方法是 HyperlinkListener 接口定义的方法。当用

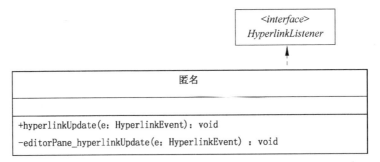

图 16-9　匿名类的 UML 图

户在 Browser 对象的 editorPane 编辑面板中单击超链接时,执行该方法。在该方法中,调用了私有方法 editorPane_hyperlinkUpdate（HyperlinkEvent）。

私有方法 editorPane_hyperlinkUpdate（HyperlinkEvent）的工作过程参见图 16-1。

2) 代码

请参考 Browser 类的源代码 Browser.java 中的匿名私有内部类部分。

### 16.4.4　"生成配置文件"功能实现

根据 16.3.3 节中的实现思路,该功能的实现需要一个类——CreateConfig 类。该类的实现依赖于外部包 dom4j-1.6.1.jar,需要将其加载到当前工程。

**1. UML 图（图 16-10）**

| CreateConfig |
|---|
| +WEB_ROOT: String |
| +PORT: String |
| +main(args[]: String): void |
| +createXml(fileName: String): void |

图 16-10　CreateConfig 类的 UML 图

图 16-10 所示的 UML 图标识出 CreateConfig 类的主要成员。

公有静态成员变量如下:

WEB_ROOT:表示页面文件的存储路径,赋值为应用程序运行时所在路径下的 webroot 目录。

PORT:表示 Web 服务器运行的端口号,在 main 主方法中由用户输入得到。

静态成员方法如下:

main 主方法中,从用户键盘接收端口号至 PORT 变量后,测试输入的端口号是否为数字字符。如为正常的数字字符端口号,则调用 createXml 方法创建 web.xml 文件。

createXml(String)方法中,创建 Document 对象,在 Document 中创建 root 根元素,之后在 root 根元素中创建 webroot 和 port 子元素,最后将公有静态成员变量 WEB_ROOT 和 PORT 的值设置给 webroot 和 port 子元素。创建 XMLWriter 输出流,将上述文档存入方法参数给出的 web.xml 文件中。

## 2. 代码(CreateConfig.java)

```java
package webServer;
import java.io.File;
import java.io.FileWriter;
import java.io.IOException;
import java.io.Writer;
import java.util.Iterator;
import javax.swing.JOptionPane;
import org.dom4j.Document;
import org.dom4j.DocumentException;
import org.dom4j.DocumentHelper;
import org.dom4j.Element;
import org.dom4j.io.SAXReader;
import org.dom4j.io.XMLWriter;
public class CreateConfig {
    public static String WEB_ROOT = System.getProperty("user.dir")
            + File.separator + "webroot";
    public static String PORT;
    public static void main(String a[]) {
        PORT = new JOptionPane()
                .showInputDialog(null, "请输入 Web 服务器使用的端口号：", "80");
        boolean flag = true;
        char ch = PORT.charAt(0);
        if (ch > '0' && ch <= '9') {
            for (int i = 1; i < PORT.length(); i++) {
                char ch1 = PORT.charAt(i);
                if (ch1 < '0' && ch > '9')
                    flag = false;
            }
        } else {
            flag = false;
        }
        if (flag) {
            createXml("web.xml");
            new JOptionPane().showMessageDialog(null, "已成功创建配置文件：web.xml", "", 0);
        } else {
            new JOptionPane().showMessageDialog(null, "错误的端口号：" + PORT, "", 0);
        }
    }
    private static void createXml(String fileName) {
        Document document = DocumentHelper.createDocument();
        Element root = document.addElement("root");
        Element webroot = root.addElement("webroot");
        Element port = root.addElement("port");
        webroot.setText(WEB_ROOT);
        port.setText(PORT);
        try {
            Writer fileWriter = new FileWriter(fileName);
            XMLWriter xmlWriter = new XMLWriter(fileWriter);
```

```
            xmlWriter.write(document);
            xmlWriter.close();
        } catch (IOException e) {
            System.out.println(e.getMessage());
        }
    }
}
```

## 16.5 Web 服务器与浏览器系统调试与软件发布

### 16.5.1 系统调试

#### 1. 生成配置文件

在 Eclipse 环境下，创建一个新的工程，在工程中加载外部包 dom4j-1.6.1.jar（加载方法：将光标移至工程名，右击，在弹出的菜单中选择 Build Path→Add External Archives，在弹出的打开文件对话框中选择 dom4j-1.6.1.jar 文件打开即可）。在该工程中用来存放源代码的路径 src 下创建两个包：appProtocol 包和 webServer 包。在 appProtocol 包中创建两个类：Request 类和 Response 类（详见 16.4.1 节）。在 webServer 包中创建两个类：CreateConfig 类（详见 16.4.4 节）和 WebServer 类（详见 16.4.2 节）。然后运行主类 CreateConfig 即可。

#### 2. Web 服务器

生成配置文件后，将主类改为 WebServer，运行主类 WebServer 即可。

#### 3. 浏览器

在 Eclipse 环境下，创建一个新的工程。在该工程中用来存放源代码的路径 src 下创建两个包：appProtocol 包和 browser 包。在 appProtocol 包中创建两个类：Request 类和 Response 类（详见 16.4.1 节）。在 browser 包中创建一个类：Browser 类（详见 16.4.3 节）。然后运行主类 Browser 即可。

### 16.5.2 软件发布

#### 1. 生成配置文件

在 Eclipse 环境下，在菜单中单击"文件"→Export，会弹出一个界面；在弹出的界面中展开 Java 后，选择 Runnable JAR file，然后单击 Next 按钮，会弹出另外一个界面；在该界面中有两个组合框，在提示有 Launch configuration 的组合框中选择工程的主类 CreateConfig，在提示有 Export destination 的组合框中给出生成的 JAR 文件的文件名（这里给出的文件名为 CreateConfig.jar）及其存放路径，然后单击 Finish 按钮。

现在，可以将生成的文件 CreateConfig.jar 复制到任何一台安装了 Java 运行环境并已生成配置文件的计算机上，读者用鼠标双击该文件的图标，即可生成配置文件。

## 2. Web 服务器

在 Eclipse 环境下,在菜单中单击"文件"→Export,会弹出一个界面;在弹出的界面中,展开 Java 后,选择 Runnable JAR file,然后单击 Next 按钮,会弹出另外一个界面;在该界面中有两个组合框,在提示有 Launch configuration 的组合框中选择工程的主类 WebServer,在提示有 Export destination 的组合框中给出生成的 JAR 文件的文件名(这里给出的文件名为 WebServer.jar)及其存放路径,然后单击 Finish 按钮。

现在,可以将生成的文件 WebServer.jar 复制到任何一台安装了 Java 运行环境并已生成配置文件的计算机上,读者用鼠标双击该文件的图标,即可运行 Web 服务器软件。

## 3. 浏览器

在 Eclipse 环境下,在菜单中单击"文件"→Export,会弹出一个界面;在弹出的界面中,展开 Java 后,选择 Runnable JAR file,然后单击 Next 按钮,会弹出另外一个界面;在该界面中,有两个组合框,在提示有 Launch configuration 的组合框中选择工程的主类 Browser,在提示有 Export destination 的组合框中给出生成的 JAR 文件的文件名(这里给出的文件名为 Browser.jar)及其存放路径,然后单击 Finish 按钮。

现在,可以将生成的文件 Browser.jar 复制到任何一台安装了 Java 运行环境的计算机上,读者用鼠标双击该文件的图标,即可运行浏览器软件。

## 16.6 综合课程设计作业

(1) 设计和实现 Https 服务器。
(2) 设计和实现 FTP 服务器和客户端。

# 第17章 Java与网络、数据库：基于B/S的用户登录管理系统

## 17.1 基于B/S的用户登录管理系统需求分析

基于B/S的用户登录管理系统需实现的功能如下：

(1) 客户端从服务器下载程序代码后，在客户端运行。运行时接收用户输入的用户名和密码，将其发送给服务器。

(2) 服务器接收到用户名与密码后，判断其是否正确。如正确，则发送"登录成功"提示信息给客户端；如不正确，则发送"必须输入用户名和密码"或"用户名或密码不正确"提示信息给客户端。

(3) 客户端显示提示信息。

此外，服务器应能同时为多个客户提供上述服务。

## 17.2 基于B/S的用户登录管理系统设计

### 17.2.1 数据库设计

根据以上需求分析，设计数据库Database1，包含一个表（表17-1）：用户。

表17-1 用户信息表（用户）结构

| 字段名 | 数据类型 |
| --- | --- |
| 用户名 | 字符串 |
| 密码 | 字符串 |

### 17.2.2 登录服务器功能设计

根据需求分析，登录服务器启动后，不断检测新上线的客户端发来的连接请求。当接收到新的连接请求后，登录服务器需并行完成以下多项工作：

(1) 继续检测新上线的客户端发来的连接请求。
(2) 继续为已连接的客户端提供服务。
(3) 与刚发送请求的客户端建立 TCP 连接后为其提供服务。为此，需循环执行以下操作：
① 接收请求；
② 解析请求；
③ 发送响应。

在②中，解析出接收到的请求，如为登录请求，则首先根据记录的信息判断该用户是否已经登录，如已经登录，则生成"重复登录"的响应信息；否则查询用户表，判断用户名和密码是否正确，如正确，则记录登录用户名，生成"登录成功"的响应信息；如不正确，则生成"必须输入用户名和密码"或"用户名或密码不正确"的响应信息。

在②中，如解析出接收到的请求为退出服务器请求（例如客户端关闭主界面窗口），则从记录的登录用户名中去除该客户端，生成"已退出服务器"的响应信息，并结束与客户端的通信循环。

主程序不断检测新上线的客户端发来的连接请求，当接收到新的连接请求后，创建子线程。在子线程中，为当前客户端提供服务。

登录服务器不需要与用户交互和显示信息，所以不需提供图形用户界面。

### 17.2.3 客户端功能设计

根据需求分析，该客户端代码是从服务器端下载，在客户端执行的。设计功能如下：

（1）从服务器下载的程序代码在客户端运行时，显示登录界面，接收用户输入的"用户名"和"密码"，用户单击"登录"按钮后，系统生成登录服务器的请求，通过网络将登录请求发送至服务器。

（2）接收服务器发来的响应信息，显示响应信息。如响应信息为"登录成功"，则显示主界面（本系统仅设计和实现了登录功能，主界面上其他功能省略）。

（3）如用户关闭主界面，系统生成退出登录服务器的请求，通过网络将退出请求发送至服务器。同样接收服务器发来的响应信息，显示响应信息后，退出客户端应用程序。

## 17.3 基于B/S的用户登录管理系统实现思路

### 17.3.1 系统实现采用分层结构模型

为避免图形用户界面中对按钮事件处理的方法中代码堆砌太多，造成程序的结构性太差，系统实现时采用清晰的分层结构模型，如图17-1所示。

在客户端，用户通过界面输入信息或单击按钮，界面层通过事件处理控制将信息封装成请求包后，发送至服务器的客户服务层。

在服务器端，客户服务层接受和解析请求，根据请求类型，调用业务层的相应函数；业务层的方法或函数将界面层所收集的数据传递给数据操作层的相应方法或函数；数据操作层的方法或函数根据业务层需求生成相应的SQL语句，完成数据库操作，将结果返回至业务层；业务层返回标记至客户服务层；客户服务层根据标记生成相应的响应信息包，发送至客户端；再由客户端界面层显示响应包中的信息给用户。

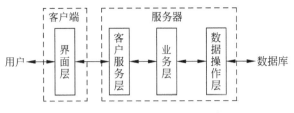

图 17-1　本系统采用的四层结构模型

### 17.3.2　客户端与服务器的应用协议

设计 Request 类和 Response 类，Request 对象封装用户名和密码信息，Response 对象封装服务器发送给客户端的字符串信息。客户端的界面层创建 Request 对象，发送给服务器的客户服务层；客户服务层生成相应的 Response 对象，反馈给客户端。

### 17.3.3　客户端的实现思路

从服务器下载小应用程序在客户端执行，具体使用 JApplet 实现。

## 17.4　基于 B/S 的用户登录管理系统实现

在基于 B/S 的用户登录管理系统中，根据以上设计和实现思路，共写了 8 个类。其中客户端的界面层两个类：LoginApplet 类、MD5 类，MD5 类用于添加用户，以及登录时对密码进行加密处理；服务器端的客户服务层两个类：LoginServer 类、LoginHandler 类；服务器端的业务层一个类：Service 类；服务器端的数据操作层一个类：DataOperator 类；界面层与客户服务层之间的通信使用的应用协议涉及两个类：Request 类和 Response 类。

以下根据系统功能分类，对其实现细节进行讲解。

### 17.4.1　建立数据库表和数据源

#### 1．建立数据库表

在 SQL Server 数据库管理系统下，建立数据库 Database1，在该数据库中创建一个表，表名为"用户"，表结构参见 17.2.1 节。

参见 17.4.3 节中，数据操作层——DataOperator 类的源代码中的 connect() 方法，在 SQL Server 数据库管理系统下，建立登录用户名"sa1"，密码"123"，默认操作的数据库为 Database1。

#### 2．建立 ODBC 数据源

在控制面板中建立名字为 Database1 的 ODBC 数据源，默认操作的数据库名为 Database1。

## 17.4.2 应用协议的实现

**1. Request 类**

Request 类封装了客户端的登录请求信息或退出服务器请求信息。
1) UML 图(图 17-2)

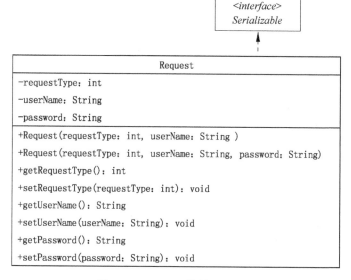

图 17-2 Request 类的 UML 图

Request 类创建的对象需在网络中传输,所以实现了 Serializable 接口。

成员变量：requestType 定义了请求类型,userName 定义了用户名,password 定义了密码。

构造方法：Request(int, String, String)在客户登录时调用,用来创建登录请求对象,第一个参数传递请求类型,值为 1,第二和第三个参数分别传递用户名和密码信息；Request(int, String)在客户退出服务器时调用,用来创建退出服务器请求对象,第一个参数传递请求类型,值为 4,第二个参数传递用户名信息。

成员方法：分别定义了三个成员变量的 get 方法和 set 方法。
2) 代码(Request.java)

```
package appProtocol;
import java.io.Serializable;
public class Request implements Serializable {
    private int requestType;
    private String userName;
    private String password;
    public Request(int requestType, String userName){
        this.requestType = requestType;
        this.userName = userName;
    }
```

```
    public Request(int requestType, String userName, String password) {
        this.requestType = requestType;
        this.userName = userName;
        this.password = password;
    }
    public String getUserName() {
        return userName;
    }
    public void setUserName(String userName) {
        this.userName = userName;
    }
    public int getRequestType() {
        return requestType;
    }
    public void setRequestType(int requestType) {
        this.requestType = requestType;
    }
    public String getPassword() {
        return password;
    }
    public void setPassword(String password) {
        this.password = password;
    }
}
```

2．Response 类

Response 类封装了当服务器收到客户端的登录请求信息时,向客户端发送的应答信息。

1) UML 图(图 17-3)

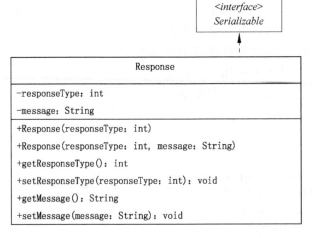

图 17-3　Response 类的 UML 图

Response 类创建的对象需在网络中传输,所以实现了 Serializable 接口。

成员变量：responseType 定义响应类型，message 定义服务器发送给客户端的响应信息。

2) 代码（Response.java）

```java
package appProtocol;
import java.io.Serializable;
public class Response implements Serializable {
    private int responseType;
    private String message;
    public Response(int responseType) {
        this.responseType = responseType;
    }
    public Response(int responseType, String message) {
        this(responseType);
        this.message = message;
    }
    public int getResponseType() {
        return responseType;
    }
    public void setResponseType(int responseType) {
        this.responseType = responseType;
    }
    public String getMessage() {
        return message;
    }
    public void setMessage(String message) {
        this.message = message;
    }
}
```

### 17.4.3 登录服务器的实现

根据以上设计和实现思路，登录服务器编写 4 个类。

客户服务层：LoginServer 类（主线程）和 LoginHandler 类（子线程）。

业务层：Service 类。

数据操作层：DataOperator 类。

**1. 客户服务层（主线程）——LoginServer 类**

1) UML 图（图 17-4）

PORT 表示端口号，具有 public static final 属性，赋值 8000。登录服务器在该端口地址上等待客户端的连接请求。

MAX_QUEUE_LENGTH 表示最大队列长度，也具有 public static final 属性，赋值 100。

| LoginServer |
|---|
| +PORT{immutable}：int = 8000 |
| +MAX_QUEUE_LENGTH{immutable}：int = 100 |
| +start()：void |
| +main(args[]：String)：void |

图 17-4  LoginServer 类的 UML 图

在 start()方法中,服务器创建 ServerSocket 对象,该对象通过调用其 accept()方法在 8000 端口上监听。如监听到客户端,与之连接后返回 Socket 对象,start()方法创建和启动以 LoginHandler 对象为线程体的子线程,与该客户通信。start()方法又回到循环开始,继续监听。

2) 代码(LoginServer.java)

```java
package loginServer;
import java.net.ServerSocket;
import java.net.Socket;
import java.io.IOException;
public class LoginServer {
    public static final int PORT = 8000;
    public static final int MAX_QUEUE_LENGTH = 100;
    public static void main(String args[]) {
        LoginServer loginServer = new LoginServer();
        loginServer.start();
    }
    public void start() {
        try {
            ServerSocket serverSocket = new ServerSocket(PORT, MAX_QUEUE_LENGTH);
            System.out.println("服务器已启动...");
            while (true) {
                Socket socket = serverSocket.accept();
                System.out.println("已接收到客户来自：" + socket.getInetAddress());
                LoginHandler handler = new LoginHandler(socket);
                handler.start();
            }
        } catch (IOException e) {
            e.printStackTrace();
        }
    }
}
```

3) 运行结果显示

服务器已启动...
已接收到客户来自：127.0.0.1
"admin"登录成功...
已接收到客户来自：127.0.0.1
"yu"登录成功...

## 2. 客户服务层（子线程）——LoginHandler 类

1）UML 图（图 17-5）

图 17-5 所示的 UML 图标识出 LoginHandler 类的主要成员。

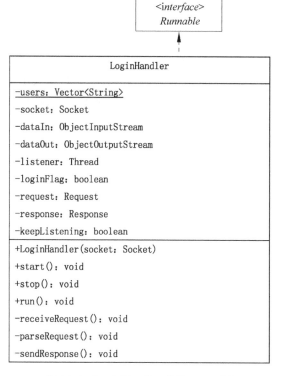

图 17-5　LoginHandler 类的 UML 图

作为子线程的线程体类，LoginHandler 类实现了 Runnable 接口。

类成员变量 users 定义为 Vector(String)集合类型，用来存储各子线程对象接收到的登录用户名。该集合不允许有重复元素，用来限制同一个用户名的重复登录。

成员变量 listener 引用子线程对象，dataIn 与 dataOut 分别为子线程对象与客户端通信时使用的输入输出流变量，request 和 response 为子线程对象与客户端交互时使用的协议数据单元。loginFlag 为登录标记，控制当用户在未登录条件下发送其他请求时，提示其"未登录"。

构造方法 LoginHandler(Socket)创建子线程的线程体对象，利用参数接收从主程序传递的 Socket 对象赋值于成量变量 socket；start()方法从 Socket 对象获取输入输出流对象，并利用线程体对象做参数构造出子线程对象 listener，启动 listener 子线程；stop()方法执行与 start()方法相反的操作；receiveRequest()方法和 sendResponse()方法分别用来接受请求和发送响应，且均为私有方法；parseRequest()方法解析接收到的请求对象类型，执行相应的操作后，生成响应对象，该方法也为私有方法。

在线程体的 run()方法中，使用布尔变量 keepListening 控制，循环执行接收请求、解析请求、生成响应、发送响应的操作。当客户端请求退出服务器时，parseRequest()方法将

keepListening 变量设置为 false,循环被中断,此时调用 stop()方法断开与该客户端的 TCP 连接后,线程体结束执行。

在 parseRequest()方法中,如解析出 requestType 为 1,则为登录请求,此时从请求对象解析出用户名与密码,首先判断类成员变量 users 集合中是否已经有此用户名,如已有此用户名,则为重复登录,这时需生成信息为"重复登录"的 Response 响应对象,将 keepListening 变量设置为 false 结束循环,断开与客户端的 TCP 连接;否则调用 Service 类的 login(String,string)方法,完成登录操作,如接收 login(String,string) 方法返回值大于 0,则生成信息为"登录成功"的响应对象。

在 parseRequest()方法中,如解析出 requestType 为 4,则表示退出服务器请求,此时将该用户名从 users 集合中去除,生成信息为"您已经从服务器退出"的 Response 响应对象,并将 keepListening 变量设置为 false,以结束循环,终止子线程。

2) 代码(LoginHandler.java)

```java
package loginServer;
import java.net.*;
import java.io.*;
import java.util.*;
import appProtocol.Request;
import appProtocol.Response;
public class LoginHandler implements Runnable {
    private static Vector<String> users = new Vector<String>();
    private Socket socket;
    private ObjectInputStream dataIn;
    private ObjectOutputStream dataOut;
    private Thread listener;
    private boolean loginFlag = false;
    private Request request;
    private Response response;
    private boolean keepListening = true;
    public LoginHandler(Socket socket) {
        this.socket = socket;
    }
    public synchronized void start() {
        if (listener == null) {
            try {
                dataIn = new ObjectInputStream(socket.getInputStream());
                dataOut = new ObjectOutputStream(socket.getOutputStream());
                listener = new Thread(this);
                listener.start();
            } catch (IOException e) {
                e.printStackTrace();
            }
        }
    }
    public synchronized void stop() {
        if (listener != null) {
            try {
```

```java
                listener.interrupt();
                listener = null;
                dataIn.close();
                dataOut.close();
                socket.close();
            } catch (IOException e) {
                e.printStackTrace();
            }
        }
    }
    public void run() {
        try {
            while (keepListening) {            // 监听该客户端
                receiveRequest();              // 接收请求
                parseRequest();                // 解析请求
                sendResponse();                // 发送响应
                request = null;                // 清除请求变量
                if(!keepListening)
                    stop();
            }
        } catch (ClassNotFoundException e) {
            e.printStackTrace();
        } catch (IOException e) {
            stop();
            System.err.println("与客户端通信出现错误...");
        }
    }
    private void receiveRequest() throws IOException, ClassNotFoundException {
        request = (Request) dataIn.readObject();
    }
    private void parseRequest() {
        if (request == null)
            return;
        response = null;
        int requestType = request.getRequestType();
        String userName = request.getUserName();
        String password = request.getPassword();
        if(requestType == 1 && users.indexOf(userName)!= -1){
            response = new Response(1, userName + "该用户已登录,您不能重复登录!");
            keepListening = false;
            System.out.println("\"" + userName + "\"重复登录,拒绝...");
            return;
        }
        if (requestType != 1 && !loginFlag) {
            // 请求类型不为1,不是登录请求,且该客户端还未登录
            response = new Response(1, userName + ",您还未登录!");
            return;
        }
        switch (requestType) {                 // 测试请求类型
        case 1:                                // 客户端登录
            int id = Service.login(userName, password);
```

```
                if (id == 0) {
                    response = new Response(1, "用户名或密码不正确!");
                    break;
                }
                if (id < 0) {
                    response = new Response(1, "查询出错!");
                    break;
                }
                users.add(userName);
                response = new Response(1, userName + ",您已经登录成功!");
                System.out.println("\"" + userName + "\"登录成功...");
                loginFlag = true;
                break;
            case 2:
            case 3:
            case 4:// 客户端请求退出服务器
                users.remove(userName);
                response = new Response(1, userName + ",您已经从服务器退出!");
                keepListening = false;
                System.out.println("\"" + userName + "\"从服务器退出...");
        }
    }
    private void sendResponse() throws IOException {
        if (response != null) {
            dataOut.writeObject(response);
        }
    }
}
```

### 3. 业务层——Service 类

1) UML 图(图 17-6)

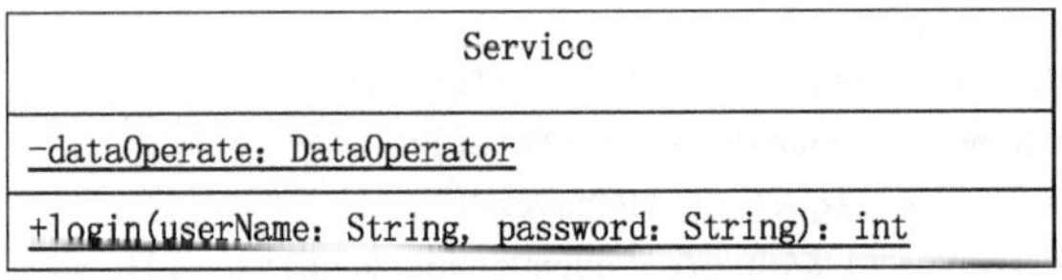

| Service |
| --- |
| -dataOperate: DataOperator |
| +login(userName: String, password: String): int |

图 17-6　Service 类的 UML 图

　　静态成员变量 dataOperate 引用 DataOperate 对象,通过该对象调用数据库操作的方法。
　　在 login（String，String）方法首次执行时,依次调用了 DataOperator 对象的 loadDatabaseDriver()方法、connect()方法和 userQuery(userName,password)方法,分别完成"加载 SQL Server 数据库的 JDBC 驱动程序"、"连接 Database1 数据库"、"查询用户表、核对用户名和密码是否正确"的操作。如 login（String，String）方法返回值大于 0,则登录成功,且返回值即为用户表中存储的用户标识号。login 方法后续执行时,只调用 DataOperator 对象的 userQuery 方法。

2）代码（Service.java）

```java
package loginServer;
public class Service {
    private static DataOperator dataOperate = new DataOperator();
    public static int login(String userName, String password) {
        boolean connectFlag = false;
        if(!connectFlag) {
            dataOperate.loadDatabaseDriver();
            dataOperate.connect();
            connectFlag = true;
        }
        return dataOperate.userQuery(userName, password);
    }
}
```

### 4．数据操作层——DataOperator 类

1）UML 图（图 17-7）

| DataOperator |
|---|
| -con: Connection |
| -pstmt: PreparedStatement |
| -sql: String |
| +loadDatabaseDriver()：void |
| +connect()：void |
| +userQuery(userName：String, password：String)：int |

图 17-7　DataOperator 类的 UML 图

2）代码（DataOperator.java）

```java
package loginServer;
import java.sql.*;
import java.util.Vector;
public class DataOperator {
    private Connection con;
    private PreparedStatement pstmt;
    private String sql;
    public void loadDatabaseDriver() {
        try {
            Class.forName("sun.jdbc.odbc.JdbcOdbcDriver");
        } catch (ClassNotFoundException e) {
            System.err.println("加载数据库驱动失败!");
            System.err.println(e);
        }
    }
    public void connect(){
        try {
```

```
                String connectString = "jdbc:odbc:Database1";
                con = DriverManager.getConnection(connectString, "sa1", "123");
            } catch (SQLException e) {
                System.err.println("数据库连接出错!");
                System.err.println(e);
            }
        }
        public int userQuery(String userName, String password){
            //查询用户表,核对用户名和密码是否正确
            try {
                sql = "SELECT id from 用户 WHERE 用户名 = ? AND 密码 = ?";
                pstmt = con.prepareStatement(sql);
                pstmt.setString(1, userName);
                pstmt.setString(2, password);
                ResultSet rs = pstmt.executeQuery();
                if (rs.next()) {
                    int id = rs.getInt(1);
                    return id;
                }
                return 0;
            } catch (SQLException se) {
                System.err.println("查询用户表出错!");
                System.err.println(se);
                return -1;
            }
        }
    }
```

## 17.4.4　JApplet 的实现——LoginApplet 类

### 1. 效果图(图 17-8)

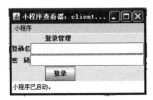

图 17-8　LoginApplet 运行效果

### 2. UML 图(图 17-9)

图 17-9 所示的 UML 图标识出 LoginApplet 类的主要成员。

LoginApplet 类是 JApplet 类的子类,并实现了 ActionListener 接口。

init()方法是小应用程序生命周期中运行的起始方法,负责完成登录窗口的初始化,并为 login 按钮注册动作事件监听器。

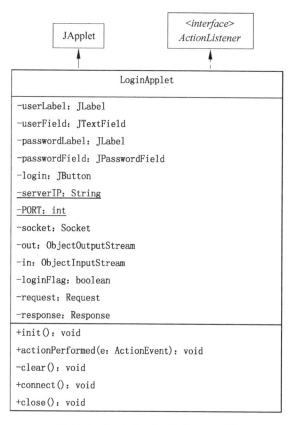

图 17-9　LoginApplet 类的 UML 图

　　actionPerformed(ActionEvent)是 ActionListener 接口定义的方法。在 LoginApplet 窗口中,当用户输入了用户名和密码,单击"登录"按钮(用成员变量 login 定义)时,执行该方法。在该方法中,取得用户在 userField 和 passwordField 输入的用户名与密码(密码需经过 MD5 类中的方法加密),封装成请求类型为 1 的 Request 对象,赋值于 request 成员变量,然后调用 connect()方法与服务器建立 TCP 连接(服务器的 IP 地址与端口号存储于静态成员变量 serverIP 和 PORT 中),得到 Socket 对象赋值于 socket 成员变量,并从 socket 中获取输入输出流对象分别赋值于 out 和 in 成员变量,通过 out 发送 Request 对象至服务器,完成登录操作。之后通过输入流对象 in,使用 Response 类型的成员变量 response 接收服务器发来的响应,将响应信息显示于弹出的窗口。如响应信息为"登录成功",则创建主界面 mainI 窗口对象,并为主界面窗口注册窗口监听器(使用匿名类对象,在用户关闭主界面窗口时,封装请求类型为 4 的 Request 对象,赋值于 request 成员变量,通过输出流对象 out 发送至服务器,完成退出服务器操作。当收到"已退出服务器"的响应信息后,调用 close()成员方法断开与服务器的 TCP 连接),最后显示主界面。

　　成员变量 loginFlag 是登录标记,初始化为 false,登录成功后被赋值为 true,此变量是为防止在同一界面中重复登录而定义的。

　　成员方法 clear()是在收到不成功登录(如"用户名密码错误"等)的响应信息时,重新输入用户信息前被调用的,其用途是清除单行文本框中显示的信息。

### 3. 代码（LoginApplet.java）

```java
package client;
import java.awt.Container;
import java.awt.GridLayout;
import java.awt.event.*;
import java.io.IOException;
import java.io.ObjectInputStream;
import java.io.ObjectOutputStream;
import java.net.InetAddress;
import java.net.InetSocketAddress;
import java.net.Socket;
import javax.swing.*;
import appProtocol.Request;
import appProtocol.Response;
public class LoginApplet extends JApplet implements ActionListener{
    private JLabel titleLabel;
    private JLabel userLabel;
    private JTextField userField;
    private JLabel passwordLabel;
    private JPasswordField passwordField;
    private JButton login;
    private static String serverIP = "localhost";
    private static int PORT = 8000;
    private Socket socket;
    private ObjectOutputStream out;
    private ObjectInputStream in;
    private boolean loginFlag = false;
    private Request request;
    private Response response;
    private String userName;
    public void init() {
        titleLabel = new JLabel("登录管理");
        userLabel = new JLabel("登录名");
        userField = new JTextField(25);
        passwordLabel = new JLabel("密 码");
        passwordField = new JPasswordField(25);
        passwordField.setEchoChar('*');
        login = new JButton("登录");
        login.addActionListener(this);
        Box box1 = new Box(BoxLayout.X_AXIS);
        Box box2 = new Box(BoxLayout.X_AXIS);
        box1.add(userLabel);
        box1.add(userField);
        box2.add(passwordLabel);
        box2.add(passwordField);
        Box box = new Box(BoxLayout.Y_AXIS);
```

```java
            box.add(titleLabel);
            box.add(box1);
            box.add(box2);
            box.add(login);
            Container c = getContentPane();
            c.add(box);
        }
        public void actionPerformed(ActionEvent e) {
            if (loginFlag){
                String hint = "重复登录!";
                JOptionPane.showMessageDialog(this, hint, "警告",
                        JOptionPane.WARNING_MESSAGE);
                clear();
                return;
            }

            userName = userField.getText().trim();
            String password = MD5.GetMD5Code(passwordField.getText().trim());
            if (userName.length() == 0 || password.length() == 0) {
                String hint = "必须输入用户名和密码!";
                JOptionPane.showMessageDialog(this, hint, "警告",
                        JOptionPane.WARNING_MESSAGE);
                clear();
                return;
            }
            try {
                request = new Request(1, userName, password);
                connect();
                out.writeObject(request);
                response = (Response) in.readObject();
            } catch (Exception ex) {
                JOptionPane.showMessageDialog(this, "无法连接或与服务器通信出错", "警告",
JOptionPane.WARNING_MESSAGE);
                clear();
                return;
            }
            String message = response.getMessage();
            if (message != null
                    && message.equals(request.getUserName() + ",您已经登录成功!")) {
                loginFlag = true;
            }
            JOptionPane.showMessageDialog(null, message, "信息提示",
                    JOptionPane.PLAIN_MESSAGE);
            if (!loginFlag) {
                clear();
                return;
            }
            clear();
```

```java
            JFrame app = new MainI();
            app.addWindowListener(new WindowAdapter() {
                public void windowClosing(WindowEvent e) {
                    try {
                        request = new Request(4, userName);
                        out.writeObject(request);
                        response = (Response) in.readObject();
                        JOptionPane.showMessageDialog(null, response.getMessage(),
                                "信息提示", JOptionPane.PLAIN_MESSAGE);
                    } catch (Exception ex) {
                         JOptionPane.showMessageDialog(LoginApplet.this, "与服务器通信出错",
"警告", JOptionPane.WARNING_MESSAGE);
                        clear();
                        return;
                    }
                    close();
                    System.exit(0);
                }
            });
            app.setVisible(true);
            setVisible(false);
    }
    private void clear(){
        userField.setText("");
        passwordField.setText("");
    }
    public void connect() throws IOException {
        InetAddress address = InetAddress.getByName(serverIP);
        InetSocketAddress serverSocketA = new InetSocketAddress(address, PORT);
        socket = new Socket();
        socket.connect(serverSocketA);
        out = new ObjectOutputStream(socket.getOutputStream());
        in = new ObjectInputStream(socket.getInputStream());
    }
    public synchronized void close() {
        try {
            in.close();
            out.close();
            socket.close();
        } catch (IOException e) {
            e.printStackTrace();
        }
    }
}
```

## 17.5 基于B/S的用户登录管理系统调试与软件发布

### 17.5.1 系统调试

**1. 登录服务器**

在 Eclipse 环境下，创建一个新的工程。在该工程中用来存放源代码的路径 src 下创建两个包：appProtocol 包和 loginServer 包。在 appProtocol 包中，创建两个类：Request 类和 Response 类（详见 17.4.2 节）。在 loginServer 包中，创建 4 个类：LoginServer 类、LoginHandler 类、Service 类和 DataOperator 类（详见 17.4.3 节）。然后运行主类 LoginServer 即可。

**2. 客户端**

在 Eclipse 环境下创建一个新的工程。在该工程中用来存放源代码的路径 src 下创建两个包：appProtocol 包和 client 包。在 appProtocol 包中创建两个类：Request 类和 Response 类（详见 17.4.2 节）。在 client 包中创建三个类：LoginApplet 类（详见 17.4.4 节）、MainI 类和 MD5 类。然后运行类 LoginApplet 即可。

MD5 类封装了密码加密算法，其源代码详见 14.5.1 节。因本章系统中 MD5 类被定义在 client 包中，所以需在该类源代码的第 1 行插入一条语句：package client;。

MainI 类用于读者扩展本系统功能，表示登录成功后进入主界面。此处该类简单定义如下：

```
package client;
import java.awt.*;
import javax.swing.*;
public class MainI extends JFrame {
    public MainI() {
        super("主界面");
        setSize(200,150);
    }
}
```

### 17.5.2 软件发布

**1. 登录服务器**

在 Eclipse 环境下，在菜单中单击"文件"→Export，会弹出一个界面；在弹出的界面中，展开 Java 后，选择 Runnable JAR file，然后单击 Next 按钮，会弹出另外一个界面；在该界面中，有两个组合框，在提示有 Launch configuration 的组合框中选择工程的主类 LoginServer，在提示有 Export destination 的组合框中给出生成的 JAR 文件的文件名（这里给出的文件名为 LoginServer.jar）及其存放路径，然后单击 Finish 按钮。

现在，可以将生成的文件 LoginServer.jar 复制到任何一台安装了 Java 运行环境的计算机上，读者用鼠标双击该文件的图标，即可运行登录服务器软件。

### 2. 客户端

编译 appProtocol 包中的 Request 类和 Response 类，以及 client 包中的 LoginApplet 类。
编写加载小应用程序的 Html 文件：

```
loginApplet.html
    < applet code = "LoginApplet" width = 200 height = 100 >
    </applet >
```

将 loginApplet.html 文件保存到和编译后的 LoginApplet.class 所在的相同目录中，如 F:\用户登录\client 中。

将 LoginApplet.class 所在目录作为 Web 服务目录，发布到 Tomcat 服务器中，虚拟目录名为 client。这样其他用户可以在其计算机中打开浏览器，在 URL 地址栏中输入 Tomcat 服务器所在主机的 IP 地址、虚拟目录名，从而访问包含有 LoginApplet 小应用程序的网页 loginApplet.html，到时会下载 LoginApplet.class 到客户端的浏览器中执行。例如 http://192.168.1.152/client/loginApplet.html。

## 17.6 综合课程设计作业

基于 B/S 的用户登录管理信息系统扩展：

### 1. 增设"用户注册"管理功能

在本系统已有功能的基础上，设计和实现"用户注册"功能。

### 2. 设计和实现一款小型网络版管理信息系统，题目自选。

在本系统基础上，扩展主界面功能，自选题目，设计和实现一款网络版的管理信息系统。如班级学生管理系统、家庭收支管理系统、图书管理系统、餐馆管理信息系统等。